AF352882

Fungal Architectures

Fungal Architectures

Editors

Andrew Adamatzky
Han A. B. Wösten
Phil Ayres

MDPI • Basel • Beijing • Wuhan • Barcelona • Belgrade • Manchester • Tokyo • Cluj • Tianjin

Editors

Andrew Adamatzky
Unconventional Computing
Laboratory
University of the West
of England
Bristol
United Kingdom

Han A. B. Wösten
Department of Biology
Utrecht University
Utrecht
Netherlands

Phil Ayres
Centre for Information
Technology and Architecture
Royal Danish Academy
Copenhagen
Denmark

Editorial Office
MDPI
St. Alban-Anlage 66
4052 Basel, Switzerland

This is a reprint of articles from the Special Issue published online in the open access journal *Biomimetics* (ISSN 2313-7673) (available at: www.mdpi.com/journal/biomimetics/special_issues/ Fungal_architectures).

For citation purposes, cite each article independently as indicated on the article page online and as indicated below:

LastName, A.A.; LastName, B.B.; LastName, C.C. Article Title. *Journal Name* **Year**, *Volume Number*, Page Range.

ISBN 978-3-0365-7347-2 (Hbk)
ISBN 978-3-0365-7346-5 (PDF)

Cover image courtesy of Phil Ayres

Contents

About the Editors

Andrew Adamatzky

Andrew is a Professor of Unconventional Computing and Director of the Unconventional Computing Laboratory, Department of Computer Science, University of the West of England, Bristol, UK. He conducts research in molecular computing, reaction-diffusion computing, collision-based computing, cellular automata, slime mould computing, massive parallel computation, applied mathematics, complexity, nature-inspired optimisation, collective intelligence and robotics, bionics, computational psychology, non-linear science, novel hardware, and future and emergent computation. He has authored seven books, the most notable being *Reaction-Diffusion Computing, Dynamics of Crow Minds*, and *Physarum Machines*, and he has edited 22 books in computing, the most notable being *Collision-Based Computing, Game of Life Cellular Automata*, and *Memristor Networks*. He has also produced a series of influential artworks published in the atlas *Silence of Slime Mould*. He is the founding Editor-in-Chief of "J of Cellular Automata" and "J of Unconventional Computing", and the Editor-in-Chief of "J Parallel, Emergent, Distributed Systems" and "Parallel Processing Letters".

Han A. B. Wösten

Han holds the position of chair in Microbiology within the Department of Biology at Utrecht University.

Han has over 30 years of experience researching and teaching microbiology, with a focus on fungal growth and development. He combines fundamental questions on how fungi disperse and colonize substrates in nature with applied research in how to implement this knowledge to enable the switch to a sustainable economy. In particular, he is interested in the use of fungi to create novel foods, to degrade toxic pollutants, and to use fungi to produce sustainable materials such as construction materials and leather and textile-like materials.

Han's research focus has been pursued in the context of projects funded by Utrecht University, the Netherlands Organization for Scientific Research, Novo Nordisk Fonden, and three EU-funded projects: *Fungal Architectures, MY-FI*, and *Fungateria*.

Phil Ayres

Phil is Professor of Bio-hybrid Architecture at the Royal Danish Academy, the chair for Biohybrid Architecture—situated within the Centre for Information Technology and Architecture (CITA), Institute for Architecture and Technology (IBT).

Phil has over 20 years of experience researching and teaching within architecture, with a sustained focus on the critical and creative use of technology to drive innovation within design and production practices contributing to the built environment. His practice-based research include architectural design, digital technologies, material systems, bio-technologies, and bio-fabrication.

Phil's current research focus lies within the conceptualisation, design, and production of novel bio-hybrid architectural systems that couple technical and living complexes, together with the development of complimentary design environments. His research focus has been pursued in the context of two EU-funded Future and Emerging Technology (FET) projects, *Flora Robotica* and *Fungal Architectures* and continues with the recently granted EU-funded EIC Pathfinder project *Fungateria*, for which Phil acts as coordinator.

Preface to "Fungal Architectures"

As one of the primary consumers of environmental resources, the building industry faces unprecedented challenges in needing to reduce the environmental impact of current consumption practices. This applies to both the construction of the built environment and resource consumption during its occupation and use. Where incremental improvements to current practices can be realised, the net benefits are often far outstripped by the burgeoning demands of rapidly increasing population growth and urbanisation. Against the backdrop of this grand societal challenge, it is necessary to explore approaches that envision a paradigm shift in how materials are sourced, processed, and assembled to address the magnitude of these challenges in a truly sustainable way, and which can even provide added value.

In the European-Commission-funded *FUNGAR: Fungal Architectures* Project, we proposed to develop a structural substrate by using live fungal mycelium, to functionalise the substrate with nanoparticles and polymers to make mycelium-based electronics, implement sensorial fusion and decision making in fungal electronics, and to develop monolithic buildings from the functionalized fungal substrate. We envisaged that fungal buildings will grow, build, and repair themselves subject to the substrate supplied, adapt naturally to the environment, and will sense all that humans can sense. One of the tasks of the FUNGAR project was to scout emerging technologies and development trends surrounding the cultivation and preparation of mycelium composites.

The papers cover a wide range of subjects, including the functional modifications of mycelium with inorganic particles, modifying and assessing the mechanical properties of mycelium composites, strategies for improving the flexural behaviour of composites, beehives from fungal materials, bio-welding and the reinforcement of composites, bioreactors for fungal production, the co-production of composites by fungi and bacteria, growing large-scale mycelium structures, sound absorption by composites, and the geometrical parameterisation of fungi.

This informative compendium of the techniques, methods, and insights on growing fungal material appeals to readers from all walks of life, from high school pupils to university professors, from mathematicians, computer scientists, and engineers to chemists, from craft practitioners to industrial producers of fungal materials, and from biologists to architects and artists.

Andrew Adamatzky, Han A. B. Wösten, and Phil Ayres
Editors

Review

Challenges and Opportunities in Scaling up Architectural Applications of Mycelium-Based Materials with Digital Fabrication

Selina Bitting [1,†], Tiziano Derme [2,†], Juney Lee [1,*], Tom Van Mele [1], Benjamin Dillenburger [2] and Philippe Block [1]

[1] Block Research Group, Institute of Technology in Architecture, ETH Zurich, Stefano-Franscini-Platz 1, HIB E 45, 8093 Zurich, Switzerland; bitting@arch.ethz.ch (S.B.); van.mele@arch.ethz.ch (T.V.M.); block@arch.ethz.ch (P.B.)

[2] Digital Building Technologies, Institute of Technology in Architecture, ETH Zurich, Stefano-Franscini-Platz 1, HIB E 23, 8093 Zurich, Switzerland; derme@arch.ethz.ch (T.D.); dillenburger@arch.ethz.ch (B.D.)

* Correspondence: juney.lee@ethz.arch.ch

† These authors contributed equally to this work.

Abstract: In an increasing effort to address the environmental challenges caused by the currently linear economic paradigm of "produce, use, and discard", the construction industry has been shifting towards a more circular model. A circular economy requires closing of the loops, where the end-of-life of a building is considered more carefully, and waste is used as a resource. In comparison to traditional building materials such as timber, steel and concrete, mycelium-based materials are renewable alternatives that use organic agricultural and industrial waste as a key ingredient for production, and do not rely on mass extraction or exploitation of valuable finite or non-finite resources. Mycelium-based materials have shown their potential as a more circular and economically competitive alternative to conventional synthetic materials in numerous industries ranging from packaging, electronic prototyping, furniture, fashion to architecture. However, application of mycelium-based materials in the construction industry has been limited to small-scale prototypes and architectural installations due to low mechanical properties, lack of standardisation in production methods and material characterisation. This paper aims to review the current state of the art in research and applications of mycelium-based materials across disciplines, with a particular focus on digital methods of fabrication, production, and design. The information gathered from this review will be synthesised to identify key challenges in scaling up applications of mycelium-based materials as load-bearing structural elements in architecture and suggest opportunities and directions for future research.

Keywords: mycelium; architecture; structural design; computational design; digital fabrication; additive manufacturing; subtractive manufacturing; circular economy

Citation: Bitting, S.; Derme, T.; Lee, J.; Van Mele, T.; Dillenburger, B.; Block, P. Challenges and Opportunities in Scaling up Architectural Applications of Mycelium-Based Materials with Digital Fabrication. *Biomimetics* 2022, 7, 44. https://doi.org/10.3390/biomimetics7020044

Academic Editors: Andrew Adamatzky, Han A.B. Wösten and Phil Ayres

Received: 15 March 2022
Accepted: 31 March 2022
Published: 14 April 2022

Publisher's Note: MDPI stays neutral with regard to jurisdictional claims in published maps and institutional affiliations.

1. Introduction

As a consequence of its current linear economic model of "produce, use, and discard", the construction industry is a significant contributor to global greenhouse gas emissions, destruction of natural habitat and production of industrial waste [1,2]. The industry's reliance on a select few non-renewable materials, such as concrete and steel, puts environmental pressure on finite natural resources, which could eventually lead to their permanent depletion [3]. While the industry is increasingly shifting towards more renewable building materials, the long-term environmental impact of this accelerating growth in demand and rate of regeneration remains to be seen. This shift is spearheaded by mass engineered timber (MET) and engineered bamboo composite (EBC) that depend on additional natural resources such as soil and water, which consequently experience increases in demand [4]. The transition from extracting non-renewable resources to harvesting renewable ones alone does not necessarily guarantee that the resultant material has a fully circular life cycle. For example, synthetic adhesives are fundamental for the production of MET and alternative bio-based solutions still largely rely on the usage of chemicals. This hinders the commercialisation of fully sustainable, circular wood products, which in turn impedes a transition to more circular and environmentally conscientious material sourcing [5].

Similar trends can be observed with the emergence of bio-resins and bio-plastics. These serve as alternatives to their more traditional, petroleum-based counterparts. However, the production of these alternatives typically relies on a single commodity feedstock, such as corn or sugarcane. This causes a rise in demand for these feedstocks for industrial use, creating competition with existing stock for food supply and instigating complex socio-economic policy problems [6]. There are similar consequences for the rapidly growing demand for MET, which have the potential to intensify the current volume and rate of deforestation across the world. While this rising demand can be potentially addressed with a fast-growing and high-yield material such as bamboo in certain contexts, hardwoods such as oak that are used for the production of most commercially available MET have the highest potential for contributing to further deforestation [7]. Therefore, there is a considerable need to find new alternative materials that are not just naturally cultivated and harvested, but also produced with processes that repurpose waste streams and improve the reusability and recyclability at the end of their life cycle.

1.1. Mycelium as Pilot Material for Circular Construction

A more circular approach to sustainable materials considers using industrial waste such as silica fly ash or lignocellulosic agricultural waste such as rice hulls, saw dust, corn cobs or soybean stalks as ingredients for the production of new bio-based materials [8]. Most of these approaches focus on repurposing low-value waste material into admixtures for concrete. However, there are also processes that transform industrial by-products and agricultural waste into completely new, high-value materials.

Recent developments of renewable composites based on mycelium have demonstrated a tremendous amount of potential in repurposing industrial waste streams into a viable resource for producing more sustainable and circular materials (Figure 1). Mycelium is the vegetative part of fungi that consists of a dense network of micro-filaments called hyphae [9,10] that have the capacity to bind food, agricultural and industrial waste that have very little or no commercial value and convert them into higher-value composite materials with a wide range of potential applications. The success of startups specialising in mycelium-based research, production and applications has also shown how this material can be introduced into the current market to have an immediate economic impact and reward [11–13].

In addition to using low amounts of energy during production and having a high profile of biodegradability, mycelium-based materials are highly customisable throughout the growing and manufacturing processes. This enables production of mycelium-based materials with various properties, which can satisfy varying criteria from different disciplines, and be suitable for different applications [14]. In the construction industry, mycelium-based materials have already gained traction in non-structural applications, such as for thermal and acoustical panels, thanks to their naturally high insulative properties as well as their resistance to fire [15].

1.2. Problem Statement

Despite their advantages, applications of mycelium-based materials as load-bearing structural elements have been limited, primarily due to their low mechanical properties. Considering that the structural mass is the predominant contributor to the total embodied carbon of a building, structural applications of low-carbon materials such as mycelium have the potential to significantly improve a building's environmental performance [18]. However, their comparatively low structural load capacity means that applications of mycelium-based materials need to be closely coupled with structurally informed geometry and appropriate digital fabrication methods.

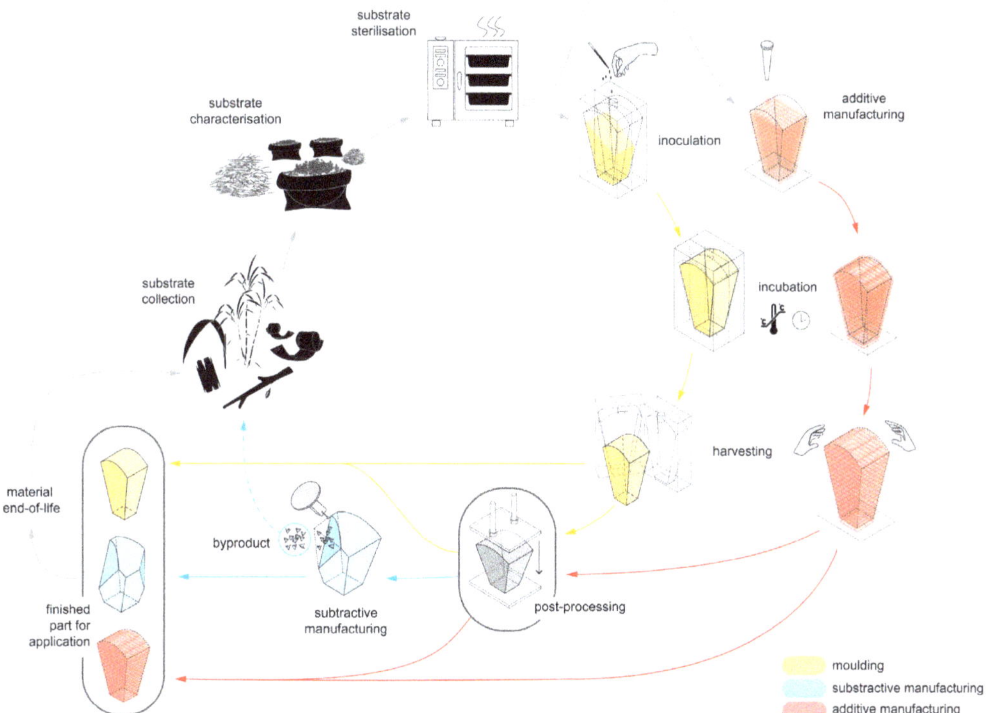

Figure 1. Circularity and workflow of current manufacturing processes applied to mycelium-based materials, adapted from [15–17].

The underutilisation of mycelium-based materials for large-scale architectural applications is also caused in large part by the monopoly of mycelium-related patents in the industry. These patents belong to a select few of the previously mentioned startups and prevent the distribution of knowledge to successfully mass-produce high-quality, standardised mycelium-based materials. The lack of generalised knowledge also leads to a lack of awareness by the public about the existence of these materials and a lack of confidence in large-scale applications beyond packaging alternatives and consumer products [19].

Furthermore, there is a disconnect between the industry and academia. Publications on new research and applications of mycelium-based materials tend to withhold information and data regarding the fungal species used, incubation parameters, substrate compositions, or detailed fabrication procedures, as a vast majority of the authors are affiliated with commercial companies [20].

There has been a recent surge of publications reviewing research and applications of mycelium-based materials in architecture, and while many have similarities, there is also a broad range of stances and scopes [4,15,19–26]. Most of these reviews focus on applications of mycelium-based materials for bespoke architectural prototypes and installations. However, this is a narrow scope that can lead to the exclusion of emerging innovations from other disciplines, such as mycology, microbiology, and material science as well as the fashion and textile industry. Additionally, the reviews tend to focus mostly on mycelium-bound composite materials, while lesser-known or less-documented implementations using pure mycelium have the potential to open new avenues of research and design applications in architecture.

1.3. Objectives

The main aim of this paper is to review the latest developments in research and applications of mycelium-based materials across multiple disciplines, including those that are not typically associated with the construction industry, with a focus on digital fabrication techniques. This review will investigate different typologies of mycelium-based materials at varying scales of interest, starting from small-scale commercial products and working up to larger architectural applications. The intent is to understand the scales at which mycelium-based materials tend to be implemented in the current state of the art. The key challenges that are hindering the scaling up of their applications then can be identified, along with the opportunities to overcome those challenges. Speculative strategies for scaling up applications of mycelium are then proposed in later sections, focusing on appropriate selection of material typology in conjunction with relevant fabrication and construction methods.

1.4. Contributions and Outline

The paper is organised as follows.

In Section 2, we present a review of the latest advancements and applications of mycelium-based materials across a wide range of scales and disciplines. Methods of growing and processing different mycelium material typologies are discussed, followed by an overview of typical production and digital fabrication methods that are often used in literature. Finally, various applications of mycelium-based materials are examined, both at the scale of smaller commercial products and larger architectural prototypes and installations.

In Section 3, we synthesise the information gathered from the state of the art to identify the key challenges in scaling up applications of mycelium-based materials in architecture as load-bearing structural elements. For each challenge, we suggest opportunities and outline directions for future research with a focus on relevant computational structural design and digital fabrication techniques.

2. State of the Art

This section on the state of the art is organised in three subsections: mycelium material typologies, fabrication methods, and applications.

2.1. Mycelium Material Typologies

There are two primary methods for developing engineered mycelium-based materials: pure mycelium materials (PMM) and mycelium-bound composites (MBC). The first refers to creating a mycological biopolymer by harvesting a liquid culture of pure mycelium. The second refers to a bio-composite wherein the hyphal network of the mycelium binds lignocellulosic substrates. In both approaches, the air contained within the mycelial network or between the loosely packed substrates and mycelia matrix results in a low-density material with foam-like properties [15].

PMM and MBC have material characteristics similar to polystyrene and polyurethane foams, respectively [19] (see Figure 2). These materials, without post-processing, are primarily suitable for non-structural applications and, particularly in the case of MBC, can be used as excellent thermal and acoustic insulators. The mechanical properties of these materials, described by several authors [15,27–29], vary depending on the fungal strain chosen for the inoculation, the origin and type of the substrate, and the growth conditions [30].

Substrates play a fundamental role in the final material properties. Their initial weight, porosity, and nutritive profile determine the consequent thermodynamic, physical, and mechanical behaviour of the biomaterial. As reported by Islam et al. [10] and Girometta [15], the balance between fungal and plant biomass can drastically determine the density of the composite, especially in the case of MBC. Implementations which prioritise material strength typically value higher density. However, in addition to density, rendering the material inert and post-processing the material have the possibility to increase or decrease the mechanical strength of both PMM bio-polymers and MBC bio-composites.

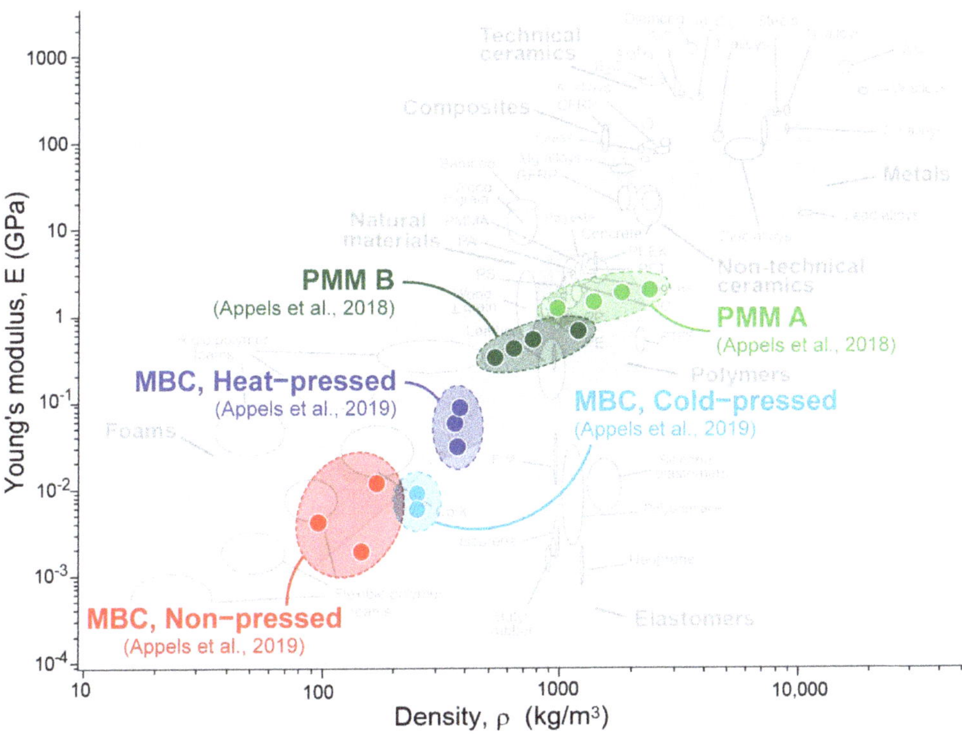

Figure 2. A comparison of mycelium-based materials to standard materials, with data adapted from [30,31]. Underlying Ashby plot created using CES EduPack 2019, ANSYS Granta © 2020 Granta Design.

2.1.1. Pure Mycelium Materials

In accordance with the definition described by Vanderlook et al. [16], PMM are mycelial bio-polymers that do not contain any lignocellulosic substrates. They are characterised solely by the biological properties of the fungal species, source of nutrients, and growing conditions. The notable feature of these materials is their reliance on the use of liquid mycelium cultures [32]. The basic method for producing these mycelial bio-polymers starts with several containers, each of which defines the cavity that contains a soft scrim, the nutritive substrate, and the desired fungal strain. These containers are subsequently placed into a closed incubation chamber with directed airflow at target humidity level and temperature [33–35].

The use of this specific culturing method allows for the rapid and consistent production of mycelial biomass and a precise control of the growth process. Thanks to its flexible foam-like properties, PMM can be produced and post-processed to be suitable for the textile, footwear and paper making industry or as green alternatives to polymeric foams such as expanded polystyrene and materials where flexibility is preferred over rigidity [36] (Figure 2). Within the various steps of growing and processing these mycelium-based materials, technology and fermentation can have a crucial influence on the resultant material as well as in scaling up its production. For example, the use of bioreactors fulfils the need for a controlled environment with the potential of producing large quantities of material [37]. Research from the VTT Technical research centre in Finland showed the potential of mass producing the mycelial bio-polymer through the creation of a continuous PMM production using bioreactor fermentation [38].

2.1.2. Mycelium-Bound Composites

MBC are bio-composites characterised by the growth of a fungal strain on substrates comprising discrete particles of organic or inorganic origin [30]. During the growth process, mycelium digests the nutrients contained in the substrate, penetrating the fibres and developing a tight network of hyphae. These hyphae bind the substrates together, ultimately transforming the mixture into a self-supporting, consolidated mycelial mass [15,39]. MBC can be produced and post-processed to create tough, pliable material with mechanical properties similar to those of wood and cork [30] (Figure 2).

The balance between the mass of the substrate and the mycelial mass enables the creation of versatile composites with tuneable density and porosity. Most of the research investigating the physical properties of MBC explore the influence of various types and sizes of substrate fibres, as well as the type of inoculation (grain spawn or liquid culture) [40,41]. The substrates used to maximise the growth of mycelium consist of lignocellulosic materials coming from agricultural crop waste such as, but not limited to, cotton, corn, flax, hemp, and wheat. These substrates, depending on their nutritional profile, affect the density of the material, as a higher proportion of grains typically corresponds to a higher material density [24,42]. Depending on the desired material profile and field of application, the use and customisation of substrates can be tailored to achieve the desired properties in the final material.

As suggested in the patent of Schaak Damen [43] from the company Ecovative, the production of MBC can also implement a more complex process that utilises a bacterial species in combination with a fungal species, non-nutritional substrates, and additional nutritive materials. In this approach, the bacteria, through its metabolic process, provides mechanical properties to the bio-composite material, and the fungal species binds the bio-composite.

2.2. Fabrication Methods

The simplest method for fabricating mycelium-based materials is to grow it in a mould with a predefined geometry. There are two primary fabrication methods that can be applied afterwards: subtractive and additive manufacturing.

Subtractive manufacturing uses pre-grown MBC components or panels, which are then altered via cutting, milling or some other subtractive process. This transforms MBC into a multitude of products and architectural building components. Subtractive processes can be applied to mycelium materials that are either alive or rendered inert, and are typically associated with discretised forms, rigid blocks, and panelised mycelium-based products.

Alternatively, additive manufacturing describes the methods that are applied prior to either MBC or PMM during the growth phase. This fabrication methodology needs to be carefully integrated into the production pipeline as it can directly influence the growing and harvesting processes of the material (i.e., preparation, inoculation, and incubation). This results in different products with different material characteristics. Typically, additive manufacturing is associated with the production of low-density mycelium products, or large in situ architectural components.

2.2.1. Moulds

The methodology typically used to produce MBC includes three main steps that come after the preparation and the sterilisation of the selected substrate: (1) inoculation of the substrates in a controlled environment; (2) incubation of the material for a period ranging from 7 to 30 days, depending on the fungal species under specific humidity, temperature, and air conditions; and (3) transfer of the colonised mycelium material outside the controlled environment and exposure to high heat to halt and terminate the growing process. Due to the high sensitivity to humidity, temperature, and air quality, the first two steps often rely on containing the mixture in pre-defined moulds. These moulds can be produced with customised geometry, then sealed to maintain a constant controlled environment for the fungal propagation [40]. Reproducibility of laboratory facilities and mass customisation are crucial steps for large-scale production and application of MBC produced with moulds [44].

2.2.2. Subtractive Manufacturing

Subtractive manufacturing (SM) refers to processes that involve removal of material. In a subtractive manufacturing workflow using mycelium-based materials, modules of prefabricated MBC are milled, wire-cut or undergo some other subtractive process. These modules can be products such as insulation, packaging or acoustical insulation, or they can be assembled into larger structures or pavilions. Subtractive methods in combination with MBC are not widely implemented due to several disadvantages, such as the need to sterilise equipment and the inconsistent fibrous nature of the material being difficult to work with. However, methods such as milling [45,46] and wire-cutting [17] have recently shown promise in spite of them. The main advantage of utilising this fabrication method is the elimination of the necessity for complex moulds, as geometric complexity can be added by machining away material instead (Figure 3).

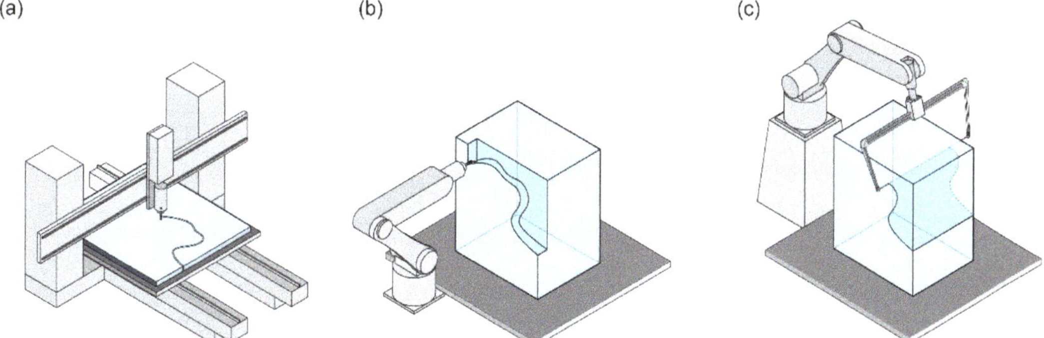

Figure 3. Typical subtractive manufacturing techniques: (**a**) milling (cutting) flat, panel-like specimen; (**b**) milling (carving) volumetric specimen; and (**c**) wire-cutting volumetric specimen.

Milling

Milling is a method that can be applied not only to panel-like geometries but also to more volumetric ones (Figure 3a,b). Milling materials such as medium-density fibreboard (MDF), oriented strand board (OSB), or polystyrene produces dust and debris as a by-product, which is typically discarded as waste. The circular nature of mycelium-based materials would allow these by-products to be re-used. MBC can be milled before or after autoclaving or applying a post-processing method such as pressing. However, there are some risks and complications in doing so. In the case where mycelium is milled while the mycelium hyphae are still growing, additional considerations must be made to prevent contamination, such as sterilising all equipment and milling in a closed, controlled environment. If rendered inert, contamination is no longer a risk.

Examples of milling MBC after autoclaving are exhibited in Figure 4 and show how the substrates in the MBC have ripped up during the milling process. This results in an undesirable surface finish and inadequate geometric precision. Furthermore, because MBC are anisotropic materials [47] with a structural performance that is stochastic [48], this method of fabrication then has the potential to aggravate the structural performance and material characteristics of the block in an unpredictable manner. Some fungal strains, for example, develop a skin (seen in white in Figure 4), and this skin increases the material's compressive strength as well as its water repellence [32]. The removal or harming of this skin is therefore disadvantageous to the mechanical properties of MBC.

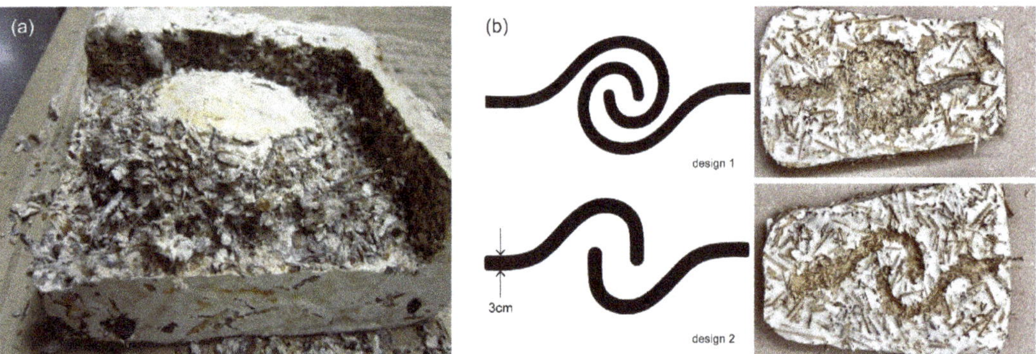

Figure 4. Two examples of MBC grown in a mould, rendered inert, and CNC milled thereafter Reproduced and reprinted with permission from: (**a**) Peek; 2021 [45]; and (**b**) Lazaro et al., Proceedings of the UbiComp/ISWC; 2019 [46].

Wire-cutting

While wire-cutting (Figure 3c) also produces excess material, it is quicker than milling, which is an advantage. Elsacker et al. [17] highlights the potential of this subtractive method in combination with MBC, where both living and inert MBC blocks are cut with an abrasive wire cutter. When a living block is cut, both halves are incubated further after cutting to allow the cut surface to grow into a thickened outer coating or "skin", which is advantageous from an architectural perspective in terms of aesthetics as well as the material properties. This shows the advantage of designing geometries with wire-cutting, as the excess material can be planned to become another inverse module to be used in construction. These blocks are later used as an insulative formwork for concrete, which is elaborated upon in Section 2.3.1. The advantage of this approach is in not creating excessive amounts of dust and the potential to incorporate both the "positive" and the "negative" of the cut specimen.

2.2.3. Additive Manufacturing

In the past decade, additive manufacturing (AM) has been widely used to produce three-dimensional products following a layer-by-layer, gradual material deposition method. The growing demand for the use of this technology refers directly to its advantages of being a form of rapid prototyping with the capacity of producing geometrically complex parts without added cost [49]. This led to the expansion of the technology towards the use of numerous materials across different industries from the automobile, aerospace, construction, medical, and food industries, to many others. AM can be divided into various categories that depend on the type of process and material state. These include liquid-based, solid-based, powder-based, and gas-based processes [50] as well as new subcategories that have emerged from the combinations of two or more different types of AM such as slurry-based 3D printing.

AM applied to mycelium-based materials refers to the creation of mycelium-based composites that integrate lignocellulosic substrates with the inoculation of a fungal strain during or after the manufacturing process. Due to their new and unique processes, various applications of AM to mycelium-based materials represent extensions of current advancements into bio-printing. This technique is mainly used in tissue engineering and regenerative medicine, and consists of manufacturing with living microorganisms, scaffolds, and the transformation of materials into complex living tissue [51]. While the differentiation between current approaches of AM is somewhat flexible, AM approaches that are used with mycelium-based materials can be largely categorised into three main areas of investigation: substrate core deposition; filament-based scaffolds; and bio-inks (Figure 5).

Substrate core deposition

This fabrication methodology can be associated with Direct Ink Writing (DIW), a 3D printing technique in which a paste-like material is deposited in a layer-by-layer fashion. Most of the experimentations found in literature use this methodology to develop an extrudable paste made of lignocellulosic material for the fabrication of a scaffolding structure. The deposition process is usually followed by the inoculation of the substrate core scaffold with a selected fungal strain. These approaches often result in small-scale vase-like structures, as seen in the work of Fraunhofer UMSICHT [52]. In the case of the experimentation developed at the IAAC in Barcelona [53], the use of clay as the main substrate characterised the additive manufacturing process. The resulting lattice structure was printed and designed to be ideal for the post-inoculation process.

Filament-based scaffolds

This fabrication approach is an extension of widely available fused deposition modelling (FDM) printers, capitalising on the research of specific filaments that contain sufficient nutrients to enhance mycelium interfacial bonding. The filament is printed first and thereafter inoculated with the desired fungal strain. These opportunities came along with the commercialisation of the first wooden filaments on the market, such as Laywoo-D3 or Growlay [54]. Due to their composition and capacity to absorb water, these filaments may serve as a means for the growing plants, or in this case, fungi. Investigations of this application are mostly academic and emphasise the role of 3D printed scaffolds. Notable experiments in this area come from the PhD research of Nicole Alima [55] on bio-scaffolds, and from the Myco Mensa table developed by Richard Beckett [56]. These experimentations are characterised using generative design tools and less on the definition of material properties and performance.

Bio-inks

This material typology integrates an organic substrate, a carrier, and living cells directly into an extrudable paste or mixture with shape retention properties. The carrier is normally a biopolymeric gel that acts as a molecular scaffold [57]. Promising investigations in this area come from Eugene Soh et al. [58] and the group of Professor H. La Ferrand from Nanyang Technological University of Singapore. The publication suggests an experimental framework that integrates (a) the direct experiment with new substrates such as bamboo fibres, (b) with the selection of a specific fungal strain (i.e., *Ganoderma lucidum*), and (c) the use of additives, such as chitosan-based solutions, are used to increase the workability and buildability of the extrusion paste. In this research, the carrier acts as a provisional binder for the substrate and combines the fabrication and inoculation into one single step. Goidea et al. [59] use a combination of clay and organic substrates for the creation of an extrudable inoculated paste. These experimentations are interested in the production of large-scale, load-bearing demonstrators. Similar is the case in 2016 of the project Bio Ex-Machina conducted by Officina Corpuscoli, Utrecht University, and the European Space Agency's Advanced Concepts Team (ESA/ACT) [60].

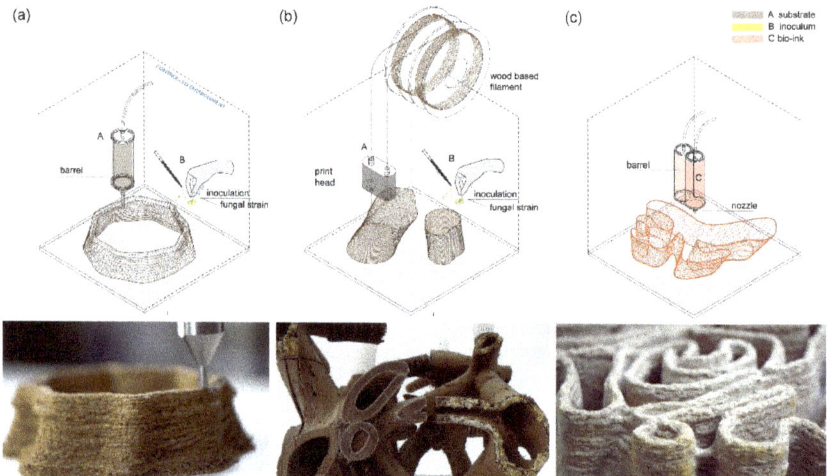

Figure 5. Current additive manufacturing approaches for mycelium-bound composites: (**a**) substrate core deposition [52] (Julia Krayer and Fraunhofer UMSICHT © 2022); (**b**) filament-based scaffolds [55] (reprinted with permission from Alima et al., Proceedings of the 2021 DigitalFUTURES; 2022); and (**c**) bio-inks [59] (reprinted with permission from Goidea et al., Pulp Faction: 3d Printed Material Assemblies through Microbial Biotransformation; published by UCL Press, 2020).

2.3. Applications

The applications of mycelium-based materials to be reviewed are sorted into two categories based on the scale of the application: products (small) and architectural projects (large). Some examples of implementations that are in the products category are items aimed at consumers, such as furniture, jewellery, clothing, or building components such as interior finishing products and insulation. The architectural projects section showcases assemblies composed of mycelium-based materials that are larger in scale than a singular module or piece of furniture. The products and architectural applications are summarised and categorised in Figure 6, according to the typology of mycelium-based material used and the fabrication method applied.

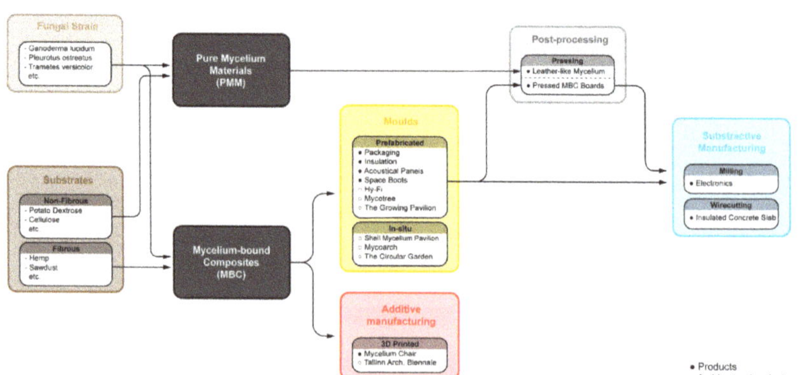

Figure 6. Overview of mycelium-based material typologies and applications.

2.3.1. Products

The following products are small-scale applications of the mycelium material typologies presented in Section 2.1.

Fashion

There are multiple types of mycelium-based textiles that are relevant to the fashion industry, applicable to items such as clothing and shoes. For example, flexible mycelium foams can be used as an inner insulation layer to fabric as an alternative to down feathers [61]. The foam is grown by filling macroscopic void spaces with elements such as agar beads, which are incorporated into the mycelial matrix during the growth process and then removed later through exposure to high heat. This does not damage the mycelial network itself but does destroy the filler elements and results in a flexible foam-like material. This product is patented by Ecovative [62], and demonstrates how influential post-processing of PMM is on the resulting material. The production of a leather-like mycelium material is an elaboration of this process.

The development of mycelium-based substitutes for leather utilises liquid culture to produce sheets that are then manipulated to be tanned and turned into bags, clothes, etc. The rolls are produced as flexible mycelium foams, then pressed and further manipulated to have leather-like material properties. One company that is utilising this technology is Bolt Threads, which engineered Mylo in tandem with Ecovative, who are working to bring these products to the consumer market. The process used to produce Mylo results in an non-biodegradable product that is not plastic-free and is certified 50–85% bio-based [12]. Companies such as Adidas, Lululemon, and Stella McCartney have designed numerous products with Mylo, although in most cases they are not yet sold in stores.

Alternatively, a recent article by E.R. Kanishka B. Wijayarathna explores a fully biodegradable mycelium-based leather alternative, which is made of fungal biomass cultivated on bread waste. Glycerol and a bio-based binder were used in the treatment of this leather-like material, which was successfully used to create a prototype phone pouch as well as a coin wallet. Sheets of the untreated material with only glycerol post-treatment resulted in a tensile strength of 7.7 MPa, whereas sheets of the untreated material and tannin-treated material with both glycerol and binder treatments led to tensile strengths of 7.1 MPa and 6.9 MPa, respectively [63]. These two variations of mycelium-based leather substitutes demonstrate the influence that post-processing methods have over material characteristics.

A more futuristic application of mycelium is seen in the proposed design for mycelium space boots. The intent is to reimagine a moon boot into something that could be manufactured entirely aboard a space shuttle. The final design achieves this, with basically only human sweat and a few fungus spores, ideal for a seven-month trip to Mars with limited space. This is an implementation that displays how design goals have a great influence on mycelium products, as well as how it is possible to push the boundaries of what is possible with these materials [64].

Furniture

One of the most notable and talked-about products utilising mycelium is the chair designed in 2003 by Eric Klarenbeek [65]. This project is one of the first implementations of 3D printing technologies with MBC. The process of fabrication is characterised by the creation of a hollow 3D printed scaffold acting as temporary mould for the substrate deposition. On the contrary, Myco Mensa, a table designed by Richard Beckett [56], integrates robotic fabrication and generative design approaches with the direct deposition of organic matter. Another product is Mycosella, a project presented in 2019 from a graduate from Newcastle University, which investigates the production of a series of stools and chairs that integrate MBC with other materials (i.e., wood, steel) in different manners [66].

Packaging

One of the first mycelium-based products to reach the consumer market is Ecovative's mycelium packaging. This product utilises a growth mixture of hemp hurd and a fungal strain [61]. After use, mycelium packaging can be broken up and composted without any additional steps to naturally decompose. While this foam alternative is a significant improvement from conventional synthetic packaging material in terms of environmental impact, the main disadvantage is the increased weight and bulk. Where conventional polystyrene foam has an approximate density of 10–50 kg/m^3, an alternative mycelium packaging can have a density between 55–210 kg/m^3. Since shipping costs are typically determined by weight, mycelium packaging is not yet able to fully replace the traditional polystyrene foams [19].

Electronics

Researchers, makers, and hobbyists who prototype with electronics rely largely on plastic parts to do so. The breadboards, enclosures, battery holders, buttons, and wires tend to all be made of materials that end up in a landfill [46]. Utilising mycelium provides a biodegradable alternative. These mycelium-based alternatives can be made of MBC blocks grown in moulds, which are then laser cut, laser engraved, or milled in order to create the grooves and housings needed to attach additional electronic elements and components. The fact that mycelium is not a conductor of electricity, and has a high fire-resistance rating, makes it beneficial for these materials to be used in applications where such characteristics are sought after.

Building components

Using MBC as insulation capitalises on the inherent thermal and fire-resistance properties of mycelium-based material. The company Biohm's mycelium insulation panels are able to achieve a thermal conductivity of 0.024 W/m·K This surpasses the values that can be achieved by market leading but unsustainable materials such as glass fibre (0.032–0.044 W/m·K), mineral wool (0.032–0.044 W/m·K), expanded polystyrene (0.036 W/m·K) and extruded polystyrene (0.029–0.036 W/m·K) [13]. Furthermore, the cost of manufacturing is low enough to compete with standard materials, which is a unique advantage in this implementation of mycelium.

The high insulative properties are also utilised in Elise Elsacker's concrete slab system [17], introduced in Section 2.2.1 on subtractive manufacturing, which implements wire-cutting and mycelium blocks to cut a formwork for concrete that is left in place to become insulation. After the blocks are cut, beeswax is added to the surface of the mycelium to help prevent the concrete from fully penetrating the porous MBC shape. Any formwork not left in place is compostable, thus capitalising on the circularity of mycelium-based materials.

Another inherent material property of MBC is their favourable acoustical properties. The acoustical performance varies from product to product, but several companies and artists experiment with acoustical panelling systems made from mycelium-based materials. According to Pelletier et al. in "An evaluation study of pressure-compressed acoustic absorbers grown on agricultural by-Products", an acoustical ceiling tile with a density of 0.71 g/cm^3 and an MBC tile with a density of 0.42 g/cm^3 have an acoustical attenuation of 7.6 and 7.1, respectively [67]. Therefore, the MBC panels are a comparable alternative to commercial acoustical tiles.

Compressed mycelium panels are products that have yet to reach the consumer market, but they are a thoroughly researched variation of MBC. These panels are MBC that have been cold- or heat-pressed [4]. Once pressed, these materials become much denser as the air that once made the material light and airy is now lost in favour of a condensed material with improved structural properties. Cold-pressed MBC have a higher tensile strength and elastic modulus in comparison to non-pressed MBC, and result in a 2-fold density increase. Heat-pressed MBC further improve the tensile strength and elastic modulus, as well as having a more than 3-fold increase in density [30]. Compressed MBC, which have been heat-pressed, have panel-like characteristics and a compressive strength comparable to timber products such as oriented strand board (OSB) or medium density fibreboard (MDF). This is due to the softening of the lignin, which reacts to form new cross-links that increase the material strength [68]. In addition to the improved structural characteristics, heat-pressing also results in a decreased variation of density throughout panels and a consistent thickness.

MBC-based floor tiles made by the Italian company Mogu consist of a topcoat of a water-based paint, a cover layer of 67% bio-based polyurethane with oyster shells, and a bio-composite core layer of 100% bio-based high density fibreboard [69]. The product is made of mycelium in addition to other materials, does not contain petroleum-based plastics, and incorporates recycled products. It is possible to biodegrade the tiles at the end of their lifecycle, so long as correct industrial conditions are provided. This serves as an example of a product in which mycelium is developed in a way that it can be used in combination with different materials.

2.3.2. Architectural Projects

Architectural projects are constructions composed of more than one module of a mycelium-based material, or projects that are at a scale that is large enough to be considered an occupiable space. The following projects are categorised into in situ and prefabricated projects, as this differentiates between the fabrication techniques and geometries associated with either method of constructing with mycelium-based materials.

In Situ

Shell Mycelium Pavilion (2016)

The Shell Mycelium Pavilion by BEETLES 3.3 and Yassin Areddia Designs [70] used a triangular timber framework as the main structural geometry. The cavities of this geometry were filled with mycelium and then covered with coir pith consisting of coconut husk fibres. After some time, the top layer dried up and died, creating a protective layer over the growing mycelium. This project combined prefabricated modules and on-site assembly and growth of mycelium conducted in a non-sterile environment.

Monolithic Mycelium Experiments (2017–2018)

The monolithic mycelium experiments led by Jonathan Dessi-Olive [71] were a series of studies done to adapt mycelium to an in situ style implementation. Version 2 was an arch that used a waffle structure formwork made of laser cut cardboard onto which an inoculated substrate mixture was placed, which took over and inhabited the cardboard formwork to create an arch. The design of the arch implemented RhinoVAULT [72] to find a compression-only structure. The arch was allowed to grow for just under 5 days, and then the cover was removed to allow it to air dry for several weeks [71]. In the end, the small arch was able to carry the weight of a person and demonstrated both the importance of geometry and the possibility to grow in situ in a non-sterile environment.

Voxel Bio-Welding and Jammed Bio-Welding (2019)

This project was exhibited by David Benjamin of The Living on two occasions: in 2019 at the exhibition "Broken Nature" hosted by the Triennale of Milan, and again in 2019 at Centre Pompidou's exhibition La Fabrique du Vivant [73]. On both occasions the projects showcased the creation of large-scale demonstrators using bio-welding as the dominant fabrication technique. The structure was composed of pre-compressed, loose granular blocks held in place with a temporary formwork system. The structure was inoculated and incubated in situ for a specific period. Finally, the formwork was removed and the substrate remained in place due to the binding action of the mycelium.

Tallinn Architecture Biennale Installation (2022)

This project was a winning entry to the Tallin Architecture Biennale by Simulaa [74], which proposed a structure that grows as it is on display. Waste material from the local timber industry was harvested and combined with a biodegradable polymer, then 3D printed. The printed timber structure was then inoculated with the fungal strain that then consumes the structure and slowly replaces it over time until only the mycelium remains. The project prints the timber structure according to a generative algorithm using an industrial robot, illustrating the potential in combining new technologies with natural organisms [74].

Prefabricated

Mycotecture (2009)

This project is considered by many to be the origin of combining mycelium-based materials with prefabrication methods. The inventor-artist Philip Ross used a traditional arch shape and formed a relatively small construction that explored the relationship between people and grown architecture to expose people to the wonders of fungi [75]. The project used a small arch to create a relatively small, occupiable space. The bricks themselves grew edible mushrooms, which were used to brew tea for visitors [75].

Hy-Fi (2014)

The Hy-Fi pavilion by the Living was a 12 m tall pavilion that used 10,000 mycelium bricks of the MBC. These bricks were stacked and supplemented with an additional timber structure to stand. A limited number of moulds were used to produce the bricks, as the form was generated computationally to limit the number of unique moulds. The pavilion was left standing for three months. At the end of its life, all 10,000 bricks were composted and distributed to local gardens [76].

MycoTree (2017)

The MycoTree project is a collaboration between Karlsruhe Institute of Technology (KIT), Swiss Federal Institute of Technology (ETH) Zurich, Future Cities Laboratory in Singapore and Mycotech from Indonesia. This project utilises MBC grown in a mould in combination with connection made of laminated bamboo to create a load-bearing structure [4]. The moulds were digitally fabricated using readily available sheet material, and the overall design stems from the implementation of 3D graphic statics, a form-finding method for compression-only spatial structures [77]. The strength and stability of the structure are derived from its geometry, rather than the material. This project serves as a unique example of a structural implementation of MBC and demonstrates the advantages of integrating appropriate digital fabrication techniques with structurally informed computational design methods.

The Growing Pavilion (2019)

The Growing Pavilion was designed by set designer and artist Pascal Leboucq in collaboration with Erik Klarenbeek's studio Krown Design at Amsterdam studio Biobased Creations [78]. The pavilion showcases several bio-based materials, such as wood, hemp, mycelium, cattail, and cotton, with the facade panels being made of mycelium. The fungal strain in these panels is *Ganoderma*, and the substrate is composed of aspen wood and hemp. This substrate comes from the residual flows of local farmers in the Netherlands [79]. This is a non-structural implementation of MBC and is a unique project in that the MBC are used as facade cladding with wood substructure.

The Circular Garden (2019)

The CRA (Carlo Ratti Associati), in partnership with the global energy company Eni, developed this temporary installation for the Milan Design Week in 2019 [80]. The installation was grown over a two-month period to create a series of 60 4 m tall arches. The geometry of these arches was based on an inverted catenary curve to be a compression-only structure. If added up, the arches would be nearly 1 km tall. The arches compose a series of four architectural "open rooms" throughout the garden [80]. This project combined monolithic mycelium structures with prefabrication techniques, highlighting a new approach to MBC.

2.4. Summary

Most of the project-scale implementations of mycelium show a strong trend toward using MBC grown in moulds. In most cases, the structure is discretised into smaller components to be prefabricated in a controlled environment and assembled on-site. In these projects, MBC are then used as non- or semi-structural components, requiring an exoskeleton or an auxiliary structure to provide the main stability. This can be seen in the wood frame structure of the Hy-Fi pavilion and the Growing pavilion. MycoTree is a unique project in which mycelium is used in a load-bearing capacity, achieved through the unique geometry of the structure. However, the height of the structure was a major limiting factor due to the low mechanical properties of MBC.

In projects where the mycelium is grown on site, the size of the projects is considerably smaller due to logistical difficulties and the long duration associated with the growing of mycelium in exposed environments. Additionally, the inadequate material strength fails to accommodate the increasing self-weight as the size of the structure increases. As a result, mycelium-based material applications that have had the most success in reaching the construction industry are thermal or acoustical insulation panels. These applications demonstrate that the jump from product to project has not yet seen success in becoming a marketable construction method.

While the compostable nature of mycelium-based materials is an advantage in terms of circularity, it can also present several challenges. While these materials themselves are entirely compostable, some MBC packaging materials take only approximately 40 days to biodegrade [81]. This means that these materials can degrade quickly over time, which reduces their shelf life.

3. Challenges in Scaling up Mycelium

The following challenges are based on a review of the advantages and disadvantages of the previously discussed products and projects. Opportunities for addressing these challenges are speculated and proposed, and how they could broaden the potential of mycelium-based materials in large-scale architectural applications is discussed.

3.1. Structurally-Informed Design

The primary reason that mycelium-based materials have not been used as load-bearing elements is their inadequate structural properties. For example, the Living implemented a method of stacked bricks to construct the Hy-Fi pavilion. This construction technique is historically combined with masonry, as it is a relatively simple yet effective way to construct. However, according to Granta CES EduPack, a common clay brick has a compressive strength of 69–140 MPa [82], while MBC grown in a mould have a compressive strength of 0.35–0.75 MPa [83]. This shows the difference in strength between these two materials and highlights how these construction techniques for conventional materials are not appropriate or directly applicable to the construction of large, self-supporting structures made of mycelium-based materials.

The development and engineering of conventional materials such as MET, steel, and concrete are largely based on achieving material strength by increasing their allowable stresses. Although mycelium-based materials have numerous ecological benefits including repurposing of waste streams, biodegradability, and recyclability at end-of-life, their low structural strength significantly limits their applicability in large-scale products and projects. In order to scale up structural applications of materials with low tensile and bending strength, the geometry of the structure plays an important role in activating the material in compression only. With appropriate computational design tools, stability can be achieved through geometry rather than material strength. This allows relatively weaker materials such as mycelium to be used structurally and safely in large scale architectural applications. While this does not indicate that a multi-story masonry hi-rise can be constructed with mycelium with compression-only geometries, it does indicate that these materials can achieve a higher potential if given their own architectural vernacular according to their inherent properties to achieve small-to-medium scale constructions.

3.2. Region-Specific Material Profiles

Traditional materials such as steel, wood, and concrete have standardised mechanical properties and established industry regulations that are useful resources for architects and engineers. The disparity in material characteristics between different mycelium-based materials is in large part due to their inherent specificity to the given context and how the research on these materials is conducted in different regions across the globe. While most fungal strains–typically *Ganoderma lucidum*, *Pleurotus ostreatus*, or *Trametes versicolor*–are readily available across various geographic regions, the agricultural waste streams that are typically used as substrates are highly dependent on cultural, urban, and climatic conditions of the regions. For example, while rice husk is a common substrate used in mycelium-based materials for countries located in Asia [84], this waste material is not as common in Europe. This leads to material compositions with the same fungal strains but different substrates, and thus to divergent material properties. The relationship between fungal species and type of substrate is one of the essential factors affecting the research around mycelium-based materials. The customisability of recipes and production processes can result in different material characteristics (i.e., foams, sandwich composites, etc.) and products that are specifically tailored to the unique geographical and climatic constraints, as well as region-specific building regulations. This condition opens new research questions concerning:

- Identification and selection of nutritive and non-nutritive substrates available in a specific region in relation to their sector of origin (construction, agriculture or urban by-product);
- Determination and characterisation of nutritional profiles of substrates;
- Investigation, comparison, and classification of new fungal species potentially dominant in a region; and
- Investigation of genetic engineering to maximise the performance of region-specific recipes for mycelium-based materials.

Understanding the inner dynamics of interfacial bonding between substrates and fungal species may help to create new mycelium-based materials with more carefully customised properties such as the material strength, acoustical performance, thermal values, etc. For example, the more difficult it is for the mycelium to digest the substrate, the higher is the stiffness of the composite [85]. As research on mycelium-based materials expands into more regions with different substrate options, the generalisation of the relationship between substrates and growing conditions to material properties will aid in both the development of regional profiles and in a generalised understanding of the connection between fungal strains, substrates, and material properties.

3.3. Identification of Viable Waste Streams

In addition to the selection of substrate types, identifying sizable waste streams that are suitable as substrates for MBC production remains challenging. In order to enable and support circularity within the construction industry, the repurposing of construction waste would be optimal. However, the presence of additives such as synthetic adhesives and other chemical treatments involved in the production of natural materials such as MET makes it challenging to find suitable alternatives from the construction industry to waste from agricultural and culinary industries. Specifically in the commercialisation of mycelium-based materials, chemical treatments applied post-fabrication are currently a factor affecting the re-use and compostability of the material, as seen in the Mylo product.

Another key factor to consider is the creation of waste during the manufacturing process. While using prefabricated moulds as the dominant method of fabrication for MBC simplifies the manufacturing process on a larger scale, it can also create new waste streams if not taken into consideration when designing the overall manufacturing workflow.

3.4. New Mycelium Typologies

Typically, different typologies of mycelium-based materials are used in isolation from one another and are often combined with materials such as wood. However, there is a unique opportunity in exploring applications that combine the strengths of different material typologies. For example, PMM is a potential resource for laminating lignocellulosic materials together [86], while MBC, when left alive, can grow a thickened skin and join together separate blocks as shown in this mock-up installation [87]. Combining mycelium-based material typologies could potentially lead to the creation of new bio-composites with improved thermal, acoustic, and mechanical properties that cannot be achieved by a single typology alone.

3.5. Industrialisation of Biological Processes

The field of mycelium-based materials is currently confronted with two different tendencies: (1) the one dictated by the technological paradigm of automation, and (2) the one created by the use of biotechnologies in the built environment. Specifically, the renewed interest and development in biotechnologies applied to the built environment foresees a new material culture that can potentially overcome the separation between industrial processes and natural environmental cycles. The technologies and research developed within the context of mycelium-based material research, especially when confronted with applications on a large scale, come with a series of challenges that converge with the emergent field of construction biotechnologies. This new emergent field introduced by Ivanov and Stabnikov [88], investigates new microbial-mediated construction processes and microbial production of construction materials. Despite the advancement, the chemical, manufacturing, and construction industries remain somewhat reluctant to adopt bio-based or bio-inspired practices that could replace current oil-based processes. Both civil and environmental engineering are conspicuously developing a variety of solutions taking into consideration fundamental knowledge of microbiology. However, this type of knowledge is often a general case of "applied biological science" rather than firmly weaved and developed within design and engineering problems. In this regard, there are three main considerations to consider when scaling up bio-based material systems.

Scalability

Manufacturing process for new materials with high impact and a growing market value comes with an inherent potential for large-scale industrialisation. The case of mycelium-based materials shows a biological process (similar to other industrial fermentation processes, such as cheese, beer, etc.) characterised by various complex variables that include long cycles of production, risk of contamination, and a complex multi-step manufacturing process. The lack of consistency and reproducibility of a specific material profile, therefore, prevents the scaling up of mycelium in both size and quantity to reach an industry-standard level of product certification. Companies such as Ecovative and Biohm have started to address some of these considerations using genetic engineering, optimisation of fungal strains, and introduction of additional micro-organisms in the fabrication loop.

Reproducibility

One of the key factors facilitating the development of new materials in engineering disciplines is the adoption of a comprehensive and reliable set of standards. Such levels of standardisation are difficult to achieve in the context of biological systems, which often have a high degree of uncertainty. This condition does not refer exclusively to the complexity of the biological systems themselves, but to the availability of infrastructures that have the capacity to support the development of standardised practices. Specifically, one of the main limitations to overcome is the conversion of the laboratory scale protocols into much larger and industrial-size equivalents.

Automation

The use of machines and new manufacturing technologies in highly specialised and controlled environments is an important challenge shared by industries ranging from biomedicine, chemistry, and biotechnology to architecture. Weariful and repetitive tasks that are prone to error when done by humans are already being accomplished in many laboratories and industries with robots. This includes tasks such as liquid handling, as well as additive and subtractive manufacturing. The use of machines in environments such as the ones needed to produce mycelium-based materials creates a condition that intrinsically minimises human intervention, ensuring a safe environment during the fabrication process.

3.6. Agile In Situ Setup and Applications

In large part, mass-produced mycelium-based materials that have seen commercial success consist of MBC grown in moulds. This fabrication method relies on a closed, sterile environment. Growing mycelium in situ without a controlled environment has only seen success on a small-scale, while 3D printing has similarly succeeded on a small scale and within a controlled environment. Scaling up in situ applications requires a better understanding of preventing contamination, as well as the development of techniques that do not rely on such a restrictive, closed-off space. Whereas prefabricated MBC can rely on moulds to ensure a closed environmental system, when scaling up there is a need for this controlled environment to be larger than just the mould of one module.

Implementation of new fabrication techniques with mycelium-based materials in a controlled, sterile environment at an architectural scale presents a number of challenges. For example, the use of a robotic arm for mycelium prototyping requires a cleanroom with accurate humidity and temperature control. This constraint is an opportunity to create a walk-in prototyping space able to facilitate the production of large-format prefabricated and highly customised parts. These environments can potentially be located adjacent to the construction site, ultimately reducing costs of transport.

Within this framework, it may be even possible to consider these controlled environments as new construction sites, wherein prefabricated parts and substrates are simply installed and subsequently bound together using the binding capacity of the mycelial network. By approaching the controlled environment as the construction site, it is possible to develop an in situ bio-assembly process in which building parts are naturally bound and mechanically enhanced.

3.7. Cross-Disciplinary Research

The relevance of mycelium-based materials, as demonstrated in this paper, is not limited to their original fields of microbiology and mycology. Thanks to the work of researchers across many disciplines, their relevance has expanded into architecture. These fields are practically unrelated by tradition and in practice. However, the research and development to explore bio-based materials in general have drawn them closer together. While this increases the scope of research and development being done with these materials, it also means that this information does not easily cross over into the other fields. Furthermore, whenever that information does make the jump between disciplines, it can be difficult for an architect, for example, to discern the findings of a microbiologist. Therefore, the cohesive understanding of latest findings and material developments presents its own challenge as many factors prevent the knowledge from being distributed among experts from different industries.

As research on mycelium-based materials continues to exponentially increase year after year, the accessibility of the research also increases. This indicates that the visibility of research increases, and thus the dissemination of information across fields of research will naturally increase. Active efforts to regularly collaborate across different industries can best promote the sharing of findings and development of new technologies. The conscientious integration of the advances from different fields can also further unify the industries and spur new research. In this way, several other previously mentioned challenges in scaling up applications of mycelium-based materials can be addressed, such as the generalisation of how substrates, fungal strains, and environmental conditions affect the growth of mycelium. A further integrated approach can also begin to establish new, open-source practices for goals such as mass-production to expedite the scaling up of mycelium-based materials.

4. Conclusions

In this paper, we reviewed the latest advancements in mycelium-related research and applications from academia and industry to understand the diversity of perspectives and advancements. The advantages of using mycelium-based materials in the construction industry can incentivise the birth of new protocols, consider waste as a resource, and develop new methodologies to evaluate and assess the impact of the built environment. Thus far, mycelium-based materials have been most successfully implemented at the scale of small commercial products and commodities, with a number of challenges barring the expansion of applications to architecture. Coupling a material such as MBC with appropriate computational design and digital fabrication methods, while also developing mass-production techniques and local material profile strategies, has the potential to successfully scale up architectural applications of mycelium-based materials. The challenges and research gaps we identified are cross-disciplinary in nature and will require a close collaboration on material research, integrated design methods, appropriate digital fabrication techniques, and on-site implementation strategies from microbiologists, material scientists, manufacturers, and designers alike.

Author Contributions: S.B. and T.D. contributed equally to this paper. Conceptualization, S.B., T.D. and J.L.; investigation, S.B., T.D. and J.L.; writing—original draft preparation, S.B., T.D. and J.L.; writing—review and editing, S.B., T.D. and J.L.; visualization, S.B., T.D. and J.L.; supervision, J.L., T.V.M., B.D. and P.B.; project administration, J.L.; funding acquisition, B.D. and P.B. All authors have read and agreed to the published version of the manuscript.

Funding: This research is made possible by the ETH Singapore's Future Cities Laboratory (FCL) Global Project, funded by the National Research Foundation (NRF) of Singapore. This work is a part of a collaborative research project "Urban BioCycles", with Dirk Hebel of Karlsruhe Institute of Technology and Hortense Le Ferrand of Nanyang Technological University in Singapore.

Institutional Review Board Statement: Not applicable.

Informed Consent Statement: Not applicable.

Data Availability Statement: Not applicable.

Conflicts of Interest: The authors declare no conflict of interest.

References

1. Miyatake, Y. Technology Development and Sustainable Construction. *J. Manag. Eng.* **1996**, *12*, 23–27. [CrossRef]
2. Heisel, F.; Rau-Oberhuber, S. Calculation and Evaluation of Circularity Indicators for the Built Environment Using the Case Studies of UMAR and Madaster. *J. Clean. Prod.* **2019**, *243*, 118482. [CrossRef]
3. Ding, G.K.C. Life Cycle Assessment (LCA) of Sustainable Building Materials: An Overview. In *Eco-Efficient Construction and Building Materials*; Pacheco-Torgal, F., Cabeza, L.F., Labrincha, J., de Magalhães, A., Eds.; Woodhead Publishing: Sawston, UK, 2014; pp. 38–62. ISBN 978-0-85709-767-5.
4. Javadian, A.; Ferrand, H.L.; Hebel, D.E.; Saeidi, N. Application of Mycelium-Bound Composite Materials in Construction Industry: A Short Review. *SOJ Mater. Sci. Eng.* **2020**, *7*, 1–9.
5. Todorovic, T.; Norström, E.; Khabbaz, F.; Brücher, J.; Malmström, E.; Fogelström, L. A Fully Bio-Based Wood Adhesive Valorising Hemicellulose-Rich Sidestreams from the Pulp Industry. *Green Chem.* **2021**, *23*, 3322–3333. [CrossRef]
6. Kim, Y.; Ruedy, D. Mushroom Packages An Ecovative Approach in Packaging Industry. In *Handbook of Engaged Sustainability*; Springer: Cham, Switzerland, 2019; pp. 1–25. ISBN 978-3-319-53121-2.
7. Hebel, D.E.; Heisel, F. *Cultivated Building Materials: Industrialized Natural Resources for Architecture and Construction*; Birkhäuser: Basel, Switzerland, 2017; ISBN 978-3-0356-0892-2.
8. Jędrzejczak, P.; Collins, M.N.; Jesionowski, T.; Klapiszewski, Ł. The Role of Lignin and Lignin-Based Materials in Sustainable Construction—A Comprehensive Review. *Int. J. Biol. Macromol.* **2021**, *187*, 624–650. [CrossRef]
9. Bartnicki-Garcia, S. Cell Wall Chemistry, Morphogenesis, and Taxonomy of Fungi. *Annu. Rev. Microbiol.* **1968**, *22*, 87–108. [CrossRef]
10. Islam, M.R.; Tudryn, G.; Bucinell, R.; Schadler, L.; Picu, R.C. Morphology and Mechanics of Fungal Mycelium. *Sci. Rep.* **2017**, *7*, 13070. [CrossRef]
11. MycoComposite™. Available online: https://ecovativedesign.com/mycocomposite (accessed on 12 October 2021).
12. Bolt Threads. Available online: https://boltthreads.com/ (accessed on 1 November 2021).
13. BIOHM | Mycelium Insulation. Available online: https://www.biohm.co.uk/mycelium (accessed on 9 November 2021).

14. Jones, M.; Mautner, A.; Luenco, S.; Bismarck, A.; John, S. Engineered Mycelium Composite Construction Materials from Fungal Biorefineries: A Critical Review. *Mater. Des.* **2020**, *187*, 108397. [CrossRef]
15. Girometta, C.; Picco, A.M.; Baiguera, R.M.; Dondi, D.; Babbini, S.; Cartabia, M.; Pellegrini, M.; Savino, E. Physico-Mechanical and Thermodynamic Properties of Mycelium-Based Biocomposites: A Review. *Sustainability* **2019**, *11*, 281. [CrossRef]
16. Vandelook, S.; Elsacker, E.; Van Wylick, A.; De Laet, L.; Peeters, E. Current State and Future Prospects of Pure Mycelium Materials. *Fungal Biol. Biotechnol.* **2021**, *8*, 20. [CrossRef]
17. Elsacker, E.; Søndergaard, A.; Van Wylick, A.; Peeters, E.; De Laet, L. Growing Living and Multifunctional Mycelium Composites for Large-Scale Formwork Applications Using Robotic Abrasive Wire-Cutting. *Constr. Build. Mater.* **2021**, *283*, 122732. [CrossRef]
18. Renger, B.C.; Birkeland, J.L.; Midmore, D.J. Net-Positive Building Carbon Sequestration. *Build. Res. Inf.* **2015**, *43*, 11–24. [CrossRef]
19. Jones, M.; Huynh, T.; Dekiwadia, C.; Daver, F.; John, S. Mycelium Composites: A Review of Engineering Characteristics and Growth Kinetics. *J. Bionanosci.* **2017**, *11*, 241–257. [CrossRef]
20. Attias, N.; Danai, O.; Abitbol, T.; Tarazi, E.; Ezov, N.; Pereman, I.; Grobman, Y.J. Mycelium Bio-Composites in Industrial Design and Architecture: Comparative Review and Experimental Analysis. *J. Clean. Prod.* **2020**, *246*, 119037. [CrossRef]
21. Almpani-Lekka, D.; Pfeiffer, S.; Schmidts, C.; Seo, S. A Review on Architecture with Fungal Biomaterials: The Desired and the Feasible. *Fungal Biol. Biotechnol.* **2021**, *8*, 17. [CrossRef]
22. Cerimi, K.; Akkaya, K.C.; Pohl, C.; Schmidt, B.; Neubauer, P. Fungi as Source for New Bio-Based Materials: A Patent Review. *Fungal Biol. Biotechnol.* **2019**, *6*, 17. [CrossRef]
23. Dixit, S.; Stefańska, A.; Singh, P. Manufacturing Technology in Terms of Digital Fabrication of Contemporary Biomimetic Structures. *Int. J. Constr. Manag.* **2021**, 1–9. [CrossRef]
24. Yang, L.; Park, D.; Qin, Z. Material Function of Mycelium-Based Bio-Composite: A Review. *Front. Mater.* **2021**, *8*, 737377. [CrossRef]
25. Sydor, M.; Bonenberg, A.; Doczekalska, B.; Cofta, G. Mycelium-Based Composites in Art, Architecture, and Interior Design: A Review. *Polymers* **2022**, *14*, 145. [CrossRef]
26. Manan, S.; Ullah, M.W.; Ul-Islam, M.; Atta, O.M.; Yang, G. Synthesis and Applications of Fungal Mycelium-Based Advanced Functional Materials. *J. Bioresour. Bioprod.* **2021**, *6*, 1–10. [CrossRef]
27. Gou, L.; Li, S.; Yin, J.; Li, T.; Liu, X. Morphological and Physico-Mechanical Properties of Mycelium Biocomposites with Natural Reinforcement Particles. *Constr. Build. Mater.* **2021**, *304*, 124656. [CrossRef]
28. Shakir, M.A.; Azahari, B.; Yusup, Y.; Yhaya, M.F.; Salehabadi, A.; Ahmad, M.I. Preparation and Characterization of Mycelium as A Bio-Matrix in Fabrication of Bio-Composite. *J. Adv. Res. Fluid Mech. Therm. Sci.* **2020**, *65*, 253–263.
29. Xia, X.C.; Chen, X.W.; Zhang, Z.; Chen, X.; Zhao, W.M.; Liao, B.; Hur, B. Effects of Porosity and Pore Size on the Compressive Properties of Closed-Cell Mg Alloy Foam. *J. Magnes. Alloys* **2013**, *1*, 330–335. [CrossRef]
30. Appels, F.; Camere, S.; Montalti, M.; Karana, E.; Jansen, K.M.B.; Dijksterhuis, J.; Krijgsheld, P.; Wosten, H. Fabrication Factors Influencing Mechanical, Moisture- and Water-Related Properties of Mycelium-Based Composites. *Mater. Des.* **2019**, *161*, 64–71. [CrossRef]
31. Appels, F.V.W.; Dijksterhuis, J.; Lukasiewicz, C.E.; Jansen, K.M.B.; Wösten, H.A.B.; Krijgsheld, P. Hydrophobin Gene Deletion and Environmental Growth Conditions Impact Mechanical Properties of Mycelium by Affecting the Density of the Material. *Sci. Rep.* **2018**, *8*, 4703. [CrossRef] [PubMed]
32. Karana, E.; Blauwhoff, D.; Hultink, E.-J.; Camere, S. When the Material Grows: A Case Study on Designing (with) Mycelium-Based Materials. *Int. J. Des.* **2018**, *12*, 119–136.
33. Kaplan-Bie, J.H.; Bonesteel, I.T.; Greetham, L.; McIntyre, G.R. Increased Homogeneity of Mycological Biopolymer Grown into Void Space. U.S. Patent 11,266,085, 8 March 2022.
34. Greetham, L.; McIntyre, G.R.; Bayer, E.; Winiski, J.; Araldi, S. Mycological Biopolymers Grown in Void Space Tooling. U.S. Patent 11,277,979, 22 March 2022.
35. O'brien, M.A.; Carlton, A.; Mueller, P. Methods of Mycological Biopolymer Production. U.S. Patent Application 16/773,272, 27 July 2020.
36. Gandia, A.; van den Brandhof, J.G.; Appels, F.V.W.; Jones, M.P. Flexible Fungal Materials: Shaping the Future. *Trends Biotechnol.* **2021**, *39*, 1321–1331. [CrossRef] [PubMed]
37. Shokrkar, H.; Ebrahimi, S.; Zamani, M. A Review of Bioreactor Technology Used for Enzymatic Hydrolysis of Cellulosic Materials. *Cellulose* **2018**, *25*, 6279–6304. [CrossRef]
38. Szilvay, G.; Laine, C.; Arias Barrantes, M.; Suhonen, A.; Boer, H.; Penttilä, M.; Ahokas, P. Methods of Making Non-Woven Materials from Mycelium. International Patent Application PCT/FI2020/050875, 30 December 2020.
39. Bayer, E.; McIntyre, G. Method for Producing Grown Materials and Products Made Thereby. U.S. Patent Application Publication 15/266,640, 15 September 2016.
40. Holt, G.A.; Mcintyre, G.; Flagg, D.; Bayer, E.; Wanjura, J.D.; Pelletier, M.G. Fungal Mycelium and Cotton Plant Materials in the Manufacture of Biodegradable Molded Packaging Material: Evaluation Study of Select Blends of Cotton Byproducts. *J. Biobased Mater. Bioenergy* **2012**, *6*, 431–439. [CrossRef]
41. Van Kuijk, S.J.A.; Sonnenberg, A.S.M.; Baars, J.J.P.; Hendriks, W.H.; Cone, J.W. The Effect of Particle Size and Amount of Inoculum on Fungal Treatment of Wheat Straw and Wood Chips. *J. Anim. Sci. Biotechnol.* **2016**, *7*, 39. [CrossRef]

42. Arifin, Y.H.; Yusuf, Y. Mycelium Fibers as New Resource for Environmental Sustainability. *Procedia Eng.* **2013**, *53*, 504–508. [CrossRef]
43. Schaak, D. Bio-Manufacturing Process. U.S. Patent Application 16/363052, 25 March 2019.
44. Mathias, L.; Jipa, A. The Living Column. In *MAS DFAB Thesis Publications*; ETH Zurich: Zürich, Switzerland, 2017.
45. Peek, N. Mycelium Milling. Available online: http://infosyncratic.nl/weblog/2011/02/14/mycelium-milling/ (accessed on 26 October 2021).
46. Lazaro, E.; Vega, K. From Plastic to Biomaterials: Prototyping DIY Electronics with Mycelium. In *Proceedings of the UbiComp/ISWC '19 Adjunct: Adjunct Proceedings of the 2019 ACM International Joint Conference on Pervasive and Ubiquitous Computing and Proceedings of the 2019 ACM International Symposium on Wearable Computers*; ACM: London, UK, 2019.
47. Rigobello, A.; Ayres, P. Mycelium-Based Composites as Two-Phase Particulate Composites: Compressive Behaviour of Anisotropic Designs. Available online: https://www.researchsquare.com/article/rs-943974/v1 (accessed on 30 November 2021).
48. Islam, M.R.; Tudryn, G.; Bucinell, R.; Schadler, L.; Picu, R.C. Stochastic Continuum Model for Mycelium-Based Bio-Foam. *Mater. Des.* **2018**, *160*, 549–556. [CrossRef]
49. Singh, T.; Kumar, S.; Sehgal, S. 3D Printing of Engineering Materials: A State of the Art Review. *Mater. Today Proc.* **2020**, *28*, 1927–1931. [CrossRef]
50. Bose, S.; Koski, C.; Vu, A.A. Additive Manufacturing of Natural Biopolymers and Composites for Bone Tissue Engineering. *Mater. Horiz.* **2020**, *7*, 2011–2027. [CrossRef]
51. Murphy, S.V.; Atala, A. 3D Bioprinting of Tissues and Organs. *Nat. Biotechnol.* **2014**, *32*, 773–785. [CrossRef]
52. Fraunhofer UMSICHT Projekt-Steckbrief. Available online: https://fungifacturing.de/fungi-factoring/projekt-steckbrief/ (accessed on 10 March 2022).
53. Yildirim, D. Zoetic Morphologies 3D Printed Active Wall Systems through Mycelial Biomass. Available online: https://www.iaacblog.com/programs/zoetic-morphologies/ (accessed on 10 March 2022).
54. Parthy, K. Polymer Mixture for 3D Printing for Creating Objects with Pore Structures. DE Patent Application DE102013011243A1, 8 January 2015.
55. Alima, N.; Snooks, R.; McCormack, J. Bio Scaffolds. In Proceedings of the 2021 DigitalFUTURES, Shanghai, China, 1 January 2022; pp. 316–329.
56. Myco Mensa: Mycelium Table. Available online: http://www.richard-beckett.com/myco-mensa-mycelium-table/ (accessed on 11 February 2022).
57. Ji, S.; Guvendiren, M. Recent Advances in Bioink Design for 3D Bioprinting of Tissues and Organs. *Front. Bioeng. Biotechnol.* **2017**, *5*, 23. [CrossRef]
58. Soh, E.; Chew, Z.Y.; Saeidi, N.; Javadian, A.; Hebel, D.; Le Ferrand, H. Development of an Extrudable Paste to Build Mycelium-Bound Composites. *Mater. Des.* **2020**, *195*, 109058. [CrossRef]
59. Goidea, A.; Andreen, D.; Floudas, D. *Pulp Faction: 3D Printed Material Assemblies through Microbial Biotransformation*; UCL Press: London, UK, 2020; pp. 42–49.
60. Bio Ex-Machina. Available online: https://www.corpuscoli.com/projects/bio-ex-machina/ (accessed on 10 March 2022).
61. Ecovative. Available online: https://ecovative.com (accessed on 8 February 2022).
62. Bayer, E.; Winiski, J.M.; Lucht, M.J.; Mueller, P.J.; Mcintyre, G.R.; O'brien, M.A. An Open-Cell Mycelium Foam and Method of Making Same. U.S. Patent Application 16/444,354, 26 December 2019.
63. Wijayarathna, E.R.K.B.; Mohammadkhani, G.; Soufiani, A.M.; Adolfsson, K.H.; Ferreira, J.A.; Hakkarainen, M.; Berglund, L.; Heinmaa, I.; Root, A.; Zamani, A. Fungal Textile Alternatives from Bread Waste with Leather-like Properties. *Resour. Conserv. Recycl.* **2022**, *179*, 106041. [CrossRef]
64. Ortiz, D.A. How Fungus and Sweat Could Transform Martian Exploration. Available online: https://www.bbc.com/future/article/20181031-how-fungus-and-sweat-could-transform-martian-exploration (accessed on 21 February 2022).
65. Klarenbeek & Dros Designers of the Unusual. Available online: https://www.ericklarenbeek.com/ (accessed on 10 March 2022).
66. Piórecka, N. MYCOsella—Growing the Mycelium Chair. Available online: https://issuu.com/nataliapiorecka/docs/dissertation_project_ba_architectur (accessed on 10 March 2022).
67. Pelletier, M.G.; Holt, G.A.; Wanjura, J.D.; Lara, A.J.; Tapia-Carillo, A.; McIntyre, G.; Bayer, E. An Evaluation Study of Pressure-Compressed Acoustic Absorbers Grown on Agricultural by-Products. *Ind. Crops Prod.* **2017**, *95*, 342–347. [CrossRef]
68. Bouajila, J.; Limare, A.; Joly, C.; Dole, P. Lignin Plasticization to Improve Binderless Fiberboard Mechanical Properties. *Polym. Eng. Sci.* **2005**, *45*, 809–816. [CrossRef]
69. Mogu. Available online: https://mogu.bio/ (accessed on 8 October 2021).
70. Shell Mycelium Pavillion. Available online: https://www.archdaily.com/tag/shell-mycelium-pavillion (accessed on 20 October 2021).
71. Desi-Olive, J. Tactical Mycelium 1 & 2. Available online: https://jdovaults.com/Tactical-Mycelium-1-2 (accessed on 1 March 2022).
72. Rippmann, M.; Lachauer, L.; Block, P. RhinoVAULT—Interactive Vault Design. *Int. J. Space Struct.* **2012**, *27*, 219–230. [CrossRef]
73. Sommariva, E.; Mairs, J. 5 Works of "Restorative Design" from Broken Nature: Design Takes on Human Survival. Available online: https://www.domusweb.it/en/design/gallery/2019/02/28/5-installations-to-see-at-triennales-broken-nature-exhibition-.html (accessed on 10 March 2022).

74. Simulaa. Available online: https://simulaa.com (accessed on 1 March 2022).
75. Mok, K. Mycotecture: Building With Mushrooms? This Inventor Says Yes. Available online: https://www.treehugger.com/mycotecture-mushroom-bricks-philip-ross-4857225 (accessed on 14 March 2022).
76. Hy-Fi. Available online: https://urbannext.net/hy-fi/ (accessed on 21 February 2022).
77. Heisel, F.; Lee, J.; Schlesier, K.; Rippmann, M.; Saeidi, N.; Javadian, A.; Nugroho, R.; Mele, T.; Block, P.; Hebel, D. Design, Cultivation and Application of Load-Bearing Mycelium Components: The MycoTree at the 2017 Seoul Biennale of Architecture and Urbanism. *Int. J. Sustain. Energy Dev.* **2018**, *6*, 296–303. [CrossRef]
78. Pownall, A. Pavilion Grown from Mycelium Acts as Pop-Up Performance Space at Dutch Design Week. Available online: https://www.dezeen.com/2019/10/29/growing-pavilion-mycelium-dutch-design-week/ (accessed on 25 October 2021).
79. The Growing Pavilion. Available online: https://thegrowingpavilion.com/ (accessed on 1 March 2022).
80. The Circular Garden. Available online: https://carloratti.com/project/the-circular-garden/ (accessed on 1 March 2022).
81. How Can Mushrooms Help Solve the Issue of Single-Use Plastic? Available online: https://www.seedlipdrinks.com/en-gb/journal/mycelium-101/ (accessed on 14 March 2022).
82. *Ansys Granta EduPack Software*, ANSYS Inc.: Cambridge, UK, 2022.
83. Yang, Z.; Zhang, F.; Still, B.; White, M.; Amstislavski, P. Physical and Mechanical Properties of Fungal Mycelium-Based Biofoam. *J. Mater. Civ. Eng.* **2017**, *29*, 04017030. [CrossRef]
84. Singh, B. Rice Husk Ash. In *Waste and Supplementary Cementitious Materials in Concrete*; Siddique, R., Cachim, P., Eds.; Woodhead Publishing Series in Civil and Structural Engineering; Woodhead Publishing: Sawston, UK, 2018; pp. 417–460. ISBN 978-0-08-102156-9.
85. Haneef, M.; Ceseracciu, L.; Canale, C.; Bayer, I.S.; Heredia-Guerrero, J.A.; Athanassiou, A. Advanced Materials From Fungal Mycelium: Fabrication and Tuning of Physical Properties. *Sci. Rep.* **2017**, *7*, 41292. [CrossRef] [PubMed]
86. Sun, W.; Tajvidi, M.; Howell, C.; Hunt, C.G. Functionality of Surface Mycelium Interfaces in Wood Bonding. *ACS Appl. Mater. Interfaces* **2020**, *12*, 57431–57440. [CrossRef] [PubMed]
87. Dahmen, J. Mycelium Mockup. Available online: https://sala.ubc.ca/work/mycelium-mockup (accessed on 1 March 2022).
88. Ivanov, V.; Stabnikov, V. *Construction Biotechnology*; Green Energy and Techology; Spring: Singapore, 2017; ISBN 978-981-10-1444-4.

 *biomimetics*

Article

Effect of Composition Strategies on Mycelium-Based Composites Flexural Behaviour

Adrien Rigobello *, Claudia Colmo and Phil Ayres

Centre for IT and Architecture, Royal Danish Academy, 1435 Copenhagen, Denmark; ccol@kglakademi.dk (C.C.); phil.ayres@kglakademi.dk (P.A.)
* Correspondence: arig@kglakademi.dk

Abstract: Mycelium-based composites (MBC) are a promising class of relatively novel materials that leverage mycelium colonisation of substrates. Being predicated on biological growth, rather than extraction based material sourcing from the geosphere, MBC are garnering attention as potential alternatives for certain fossil-based materials. In addition, their protocols of production point towards more sustainable and circular practices. MBC remains an emerging practice in both production and analysis of materials, particularly with regard to standardisation and repeatability of protocols. Here, we show a series of flexural tests following ASTM D1037, reporting flexural modulus and flexural modulus of rupture. To increase the mechanical proprieties, we contribute with an approach that follows the composition strategy of reinforcement by considering fibre topology and implementing structural components to the substrate. We explore four models that consist of a control group, the integration of inner hessian, hessian jacketing and rattan fibres. Apart from the inner hessian group, the introduction of rattan fibres and hessian jacketing led to significant increases in both strength and stiffness (α = 0.05). The mean of the flexural modulus for the most performative rattan series (1.34 GPa) is still close to three times lower than that of Medium-Density Fibreboard, and approximately 16 times lower in modulus of rupture. A future investigation could focus on developing a hybrid strategy of composition and densification so as to improve aggregate interlocking and resulting strength and stiffness.

Keywords: mycelium-based composite; biomaterials; natural composites

Citation: Rigobello, A.; Colmo, C.; Ayres, P. Effect of Composition Strategies on Mycelium-Based Composites Flexural Behaviour. *Biomimetics* **2022**, *7*, 53. https://doi.org/10.3390/biomimetics7020053

Academic Editor: Bernhard Schuster

Received: 15 March 2022
Accepted: 21 April 2022
Published: 25 April 2022

Publisher's Note: MDPI stays neutral with regard to jurisdictional claims in published maps and institutional affiliations.

1. Introduction

Mycelium-based composites (MBC) are a novel field of material development leveraging wood-decaying basidiomycota to bind lignocellulosic particulate media via Solid-State Fermentation (SSF) [1]. Since 2006 and the establishment of the first commercial venture for MBC products (Ecovative Design, LLC, Green Island, NY, USA) the development of these materials has been supporting packaging and insulation use cases. Both these cases are relevant with a state-of-the-art that has favoured local replication of the production process across the globe and across institutions before advancing studies of the material mechanical model for assessing alternative use cases. The variety of versatile fungi that can be used, coupled with the extensive variety of lignocellulosic substrates with respect to aggregate geometries and chemical profiles, supports a wide design space. The spread of the MBC state-of-the-art regarding only reported stiffness and strength is a strong testimony to this, while research has yet to lead to functional poles definition [1]. Furthermore, the principal advantage of MBC designs lies in their higher potential for upcycling agricultural wastes, leading to the production of resource conscious and biodegradable materials with a low environmental impact [2] that can contribute to shifting towards a circular economy.

Across the literature, three principal design strategies for modifying MBC mechanical behaviour have been identified: densification, composition, and supplementation (targeting mycelium properties, based on chemical tuning of the substrate) [1]. Mechanically,

MBC have been investigated mainly as per their compressive and flexural behaviour [1]. Densification is a strategy for stiffening the composite by increasing the density of the substrate by dense packing, cold or hot pressing. Densified specimens have been reportedly leading to an increase of flexural modulus between 27 and 72-fold, and 4 to 14.5-fold increase in flexural strength for a densification from 100–130 kg/m^3 to 350–390 kg/m^3 [3]. While staying at the lower end of MBC densities, another study reported an increase of 4.4-fold in flexural strength while increasing composite density by a factor 1.4 (from 102 to 141 kg/m^3) [4]. Composition is a second strategy for modifying MBC behaviour by adding structural components to the substrate, including, for instance, orientated fibres or textiles. Modifying particle properties is also considered an instance of the composition strategy [1]. The MBC state-of-the-art largely considers monolithic and homogeneous composites, besides a few study groups investigating jute type materials in sandwich composite reinforcement, and wood panels introduction [5–7]. Composition strategies by arming or particulate design are therefor a scarcely studied area of material development still, while the lower stiffness of the mycelial matrix (the tensile modulus of *Pleurotus ostreatus* and *Ganoderma lucidum* species is reportedly in the 4–28 MPa range [8]) as compared to, for instance, American beech wood (*Fagus grandifolia*) elastic modulus of 11.9 GPa at 12% moisture content (MC) [9], suggests that the design of the composite dispersed phase can considerably influence the final composite mechanical stress response. The significance of composition strategies over the composite compressive behaviour has been reported previously, both for aggregate size and fibre placement [10].

Three studies have been investigating the effect of composition strategy on MBC flexural behaviour. A hybrid protocol using both composition and densification techniques has been employed by using blend bacterial cellulose (BC) produced by a *Komagataeibacter xylinus* colony. The BC cellulose fibrils were mixed to the hemp fibres serving as principal substrate, and set to be colonised by a *Trametes versicolor* species [11]. This method targets an increase in aggregates binding. BC introduction did not result in a statistically significant difference. Nonetheless, pressing temperature change from 70 °C to 200 °C led to a 1.42-fold increase in stiffness and a factor 1.54 increase in flexural strength to reach 2.94 MPa. The two remaining relevant studies have studied the effect of textile lamination on top and bottom surfaces of a composite. One of them reported flexural moduli in 4.65–6.57 MPa, with moduli of rupture in 0.76–1.5 MPa, without reporting on the statistical significance of the different lamination materials used neither on the fungal species that the material was cultivated with [5]. Results of a greater stiffness have been reported with the introduction of top and bottom carbon-fibre layers, leading to a modulus of 296 MPa for a modulus of rupture of 2.9 MPa. This last study also investigated bamboo lamination and saw a stiffness increase of a factor 2.18, while flexural strength dropped to 0.31 times the carbon fibre laminated composite group. No density was reported for the specimens in this study, neither proprietary supplements to the substrate [12].

In the study reported here, we focus on the effect of composition over the flexural behaviour with the introduction of orientated fibres and hessian. Following ASTM D1037, we report on three point bending for three categories of composition: the embedding of a hessian arming at mid-thickness, hessian jacketing, and rattan arming in specimen length. Apart from the inner hessian group (BM_I), the introduction of rattan fibres and hessian jacketing led to significant increases in both strength and stiffness ($\alpha = 0.05$).

2. Materials and Methods

Referenced standards for evaluating flexural properties of MBC in the state-of-the-art are presented in Table 1. ASTM D1037 was used as the standard evaluation method [13]. We use this standard because it is the most referred set of guidelines in MBC development, covers various tests and refers to the fittest material model. We report on three point bending. To this end, four specimen groups were designed:

- Control: no fibre,
- Inner hessian: a flat layer of hessian was introduced at mid-thickness,

- Hessian jacketing: a hessian jacketing was introduced in the length,
- Rattan: five parallel rattan fibres of 5 mm diameter by 500 mm, separated by 8 mm, were introduced in the length and at mid-thickness.

The wet specimens are parallelepipeds of 520 mm × 72 mm, with a nominal thickness of 20 mm. The width and thickness were not affected by the desiccation, but the length of the dry specimens varied between reinforcement strategies and shrank on average by 3.5% in the control group, 2.7% in the inner hessian group, 1.7% in the hessian jacketing group, and 0.8% in the rattan group. The distance of the top and bottom specimen surfaces to the neutral axis of stress in flexion was of 10 mm. Six replicates were produced and tested for each of the specimen types. Load testing was performed on a Mecmesin MultiTest-dV testing bench equipped with a 2500 N load sensor, with a loading speed of 10.0 mm/min. Flexural modulus and modulus of rupture were calculated following ASTM D1037. The four specimen groups specifics are illustrated in Figure 1.

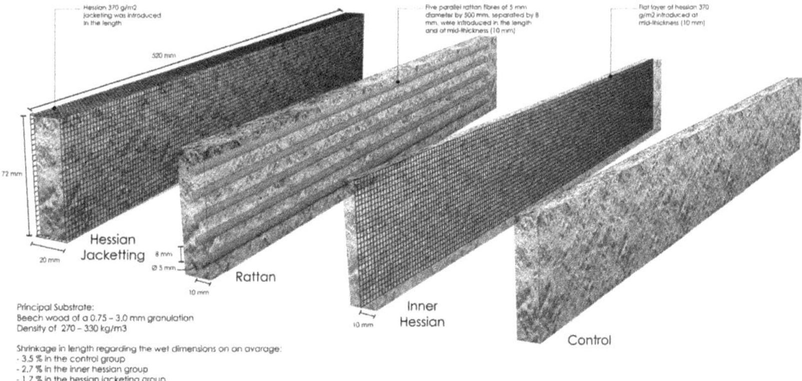

Figure 1. Fibre placement strategies (left to right): hessian jacketing (BM_H), rattan fibres (BM_R), inner hessian (BM_I), control (BM).

Table 1. Referenced standards for flexural characterisation in the MBC state-of-the-art.

Standard	Designation	Refs.
ASTM C203	Standard Test Methods for Breaking Load and Flexural Properties of Block-Type Thermal Insulation.	[14]
ASTM C393	Standard Test Method for Core Shear Properties of Sandwich Constructions by Beam Flexure	[5]
ASTM D7250	Standard Practice for Determining Sandwich Beam Flexural and Shear Stiffness	[5]
ASTM D1037	Standard Test Methods for Evaluating Properties of Wood-Base Fiber and Particle Panel Materials.	[15–17]
ISO 16978	Wood-based panels—Determination of modulus of elasticity in bending and of bending strength.	[18]
ISO 12344	Thermal insulating products for building applications—Determination of bending behavior.	[18]

2.1. Materials

A millet-grown spawn of species *G. lucidum* (reference M9726) was acquired from Mycelia BVBA (Nevele, Belgium). The spawn was stored at a constant 4 °C and 65% RH prior to being used. The principal substrate of the specimens is European beech wood (*Fagus sylvatica*) of a 0.75–3.0 mm granulation (Räuchergold type HB 750/2000, J. Rettenmaier

& Söhne GmbH + Co KG, Rosenberg, Germany), and nominal density in 270–330 kg/m^3. Longitudinal reinforcement was introduced in the BM_R specimens group by using 5 mm diameter rattan fibres (*Calamus manan*; B.V. INAPO, Bloemendaal, Netherland), and hessian was used for the BM_I and BM_H groups (*Cannabis sativa* subsp. *sativa* ; NEMO Hemp jam web 370 g/m^2, Naturellement Chanvre, Echandelys, France).

2.2. Cultivation Protocol

The principal substrates, fibres and hessian were prepared at 40% moisture content (MC) with mineralized water and sterilised at 121 °C for 15 min. The principal substrates were then mixed with 16 wt% spawn and incubated in polypropylene filtered bags (SacO2, Deinze, Belgium) for 7 days at 27 °C in the dark. Once colonised, the principal substrates were broken down and formed with the sterile fibre and hessian into alcohol cleaned aerated PETG moulds. The formed specimens were incubated for 21 days at 27 °C in the dark, then oven-dried for 48 h at 60 °C. The dried specimens were stored at 4 °C and 65% RH prior to testing. No external mycelium was cultivated on the boundaries of the specimens. No additives were used.

3. Results and Discussion

We investigated the effect of diverse reinforcements on the mechanical behaviour in flexion of MBC using four levels: control (BM), inner hessian (BM_I), hessian jacketing (BM_H), and rattan fibres in the length (BM_R). In Figure 2 we can observe the dissected specimens after testing. Isotropic controls were added to the experimental series (BM). Experimental parameters per specimen type and resulting mean density, mean flexural modulus and mean modulus of rupture are presented in Table 2. Box plots of the results for flexural modulus and modulus of rupture are presented in Figure 3. It can be noted that the mechanical failure of the specimens was related to dewetting of the principal substrate, beech wood particles, across all groups and at the level of the highest tensile stress, that is at the middle of the span on the opposite surface to the applied load. Rattan fibres did not fail nor deform plastically. Likewise, no debonding of the hessian jacketing was visible after testing. Representative failure modes are presented in Figure 4, where we can appreciate the limited fractures in the most elastic specimen groups BM and BM_I where the mycelial matrix also deformed plastically on the surface that was the most exposed to tensile stress (10 mm from the neutral axis), and the lack of external fracture in the BM_H specimen group. The BM_R group resulted in more external fractures as the aligned continuous rattan fibres were favourable to increasing strength and stiffness while its continuous interfacial bond between the matrix and the rattan fibres constrained the deformation. A future investigation may focus on displacing the fibre reinforcement towards the most stressed opposite surface to the load, distancing it from the neutral axis to improve its efficiency. A backdrop we could expect from this is the earlier formation of fractures on the surface exposed to tensile stress, while the fibre alignment with the neutral axis in this experimental series left a thicker sectional area of beech wood and mycelium complex under the fibres, which allowed for the mycelial matrix to deform elastically and plastically to a greater strain.

Table 2. Summary of specimen types parameters, resulting dried densities, and flexural properties.

Type	Fibre Composition	Mean Density (s.d.) [kg/m^3]	Mean Flexural Modulus (s.d.) [MPa]	Mean Modulus of Rupture (s.d.) [MPa]
BM	Control	232.24 (18.24)	192.71 (52.40)	0.12 (0.03)
BM_I	Inner hessian	227.22 (8.46)	197.33 (45.56)	0.11 (0.02)
BM_H	Hessian jacketing	236.75 (12.00)	375.14 (98.81)	0.18 (0.03)
BM_R	Rattan	249.48 (9.78)	1.34×10^3 (570.68)	0.61 (0.12)

Similarly to our previous investigation on compression behaviour characterisation [10], the hessian jacketing series offers a contrasting instance of the effect that the cultivation of an external mycelial skin on test specimen might lead to. The BM_H series resulted in a 1.95-fold increase in flexural modulus as compared to the control group, and 1.5-fold increase in strength. In contrast to the BM_I series which had a hessian reinforcement added at mid-thickness and along the neutral stress axis, jacketing is a particularly suitable fibre composition strategy as the low elasticity of hessian contributes to resisting the tensile stress on the surface opposite to the load and on covered side surfaces. Beech wood particles further contribute to the composite strength where subjected to compression—on the surface where the load is applied—as the elastic mycelial matrix reaches its maximal strain and beech particles of a higher stiffness interlock. The BM_R series has the highest relative standard deviation with 41.3% of the mean, while other groups have a standard deviation of a maximum of 27.2%. An extremum is reported at 2.44 GPa while Q3 is 1.35 GPa for a mean at 1.34 GPa. If not considering this extremum, the mean of the series is at 1.12 GPa for a standard deviation closer to MBC standards at 211.06 MPa (18.8%).

The results of the experimental series are plotted on an Ashby map (Figure 5) and are presented as normalised by density (Figure 6). In the latter figure we can notice the increased mechanical efficiency of the BM_R series thanks to the composition strategy. In both figures flexural reports from the published MBC state-of-the-art are plotted [3–5,11,12,18–22]. These two figures gather evidences produced with approximately ten fungal species, two studies having not disclosed the ones they used [5,20]. There are 42 data points gathered from ten journal and conference articles. These include articles reporting on strength and/or stiffness in flexion; 3 data points had no density reported [12]. Only the reports with sufficient data are rendered on the figures.

Figure 2. Fibre placement strategies in tested specimens (left to right): rattan fibres (BM_R), hessian jacketing (BM_H), inner hessian (BM_I), control (BM).

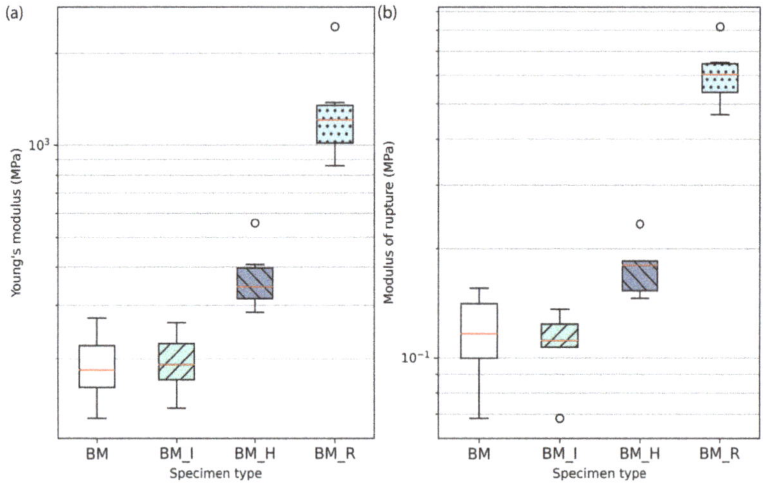

Figure 3. Flexural modulus (**a**) and modulus of rupture (**b**) box plots results.

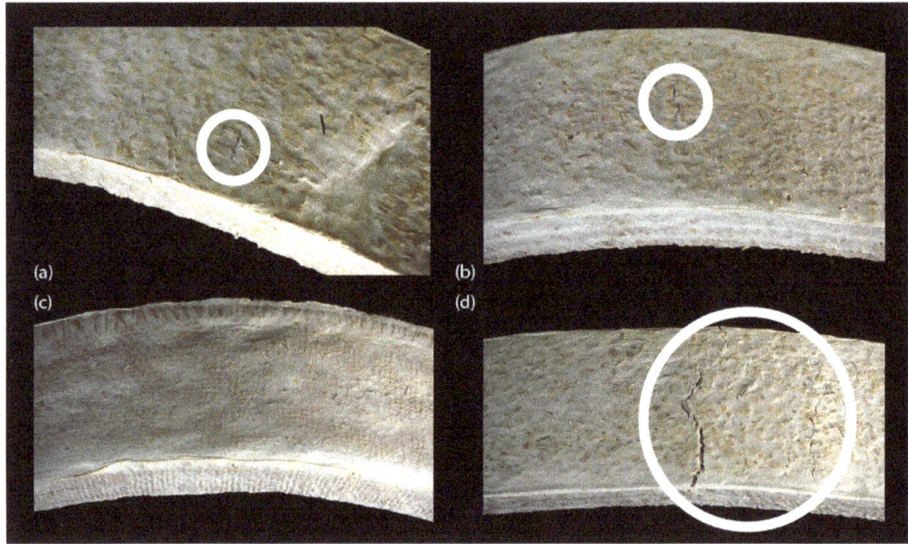

Figure 4. Representative failure modes for the (**a**) BM group, (**b**) BM_I group, (**c**) BM_H group, (**d**) and BM_R group.

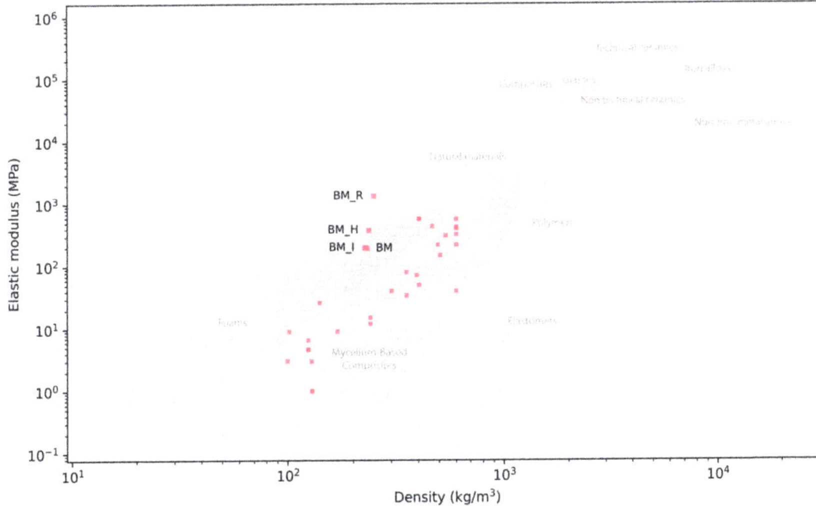

Figure 5. Flexural modulus results as a function of density. Labelled data points: results from this study; unlabelled data points: reports from the state-of-the-art.

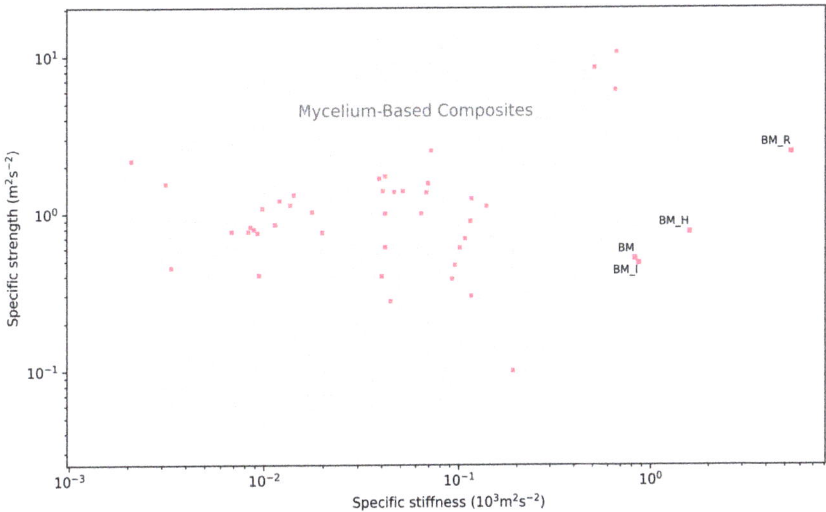

Figure 6. Specific strength results as a function of specific stiffness. Labelled data points: results from this study; unlabelled data points: reports from the state-of-the-art.

Statistical Analysis

The result distributions are two-tailed. The mean of Fisher's defined kurtosis for flexural modulus series is -0.7243 (s.d. 0.6040) and -0.5844 for modulus of rupture (s.d. 0.5019). Fisher-Pearson's skewness coefficient mean for flexural modulus is 0.5582 (s.d. 0.4907), and -0.1625 for modulus of rupture (s.d. 0.7697). The distributions are considered normal [23], but did not satisfy the Shapiro-Wilk test (modulus: $p = 1.1266 \times 10^{-5}$, modulus of rupture: $p = 3.4568 \times 10^{-5}$, $\alpha = 0.001$). Equality of variances was controlled with the Levene test; flexural modulus result variances are equal ($p = 0.0208$, $\alpha = 0.05$), such as modulus of rupture ones ($p = 0.0146$, $\alpha = 0.05$). One-way ANOVA was conducted for flexural modulus and modulus of rupture regarding reinforcement strategies (respectively $p = 1.8619 \times 10^{-6}$ and $p = 2.8678 \times 10^{-11}$). The mean values are significantly different ($\alpha = 0.001$). Using the pairwise Games-Howell test we confirm the significant difference between the rattan group and the other groups for flexural modulus and modulus of rupture ($\alpha = 0.05$). The inner hessian group was not significantly different from the control ($p = 0.900$).

4. Conclusions

The investigation presented in this paper has focused on the use of natural fibre composition in MBC as a means to modify flexural behaviour. We have demonstrated three fibre composition designs and show significant increases in stiffness and strength comparing the BM_R and BM_H series to the BM control group ($\alpha = 0.05$). The BM_I group, with a layer of hessian inserted at mid-thickness of the specimens, did not result in a significant effect. The mean of the flexural modulus for the most performative BM_R series (1.34 GPa) is still close to three times lower than that of Medium-Density Fibreboard (MDF; 4 GPa), and approximately 16 times lower in modulus of rupture (0.62 MPa, MDF: 10 MPa). Considering the higher density of MDF (750 kg/m^3) as compared to the BM_R series (249.48 kg/m^3) a future investigation could focus on developing a hybrid strategy of composition and densification so as to improve aggregate interlocking and resulting strength and stiffness. Future investigations may also focus on various soft arming positioning strategies reflecting on the most mechanically demanding areas of the specimens as per a defined load case, and use of orientated continuous fibres for a principal substrate. Composite production accuracy improvements are expected to contribute to reducing the standard deviation of results. Furthermore, reinforcements may be strategised in developing efficient or multifunctional composites, for instance, in designing the principal substrate to be thermally performative in addition to introducing reinforcements to perform structurally. We report on specimens dimensional stability after drying linked to the diverse compositions, this aspect can be investigated in the future both as a means to explore design consequences, and for production control.

Considering MBC higher vernacular potential thanks to the versatility and diversity of fungal species that can be used for cultivating, and to meet the urgent sustainability agenda, the sourcing of the raw materials and substrates for MBC should consider local and opportunistic supplies. In this study we make use of Austrian beech wood, hessian manufactured and cultivated in France, and rattan fibres which are produced in South-East Asia and West Afrika countries [24]. Beyond the interest of rattan fibres for their very consistent supply for supporting reproducibility of results and material homogeneity, future market-orientated MBC developments may focus on constraining the geographical footprint of their attached MBC supply chains so to reduce their embedded energy and global resources stress.

Author Contributions: Conceptualization, A.R.; methodology, A.R.; formal analysis, A.R. and C.C.; investigation, C.C. and A.R.; data curation, A.R.; writing—original draft preparation, A.R. and C.C.; writing—review and editing, P.A.; visualization, A.R. and C.C.; supervision, P.A.; project administration, P.A.; funding acquisition, P.A. All authors have read and agreed to the published version of the manuscript.

Funding: This project has received funding from the European Union's Horizon 2020 research and innovation program FET OPEN "Challenging current thinking" under the project *Fungal Architectures*, grant agreement No. 858132.

Institutional Review Board Statement: Not applicable.

Informed Consent Statement: Not applicable.

Data Availability Statement: The data presented in this study are available on request from the corresponding author.

Conflicts of Interest: The authors declare no conflict of interest. The funders had no role in the design of the study; in the collection, analyses, or interpretation of data; in the writing of the manuscript, or in the decision to publish the results.

References

1. Rigobello, A.; Ayres, P. Design Strategies for Mycelium-Based Composites. In *Fungi and Fungal Products in Human Welfare and Biotechnology*; Satyanarayana, T., Deshmukh, S.K., Ed.; Springer Nature: Berlin, Germany, 2022; *in press*
2. Sisti, L.; Gioia, C.; Totaro, G.; Verstichel, S.; Cartabia, M.; Camere, S.; Celli, A. Valorization of wheat bran agro-industrial byproduct as an upgrading filler for mycelium-based composite materials. *Ind. Crops Prod.* **2021**, *170*, 113742. [CrossRef]
3. Appels, F.V.W.; Camere, S.; Montalti, M.; Karana, E.; Jansen, K.M.B.; Dijksterhuis, J.; Krijgsheld, P.; Wösten, H.A.B. Fabrication factors influencing mechanical, moisture- and water-related properties of mycelium-based composites. *Mater. Des.* **2019**, *161*, 64–71. [CrossRef]
4. Cesar, E.; Montoya, L.; Barcenas-Pazos, G.M.; Ordonez-Candelaria, V.R.; Bandala, V.M. Performance of mycelium composites of Lentinus crinitus under two compression protocols. *Madera y Bosques* **2021**, *27*, e2722047. [CrossRef]
5. Jiang, L.; Walczyk, D.; McIntyre, G.; Bucinell, R.; Tudryn, G. Manufacturing of biocomposite sandwich structures using mycelium-bound cores and preforms. *J. Manuf. Process.* **2017**, *28*, 50–59. [CrossRef]
6. Sàez, D.; Grizmann, D.; Trautz, M.; Werner, A. Developing sandwich panels with a mid-layer of fungal mycelium composite for a timber panel construction system. In Proceedings of the 2021 World Conference on Timber Engineering, Santiago, Chile, 9–12 August 2021.
7. Ziegler, A.R.; Bajwa, S.G.; Holt, G.A.; McIntyre, G.; Bajwa, D.S. Evaluation of Physico-Mechanical Properties of Mycelium Reinforced Green Biocomposites Made from Cellulosic Fibers. *Appl. Eng. Agric.* **2016**, *32*, 931–938.
8. Haneef, M.; Ceseracciu, L.; Canale, C.; Bayer, I.S.; Heredia-Guerrero, J.A.; Athanassiou, A. Advanced Materials From Fungal Mycelium: Fabrication and Tuning of Physical Properties. *Sci. Rep.* **2017**, *7*, 41292. [CrossRef] [PubMed]
9. Green, D.W.; Winandy, J.E.; Kretschmann, D.E. Mechanical properties of wood. In *Wood Handbook: Wood as an Engineering Material*; General technical report FPL; USDA Forest Service, Forest Products Laboratory: Madison, WI, USA, 1999; Volume GTR-113, pp. 4.1–4.45.
10. Rigobello, A.; Ayres, P. Mycelium-Based Composites as Two-Phase Particulate Composites: Compressive Behaviour of Anisotropic Designs. *Sci. Rep.* **2021**, *in press*. [CrossRef]
11. Elsacker, E.; Vandelook, S.; Damsin, B.; Van Wylick, A.; Peeters, E.; De Laet, L. Mechanical characteristics of bacterial cellulose-reinforced mycelium composite materials. *Fungal Biol. Biotechnol.* **2021**, *8*, 18. [CrossRef] [PubMed]
12. Travaglini, S.; Dharan, C.K.H.; Ross, P.G. Mycology Matrix Sandwich Composites Flexural Characterization. In Proceedings of the American Society for Composites 29th Technical Conference, La Jolla, CA, USA, 8–10 September 2014; DEStech Publications, Inc.: Lancaster, PA, USA, 2014; pp. 1941–1955.
13. *ASTM D1037-12*; Test Methods for Evaluating Properties of Wood-Base Fiber and Particle Panel Materials. ASTM International: West Conshohocken, PA, USA 2020.
14. Holt, G.A.; Mcintyre, G.; Flagg, D.; Bayer, E.; Wanjura, J.D.; Pelletier, M.G. Fungal Mycelium and Cotton Plant Materials in the Manufacture of Biodegradable Molded Packaging Material: Evaluation Study of Select Blends of Cotton Byproducts. *J. Biobased Mater. Bioenergy* **2012**, *6*, 431–439. [CrossRef]
15. Lokko, M.L.; Rowell, M.; Dyson, A.; Rempel, A. Development of Affordable Building Materials Using Agricultural Waste By-Products and Emerging Pith, Soy and Mycelium Biobinders. In Proceedings of the PLEA 2016: The 32nd International Conference on Passive and Low-Energy Architecture, Los Angeles, CA, USA, 11–13 July 2016.
16. Sun, X.; Tang, M. Comparison of four routinely used methods for assessing root colonization by arbuscular mycorrhizal fungi. *Botany* **2012**, *90*, 1073–1083. [CrossRef]
17. Chan, X.Y.; Saeidi, N.; Javadian, A.; Hebel, D.E.; Gupta, M. Mechanical properties of dense mycelium-bound composites under accelerated tropical weathering conditions. *Sci. Rep.* **2021**, *11*, 22112. [CrossRef] [PubMed]
18. Elsacker, E.; Søndergaard, A.; Van Wylick, A.; Peeters, E.; De Laet, L. Growing living and multifunctional mycelium composites for large-scale formwork applications using robotic abrasive wire-cutting. *Constr. Build. Mater.* **2021**, *283*, 122732. [CrossRef]
19. López Nava, J.A.; Méndez González, J.; Ruelas Chacón, X.; Nájera Luna, J.A. Assessment of Edible Fungi and Films Bio-Based Material Simulating Expanded Polystyrene. *Mater. Manuf. Process.* **2016**, *31*, 1085–1090. [CrossRef]
20. Sun, W.; Tajvidi, M.; Hunt, C.G.; McIntyre, G.; Gardner, D.J. Fully Bio-Based Hybrid Composites Made of Wood, Fungal Mycelium and Cellulose Nanofibrils. *Sci. Rep.* **2019**, *9*, 3766. [CrossRef] [PubMed]
21. Sivaprasad, S.; Byju, S.K.; Prajith, C.; Shaju, J.; C R, R. Development of a novel mycelium bio-composite material to substitute for polystyrene in packaging applications. *Mater. Today Proc.* **2021**, *47*, 5038–5044. [CrossRef]
22. Ongpeng, J.; Inciong, E.; Siggaoat, A.; Soliman, C.A.; Sendo, V.B. Using Waste in Producing Bio-Composite Mycelium Bricks. *Appl. Sci.* **2020**, *10*, 5303. [CrossRef]
23. Hair, J.F.; Black, W.C.; Babin, B.J.; Anderson, R.E. *Multivariate Data Analysis*; Pearson Education Limited: London, UK, 2013.
24. Zhao, H.; Zhao, S.; Fei, B.; Liu, H.; Yang, H.; Dai, H.; Wang, D.; Jin, W.; Tang, F.; Gao, Q.; et al. Announcing the Genome Atlas of Bamboo and Rattan (GABR) project: Promoting research in evolution and in economically and ecologically beneficial plants. *GigaScience* **2017**, *6*, gix046. [CrossRef] [PubMed]

 *biomimetics*

Article

Basic Research of Material Properties of Mycelium-Based Composites

Hana Vašatko * , Lukas Gosch, Julian Jauk and Milena Stavric

Faculty of Architecture, Institute of Architecture and Media, Graz University of Technology, 8010 Graz, Austria; lukas.gosch@tugraz.at (L.G.); julian.jauk@tugraz.at (J.J.); mstavric@tugraz.at (M.S.)
* Correspondence: vasatko@tugraz.at

Abstract: The subject of this research is growing mycelium-based composites and exploring their basic material properties. Since the building industry is responsible for a large amount of annual CO_2 emissions, rethinking building materials is an important task for future practices. Using such composites is a carbon-neutral strategy that offers alternatives to conventional building materials. Yet, in order to become competitive, their basic research is still needed. In order to create mycelium-based composites, it was necessary to establish a sterile work environment and develop shaping procedures for objects on a scale of architectural building elements. The composite material exhibited qualities that make it suitable for compression-only structures, temporary assemblies, and acoustic and thermal insulation. The methodology includes evaluating several substrates, focused on beech sawdust, with two mycelium strains (*Pleurotus ostreatus* and *Ganoderma lucidum*), density calculations, compression tests, three-point flexural tests and capillary water absorption. The results of this study are presented through graphical and numerical values comparing material and mechanical properties. This study established a database for succeeding investigations and for defining the potentials and limitations of this material. Furthermore, future applications and relevant examinations have been addressed.

Keywords: mycelium; growth; bio-composites; mechanical properties; architecture; materials science

Citation: Vašatko, H.; Gosch, L.; Jauk, J.; Stavric, M. Basic Research of Material Properties of Mycelium-Based Composites. *Biomimetics* **2022**, 7, 51. https://doi.org/10.3390/biomimetics7020051

Academic Editors: Andrew Adamatzky, Han A.B. Wösten and Phil Ayres

Received: 15 March 2022
Accepted: 19 April 2022
Published: 21 April 2022

Publisher's Note: MDPI stays neutral with regard to jurisdictional claims in published maps and institutional affiliations.

1. Introduction

Contemporary problems, such as rapid population growth, the increased demand for food and housing, freshwater scarcity and inadequate waste management, are all influenced by demographic, environmental and economic factors. The current linear economic model relies on the extraction, transformation and disposition of raw materials once their life cycle ends. Meanwhile, the concept of circular economy is defined by the reuse of materials, with radical changes in the production itself taking place, thus preventing the accumulation of waste [1].

An additional fact driving this research is that embodied carbon emissions from the manufacturing of building materials and the construction sector account for 38% of annual worldwide greenhouse gas emissions in the current world climate [2]. Out of the four raw material types currently extracted—minerals, ores, fossil fuels and biomass—the major portion, totalling approximately 40%, finds its use in the construction and housing sector [3]. If the European energy strategy with net-zero emissions is to be realised by 2050 [4], it is time to reconsider present design concepts as well as building elements and materials.

One way to accomplish this is by investigating new bio-based building materials. In the last decade, bio-fabrication gained significance in architecture and has become an integral part of sustainable building strategies. The production of mycelium-based composites is a low-energy and carbon-neutral process [5] that fits into the circular economy and sustainable building concepts. The utilisation of mycelium-based composites is wide in terms of scale, functionality and application, as several architects, designers and enthusiasts have begun to use them in their designs during the past decade. Furthermore, the

33

composite has found its application as packaging material (*Ecovative*), acoustic insulation (*MOGU*) and as temporary objects exhibited in a larger scale, such as MycoTree [6] and the Growing Pavilion [7]. In comparison to building materials such as concrete or bricks, mycelium-based composites are a new term in architecture. Hence, there is still a high demand for basic research and testing in this respect. Several works have already presented some initial research by combining different substrates and mycelium strains and subsequently elaborating on some of their mechanical properties [8–12], and morphology and mechanics [13]. The work presented in this article extends the relevant references by providing an overview of the basic material properties in very specific material combinations, which include organic substrates, organic fibrous materials and inorganic materials. The introduction of inorganic and fibrous materials as substrates—such as clay, sand and soy silk fibres—contributes to the mentioned references. Since a wide variety of substrates and mycelium strains are present, as well as several decisive factors during the production process, the specification of the material properties is relevant. These preliminary experiments introduce the concept of growing mycelium-based lignocellulosic products, whose properties may subsequently be fine-tuned based on the material's intended application. The overarching purpose of this research is fabricating heterogeneous composites with a defined material distribution, which will optimise the structural properties within one geometry. This study offers the initial point of evaluating material properties that will be used in these experiments.

Broadly speaking, mycelium is the vegetative part of mushrooms, which consists of branching hyphae. Mycelial growth can be described as a hyphal penetration of a substrate, which results in unifying it into one piece. A spore inoculated on a nutrient forms a tube which experiences exponential non-photosynthetic growth [14]. Three growth phases can be differentiated after inoculating a lignocellulosic substrate: (1) the lag phase (zero to little population growth, the mycelium cells get used to their new environment), (2) the exponential phase (if the conditions remain favourable, increase in biomass takes effect, as well as the cell number—this is the optimal period for mycelial growth and continuation of this for as long as possible is desirable) and (3) the stationary phase (the population growth returns to zero, the fungal biomass remains constant and some fungal cells may begin to perish) [15,16]. The oyster mushroom (*Pleurotus ostreatus*) mycelium was the most frequently used species for this research, as this very common mushroom type shows a high contamination resistance compared to other tested specimens and is capable of consuming a variety of lignocellulosic substrates. Its by-products, water, carbon dioxide, enzymes, alcohols and carbohydrates, serve as a nutritious foundation for other organisms in nature [17].

2. Materials and Methods

The methods of this research were carried out through material experiments and investigation of mechanical properties for an application as a building material. A series of material mixtures was created, as well as samples with different geometries for material testing experiments (Figure 1). The research presented here constitutes a database for subsequent investigations conducted by the authors, as well for the overall understanding of the potentials and limitations of the material. This initial study offers a comparison between different material qualities and their fine-tuned versions according to the function of their application.

Various organic substrates as well as inorganic additives were tested. However, beech sawdust was the initial substrate for assessing all tested material properties due to its availability in the local area, which minimises transportation and processing costs. Additionally, bleached cellulose pulp was a substrate option, as it achieved solid results and demonstrated no problems during inoculation and the growing phase, with very little to no contamination. Obtaining large quantities of cellulose in a desired form, however, proved to be challenging, leading to the choice of beech sawdust as the preferred substrate. The mycelium grain spawn used for this research was bought from a local vendor. Other substrates were also tested on individual samples, e.g., sand and clay, in order to increase the plasticity while comparing compressive strength.

(a)   (b)   (c)

Figure 1. Samples composed of various materials: (**a**) cellulose pulp and clay; (**b**) beech sawdust and clay; (**c**) oak sawdust.

All samples were given abbreviations within a naming system, which is used throughout this research: *Pleurotus ostreatus* mycelium (PO), *Ganoderma lucidum* mycelium (GL), beech sawdust (FS), oak sawdust (Q), bleached cellulose pulp (CS1), shredded cardboard (SC), shredded newspaper (SN), cotton fibres (CO), soy silk fibres (SF), wheat bran (WB), straw (ST), burlap (B), clay (C) and sand (S). The nomenclature works in a way that the abbreviation of the substrate is used firstly, followed by the abbreviation of the mycelium strain, and ending with a sample number. The abbreviations and the number sequences are connected with hyphens (e.g., FS-PO-01 is the first sample of beech sawdust inoculated with oyster mushroom mycelium).

The following sections provide a detailed description of the production procedure, initial substrate exploration, together with the mechanical material testing and evaluation of basic material properties. The initial substrate exploration includes substrate compatibility and investigating density. The mechanical material testing includes evaluation of compression strength and a three-point flexural test. Finally, determining the capillary water absorption coefficient was conducted.

2.1. Production Procedure of Mycelium-Based Composites

The starting point for producing mycelium-based composites was preparing the substrate by soaking it in distilled water for 24 h. Afterwards, the excess water was drained, and the moisture content (MC) of the substrate was measured. The substrate was then put in a polypropylene microfilter bag (PP50/SEU4/V40-51, *SacO2*) and sterilised in a pressure cooker at 121 °C for 45 min to eliminate any competing microorganisms that might hinder mycelial development. The moulds were made of perforated transparent foil with a thickness of 0.5 mm in order to produce samples for material testing, whose dimensions depended on the conducted method. Once the sterilised substrate cooled down to room temperature, mycelium grain spawn was homogeneously distributed in microfilter bags while working in a still-air box. The amount of mycelium used for inoculation was 10% of the weight of the sterilised substrate. The moulds were cleaned with rubbing alcohol (ethanol, 70% solution) and filled with the inoculated substrate by manual pressing. The microfilter bags containing the closed moulds were sealed and stored in an environment protected from light sources at temperatures ranging from 22 to 24 °C. The initial phase of development occurred in the moulds, followed by the second phase after unmoulding to achieve growth of the outer protective skin of the sample. The duration of each growth phase is defined in the succeeding sections. Once the mycelium fully colonised the substrate, dehydration was initiated in order to terminate the growth.

2.2. Initial Substrate Exploration

Combinations of *Pleurotus ostreatus* and several substrates, including straw, beech sawdust, wheat bran and bleached cellulose pulp (Figure 2), were examined to evaluate breaking and shrinking, growth density, surface colour, and quality impression. The dimensions of the moulds were $10 \times 10 \times 2$ cm. Four samples were prepared for each substrate type, with two of them incorporating a piece of burlap in the centre as additional organic reinforcement. The obtained burlap fabric was woven from jute and cut in squares with dimensions of 11×11 cm and had a fabric weight of $180 \ \mathrm{g/m^2}$. The individual burlap pieces were larger than the moulds in order to visually assess the continuation of mycelial growth on the burlap. It was used simultaneously for nutrition, since jute fibres are composed of lignocellulose and for decreasing the shrinking of the composite. It was stiff and inelastic, and it had a moderate moisture regain [18]. Both burlap and the substrate were prepared according to the procedure described in the previous section. The moulds were not covered with a lid, as increased surface contact with oxygen accelerates mycelial growth [11].

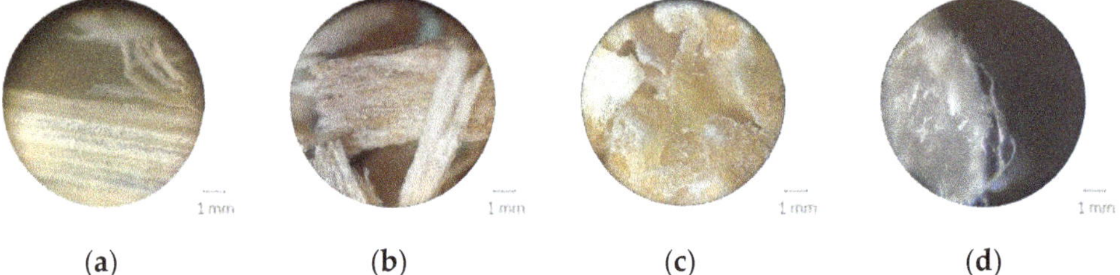

(a)   (b)   (c)   (d)

Figure 2. Substrate scale: (**a**) straw; (**b**) wood chips; (**c**) wheat bran; (**d**) bleached cellulose pulp.

In addition to successfully grown samples, a thorough growth documentation was obtained that demonstrates changes of the growth patterns and speed of the used mycelium strain as well as its preferred nutrition. The exponential growth phase [15,16] was visible after a couple of days only (Figure 3). Since the moulds were made of transparent plastic, the growth on the covered sides was noticeably reduced. After 20 days, the samples were unmoulded, flipped upside-down and placed back in the microfilter bag to achieve homogeneous surface growth on all sides. After the unmoulding, the second growth phase lasted for five days. Finally, the samples were dried over a heating source. The results of these samples are analysed in Section 3.1.

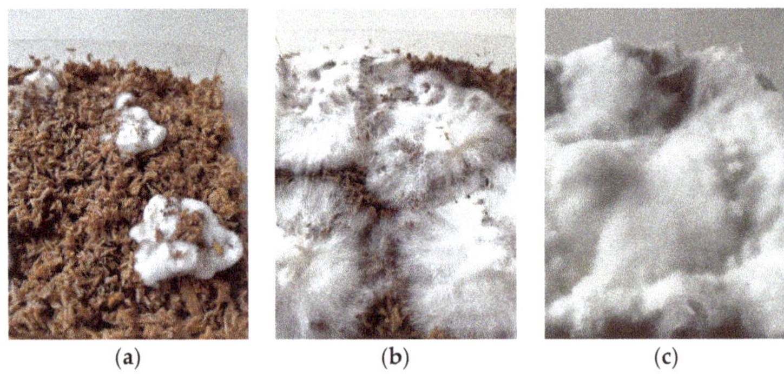

(a)   (b)   (c)

Figure 3. Close-up of the growth process of oyster mushroom mycelium on beech sawdust: (**a**) after three days; (**b**) after five days; (**c**) after 19 days.

2.3. Material Testing

Material testing included the evaluation of basic material properties and mechanical material testing. The former considered the calculation of density and capillary water absorption coefficient (Figure 4c), and the latter considered the compression strength and the three-point flexural test. The mechanical material testing was mostly executed at the Institute of Technology and Testing of Building Materials, Graz University of Technology (Figure 4a,b).

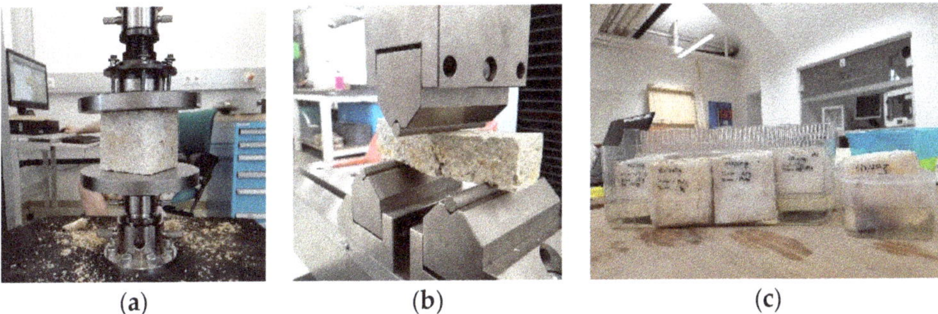
(a)   (b)   (c)

Figure 4. Setup for mechanical material testing and capillary water absorption: (**a**) compression strength; (**b**) three-point flexural test; (**c**) capillary water absorption.

The samples used to estimate the density, as well as all subsequent material tests, were prepared according to the procedures outlined in Section 2.1. For this experiment, the growth duration lasted for 14 days, i.e., until the substrate was fully colonised by mycelium. The initial growth phase lasting seven days took place in plastic moulds and the second phase took place after unmoulding. Density was measured on samples grown in dimensions of $10 \times 10 \times 10$ cm. A drying cabinet was used for sample dehydration. The temperature in the cabinet was set to 40 °C, and the drying process continued until the samples were completely dry. The documentation of weight loss and the calculation of the average water content is available in the supplementary documentation (Table S1). The method of determining the water absorption coefficient due to capillary action in hardened mortar [19] was used to determine the coefficient of water absorption of mycelium-based composites. The detailed description of the procedure and its results is provided in Section 3.5.

Both compression strength and the three-point flexural test were tested on a Shimadzu AG-X plus testing machine (Figure 4a,b). The standard used for these tests was EN 1015-11 [20], whereas the loading rate was 10 N/mm^2/sec. Compressive strength was measured on the same samples that were used for density calculations ($10 \times 10 \times 10$ cm). The failure criterion of the compression strength tests was indicated by a drop of measured compressive force. This was caused by an abrupt deformation, which led to a loss of the overall integrity of the grown test samples. The samples for the three-point flexural test had dimensions of $4 \times 4 \times 16$ cm and were placed on two linear bearings followed by a centrally applied linear load on top of those (Figure 4b). Regarding flexural tests, the failure criterion was indicated by a force drop right after achieving the peak value, which was caused by the occurrence of fractures in the sample. The graphical and numerical values of these experiments are presented in Sections 3.3 and 3.4.

3. Results

3.1. Substrate Selection

The first set of samples contained straw as a substrate. It was chopped into single pieces no longer than 3 cm in length and hand-pressed into the moulds. The MC of the substrate was 60.23%. The straw fragments were visible after unmoulding and drying (Figure 5a,b). Because the substrate had not been thoroughly compressed prior to inoculation, the airiness in the material caused the samples to break. The shrinkage factor was negligible, yet the growth speed was sufficient, as the mycelium expanded across the surface in five days.

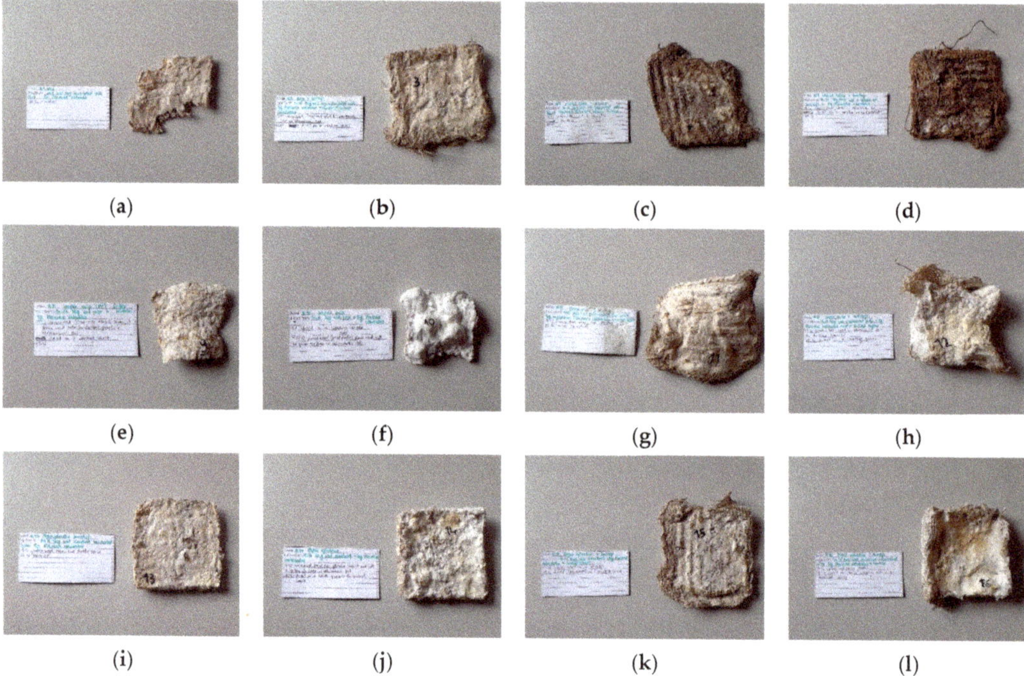

Figure 5. Samples of growth compatibility of *Pleurotus ostreatus* with straw, wheat bran, bleached cellulose pulp and beech sawdust: (**a**) ST-PO-01; (**b**) ST-B-PO-03; (**c**) WB-B-PO-03; (**d**) WB-B-PO-04; (**e**) CS1-PO-02; (**f**) CS1-PO-03; (**g**) CS1-B-PO-04; (**h**) CS1-B-PO-05; (**i**) FS-PO-20; (**j**) FS-PO-21; (**k**) FS-B-PO-22; (**l**) FS-B-PO-23.

Using pure wheat bran as a substrate was not successful—the samples without burlap became contaminated, while the other two with the burlap piece only retained their form as the fabric kept them together. The MC of the substrate was 46.75%. The remaining samples were fragile. As observed on the dark surface of the samples, the mycelium was hardly visible (Figure 5c,d). However, using wheat bran as an additive would accelerate mycelium growth [21], and it will thus be used for this purpose in further experiments.

Pieces of bleached cellulose pulp (2–6 mm diameter, MC 64.17%) were dispersed in plastic moulds, and the mycelium grew entirely within. The substrate shrank by up to 40% after drying (Figure 5e,f), making it exceedingly unpredictable in cases where specified dimensions are to be achieved. The samples containing burlap shrank up to 5% (Figure 5g,h). Additionally, the samples became deformed during the drying process. To anticipate or, at the very least, to decrease the considerable shrinkage, the cellulosic substrate should be compressed firmly prior to inoculation. The samples had a white colour and a smooth surface.

Beech sawdust was pressed manually into the moulds after inoculation but became highly porous after the drying process. The MC of the substrate was 58%. Shrinkage was 10%, which is a reliable value for future use (Figure 5i–l). Similarly, as with all samples, the density of the substrate particles was important for the stiffness of the dried product. The outer layer of the samples was light brown, and the outer skin had not developed as uniformly as it did in the cellulose pulp samples.

3.2. Density

The samples were slightly distorted after the drying process, which happened due to the inconsistent pressure from the manual filling of the moulds and the standard shrinkage factor. They were measured from edge to edge and in the centre of each side to obtain an average side length, resulting in 24 measurements per sample. Consequently, the volume was calculated followed by the density (Table 1). In addition to beech sawdust, further density measurements of other substrates were carried out, such as bleached cellulose pulp, soy silk fibres mixed with beech sawdust, cotton, cardboard, beech sawdust mixed with sand, and beech sawdust mixed with clay.

Table 1. Density comparison (density of beech sawdust samples (FS) compared to various samples).

Name	Average Side (cm)	Volume (cm^3)	Weight (g)	Density (g/cm^3)
FS-PO-01	9.26	794.67	215.00	0.27
FS-PO-02	9.29	802.20	220.00	0.27
FS-PO-03	9.42	836.12	210.16	0.25
FS-PO-04	9.41	832.80	212.52	0.26
FS-PO-05	9.35	817.40	209.67	0.26
FS-PO-06	9.42	835.01	209.28	0.25
FS-PO-average	9.36	819.70	212.77	0.26
CS1-PO-01	8.74	668.01	225.00	0.34
FS-SF-PO-01	8.96	718.92	170.00	0.24
FS-GL-01	9.30	804.36	205.00	0.25
SC-PO-01	8.93	710.93	297.00	0.42
S-FS-PO-01	9.73	919.75	215.00	0.23
S-FS-PO-02	9.47	848.38	429.00	0.51
CO-PO-01	8.82	685.35	151.00	0.22

3.3. Compression

3.3.1. Compression—Beech Sawdust

The samples whose growth was terminated after 14 days (FS-PO-05 and FS-PO-10) showed similar compressive strength, as the ones that had three additional days to grow (FS-PO-03 and FS-PO-07) (Table 2). The curves on the graph are defined by three stages: the first showing mediocre endurance, the second being the weakest stage as the sample softens, and finally, the recuperation phase, in which the curve grows more steeply than it did previously (Figure 6). Results of this kind were to be expected, taking the porosity of the material into account.

Table 2. Compression test of beech sawdust samples.

Name	Maximum Force (N)	Maximum Stress (MPa)
FS-PO-03	43,096.14	4.310
FS-PO-05	2822.56	0.280
FS-PO-07	42,275.38	4.230
FS-PO-10	11,166.73	1.120
average	24,840.20	2.490

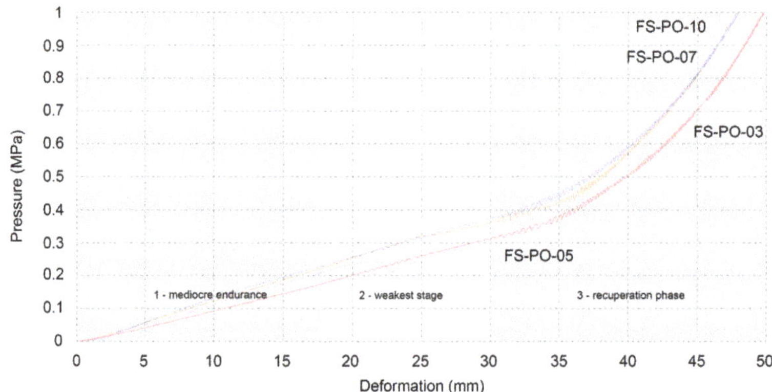

Figure 6. Compression test of beech sawdust samples.

3.3.2. Compression—Various Samples

The various samples consisted of different combinations of organic and inorganic materials. A description of each sample is provided in the following text. The order of the samples is defined by their composition similarities.

FS-SF-PO-01—The sample consisted of horizontally stacked layers of soy silk fibres between layers of sawdust. Since the sample was loaded perpendicularly to the layers of soy silk fibres, it did not break apart as those made from sawdust had. The layers of fibres behaved as elastic springs, adding a certain flexibility to the sample. The ratio of soy fibres to sawdust is 1:1, inoculated with 20% *Pleurotus ostreatus*.

CO-PO-01—Since cotton consists mostly (88–97%) of cellulose [22], this type of nutrition was viable for mycelial growth. The fibres used for the sample had a significant length (5 cm), and the task of homogenising them with the mycelium grain spawn thus proved to be challenging. The performance of this sample can be ranked between the one with soy fibres and sawdust, and that with sawdust only. CO-PO-01 exhibited a similar quality that of the sample with soy fibres, since it was compacted after testing and did not really break.

SC-PO-01—The performance of this sample was high for withstanding compressive forces. Before inoculating with 10% *Pleurotus ostreatus*, the cardboard was soaked in hot water and later torn into small pieces—35 mm in length. The sample was found to have shrunk 11% when measured by the average side length and had a density of 0.42 g/cm^3, which contributed to the best compression strength results (Table 3), as the cardboard was bound together tightly by the mycelium.

Table 3. Compression test of various samples.

Name	Maximum Force (N)	Maximum Stress (MPa)
FS-SF-PO-01	15,737.45	1.990
C-FS-GL-01	4622.05	0.510
CS1-PO-01	6429.34	0.850
FS-GL-01	6607.95	0.760
SC-PO-01	20,975.92	2.650
S-FS-PO-02	2314.20	0.260
S-FS-PO-01	692.27	0.090
CO-PO-01	6224.60	0.800

FS-GL-01—This blend consisted of the same sawdust type as the ones in the previous section, but it was inoculated with *Ganoderma lucidum*. When compared to previous results, it did not perform as well.

CS1-PO-01—The sample consisted of bleached cellulose pulp inoculated with 10% *Pleurotus ostreatus*. During the drying process, this sample shrunk significantly, as was expected after testing the substrate compatibility. The sides are reduced in size to 8.4 cm (from the original 10 cm per side). The sample thus has a density of 0.34 g/cm^3, which is higher than that of the average sample made from beech sawdust (0.26 g/cm^3). The sample was brittle, similar to the C-FS-GL-01, but withstood greater force (Figure 7).

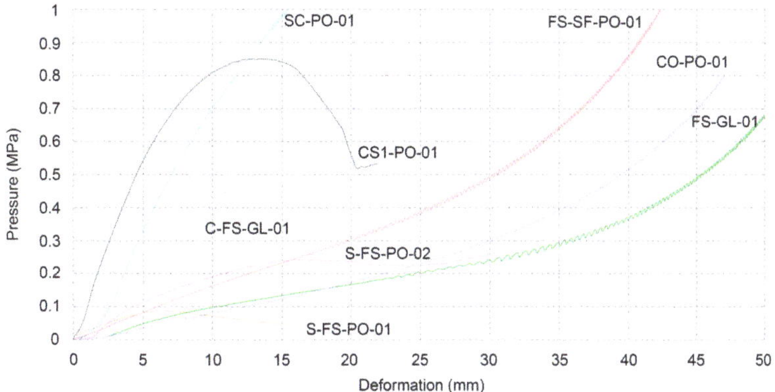

Figure 7. Compression test of various samples.

C-FS-GL-01—The mixture was made with one part of modelling clay to four parts of sawdust; the composite was inoculated with 10% of *Ganoderma lucidum* grain spawn. The organic portion was also influenced by the density of the clay—the goal was to introduce as much of the organic matter as possible in order to achieve a cohesive mycelial growth on the inside and as a result to enhance the properties of the composite. The sample showed similar brittleness as CS1-PO-01.

S-FS-PO-01 and S-FS-PO-02—The addition of sand to the mixture did not add to its mechanical strength. The two samples are differentiated by the amount of sand in the mixture: S-FS-PO-01 had equal quantities of sand and sawdust, while S-FS-PO-02 was based on a sand to sawdust volume ratio of 1:4.

3.4. Three-Point Flexural Test

3.4.1. Three-Point Flexural Test—Beech Sawdust

The first set of samples consisted of six beech sawdust samples. The numbers displayed in the table below show a wide dispersion of results (Table 4), despite them being simultaneously inoculated and incubated for the same period of time, under the same conditions. A conclusion of why the results vary so much cannot be drawn to one specific factor. However, when comparing the curve from the sample with the highest result, FS-PO-13 (Figure 8) is compared to an average result from the cellulose pulp samples, which are described in Section 3.4.2. A similar strain is not exhibited, whereby the sawdust samples can bear only half of the force that cellulose samples can.

Table 4. Three-point flexural test of beech sawdust samples.

Name	Maximum Force (N)	Maximum Stress (MPa)	Maximum Distance (mm)
FS-PO-11	61.11	0.14324	3.68
FS-PO-12	32.54	0.07626	2.40
FS-PO-13	73.02	0.17114	3.88
FS-PO-14	33.35	0.07816	2.66
FS-PO-15	52.37	0.12275	2.67
FS-PO-16	39.59	0.09280	2.85
average	48.66	0.11406	3.02

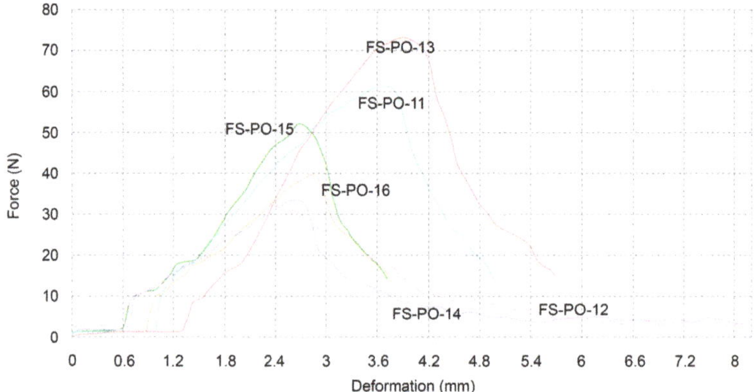

Figure 8. Three point flexural test of beech sawdust samples.

3.4.2. Three-Point Flexural Test—Various Samples

The following section describes individual composites tested in the same manner as the previous ones in Section 3.4.1. The first two samples, Q-GL-01 and FS-GL-19, were inoculated simultaneously with the same mycelium strain, *Ganoderma lucidum*, and grown under the same conditions for the same period of time. The substrate is the sole distinction between them, i.e., beech and oak sawdust. The sample inoculated with oak sawdust shows significantly better results (Table 5). It is important to note that the particles of oak sawdust were smaller (1–2 mm) compared to beech sawdust (about 3 mm). However, more samples are needed in order to make viable conclusions. This result difference will be further explored in order to relate the substrate type and mycelium strain to their specific mechanical properties.

Table 5. Three-point flexural test of various samples.

Name	Maximum Force (N)	Maximum Stress (N/mm^2)	Maximum Distance (mm)
Q-GL-01	70.38	0.16496	2.84
FS-GL-19	38.39	0.08997	4.52
SN-PO-01	276.95	0.64909	3.65
SC-PO-02	89.49	0.20973	4.23
S-FS-PO-03	29.18	0.06840	1.64
S-FS-PO-04	46.05	0.10792	2.74

The sample SN-PO-01 showed the best results (Figure 9). This sample was prepared in a similar manner as the ones made with cardboard. Sheets of newspaper were soaked in water for 24 h, which were then torn into small pieces by hand. These samples performed five times better than the average values for beech sawdust. SC-PO-02, the sample containing shredded cardboard, showed lower results than the ones containing shredded newspaper, yet it performed better compared to the sawdust composites. The cardboard and newspaper pieces were approximately 15 mm long.

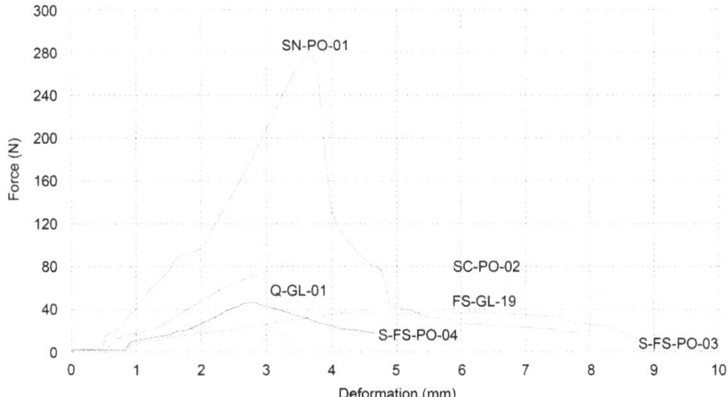

Figure 9. Three-point flexural test of various samples.

The last two pieces containing sawdust and sand in different ratios, S-FS-PO-03 and S-FS-PO-04, showed no unexpected results. Adding sand did not improve flexural strength.

Samples CS1-PO-06 and CS1-PO-07 consisted entirely of bleached cellulose pulp and exhibited excellent mechanical properties when compared to the other tested samples. Another technique was explored by varying the ratios of the two basic components, cellulose pulp and beech sawdust. The first, CS1-FS-PO-01, was made up of 30% cellulose and 70% sawdust, while the second, CS1-FS-PO-02, was made up of 70% cellulose and 30% sawdust. These were used to investigate if adding another organic component improved or degraded the qualities that were being measured. Even though the sample with more cellulose pulp performed better, the 30% sawdust in the sample reduced its total performance when compared to the samples that solely contained cellulose. Adding 30% cellulose to the sample consisting mainly of sawdust did not drastically change the result when compared to the average of the sawdust samples (Table 6, Figure 10).

Table 6. Three-point flexural test of bleached cellulose pulp compared to blend of bleached cellulose pulp with beech sawdust.

Name	Maximum Force (N)	Maximum Stress (N/mm^2)	Maximum Distance (mm)
CS1-PO-06	147.95	0.34675	3.83
CS1-PO-07	151.95	0.35614	3.92
average	149.95	0.35145	3.88
CS1-FS-PO-01	44.87	0.10516	2.09
CS1-FS-PO-02	75.17	0.17617	3.92
average	60.02	0.14067	3.01

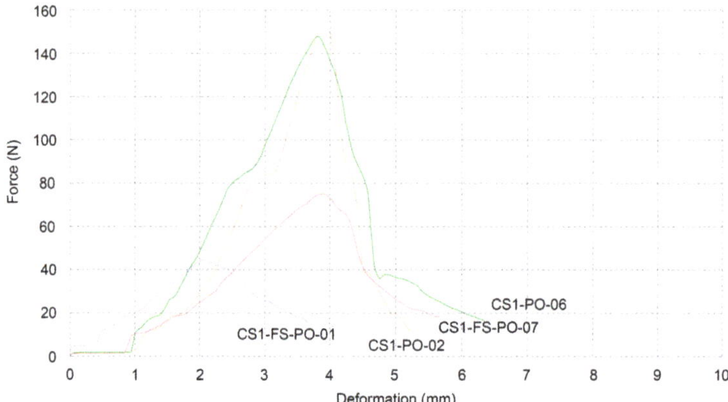

Figure 10. Three-point flexural test of bleached cellulose pulp compared to blend of bleached cellulose pulp with beech sawdust.

3.5. Capillary Water Absorption

The testing included several steps—firstly, the samples were completely dried out, and their surface was sealed with ethylene–vinyl acetate, which was applied using a standard glue gun. Once the dry weight of the samples was determined (M0), the container was filled with distilled water until the samples were immersed by 5 cm. The samples were placed in the filled container on a slant to ensure there was no air trapped beneath them, because of their uneven surfaces. The samples were weighed in a defined time range—after 10, 30, 60, 90 min, 2, 4, 8, and 24 h (Table 7). Five samples were examined, all of which had the same dimensions of $10 \times 10 \times 10$ cm. Three of them were identical, consisting of beech sawdust inoculated with *Pleurotus ostreatus*, another with the same substrate but inoculated with *Ganoderma lucidum*, and an oak sawdust inoculated with *Ganoderma lucidum*.

Table 7. Weight after certain periods of time. A water absorption coefficient greater than 2 is classified as strongly absorbent, while less than 2 is classified as water resistant, less than 0.5 is classified as water repellent, and less than 0.001 is classified as waterproof.

Name	M0: Weight Initial (g)	M1: Weight after 10 min (g)	M2: Weight after 30 min (g)	M3: Weight after 60 min (g)	M4: Weight after 90 min (g)	M5: Weight after 2 h (g)	M6: Weight after 4 h (g)	M7: Weight after 8 h (g)	M8: Weight after 24 h (g)	M8-M1	M4-M1	C (kg/(m² × min$^{0.5}$))
FS-PO-17	285	291	299	313	325	338	393	472	565	274	34	0.0034
FS-PO-18	280	286	292	302	308	311	327	254	423	137	22	0.0022
FS-PO-19	273	281	289	304	311	318	346	399	505	224	30	0.0030
Q-GL-02	240	241	244	247	248	249	258	290	577	336	7	0.0007
FS-GL-20	233	233	234	237	239	244	272	355	546	313	6	0.0006

For calculating the coefficient of water absorption, the following formula applies:

$$C = 0.1(M4 - M1)kg / \left(m^2 \times min^{0.5} \right).$$

The last column of Table 7 shows a distinct dissociation of values; the samples inoculated with *Pleurotus ostreatus* and *Ganoderma lucidum*, with the latter showing a lower water absorption than the first sample group. These findings suggest that *Ganoderma lucidum* increases water repellence in samples inoculated with sawdust. *Pleurotus ostreatus* samples show water-repellent qualities, and *Ganoderma lucidum* samples show waterproof qualities.

4. Discussion

The substrates were chosen for their performance, but availability was also an important selection criterion. Out of the four initial materials that were evaluated, beech sawdust and cellulose pulp were considered to have potential for further research. Both exhibited adequate stiffness, growth density and a satisfactory quality impression. If compressed during inoculation, straw can also be considered a viable substrate, as it creates very lightweight and porous composites.

Density was measured on samples with initial dimensions of $10 \times 10 \times 10$ cm, which was naturally decreased by mycelium digesting the substrate, compacting it and also by the drying process. In addition to these parameters, manually filling and compressing the moulds may have also influenced the density values of each sample.

The compression strength of beech sawdust composites had an average value of 2.49 MPa, with an interesting differentiation of samples whose growth lasted three days longer and exhibited higher values. Beech sawdust inoculated with *Ganoderma lucidum* did not perform as well as samples inoculated with *Pleurotus ostreatus*.

Three-point flexural tests were carried out on six beech sawdust samples with an average value of 0.11 N/mm^2. Two samples inoculated with *Ganoderma lucidum* were compared, as the sample inoculated on oak sawdust showed better results. Shredded newspaper performed well as a substrate, with a value of 0.649 N/mm^2.

Cardboard and newspaper are the materials worth considering for future experiments. Both exhibit excellent compression strength values and the highest density from purely organic samples after mycelial growth. They are also usually discarded and can be recycled in this manner. However, additional research on these two materials is necessary, since they were tested on individual samples. Cellulose pulp exhibited excellent mechanical properties, but its dimensions after drying are not as predictable due to its high shrinkage. Using sand as an additive has shown stable results while documenting shrinkage, yet it is to be considered an improper additive, since it does not enhance the mechanical properties. Adding clay to the organic substrate was beneficial for the plasticity of the samples, and it will be researched further.

Finally, capillary water absorption was tested with two mycelium strains, *Pleurotus ostreatus* and *Ganoderma lucidum*. The samples inoculated with the former strain are water repellent, while the ones inoculated with the latter exhibit waterproof qualities. There was no difference between oak and beech sawdust in terms of water absorption.

The goal of this research was to evaluate various mycelium and lignocellulosic substrate combinations. This is important if mycelium-based composites are to be introduced into the building industry—consequently, the material samples were standardised and tested to be comparable with conventional building materials in terms of potential future applications. This series of tests was used to characterise and assess their properties. The data of this research will be used by the authors in order to further develop methods of evaluating properties of mycelium-based composites for specific applications, i.e., for researching heterogeneities in mycelium-based composites.

5. Conclusions

Mycelium-based composites exhibit structural properties that open up the possibility of their implementation in the building industry. Their applications include compression-only structures, temporary assemblies, art installations [23] and materials for acoustic and thermal insulation [24]. These have already been implemented as case studies and products developed in several companies. However, their application as a widely accepted alternative to some building components and commercialisation is yet to be seen. Moulds are a viable solution for shaping the material mixtures, yet the necessity of a sterile working environment, as well as the time mycelial growth takes, are somewhat limiting factors.

Within this research, it was possible to develop a fabrication process for mycelium-based composites on a scale of architectural elements similar to masonry units. Moreover, a sterile work environment was established, and a productive shaping method developed.

The aim of this work was to gather data on several material properties and, as a result, to select the ones most suitable for composite materials depending on their application.

Another aspect that will be investigated is the correlation between growth time and the mechanical properties of mycelium-based composites, as seen in the results for different compression strength values of samples with a longer growth period. A series of samples will be made in which growth is interrupted in different samples and on numerous occasions, with a few days between each interruption. Another aspect that is planned to be looked into is the correlation between growth time, mechanical properties and weight loss of mycelium-based composites, as it has already been investigated on the decay of wood by brown-rot fungi [25]. This is important for defining the optimal growth advancements in the composite while retaining its maximal mechanical capacities.

In addition to the fine-tuning of the composite by enhancing the desired material properties, the results presented here will be used for making heterogeneous material mixtures in complex geometries, including varying mechanical requirements. The major potential that is yet to be explored is controlled material distribution within a specific element. An initial experiment to test this hypothesis was conducted by creating a series of trusses consisting of different cellulose types, each exhibiting different mechanical properties (Figure 11). After thoroughly analysing the properties of homogeneous mixtures, a heterogeneous material distribution will be implemented.

Figure 11. Truss structure made from cellulose pulp.

In addition to exploring natural coatings to prevent the degradation of organic composites caused by moisture, the waterproof qualities of *Ganoderma lucidum* can be beneficial for its future use in mycelium-based composite materials. An experimental methodology has recently been developed in one study on the subject of biodegradability of mycelium-based composites based on soil burial tests [26]. These experimental methods are crucial if mycelium-based composites will be used in exterior applications. Using a certain mycelium strain that is more resistant to the water uptake is promising but still highly dependent on the type of substrate used [10]. Yet, their rapid biodegradability is one of their qualities that make the material appealing in terms of sustainability and waste management. Consequently, the focus of application of mycelium-based composites still remains in dry interior locations.

Supplementary Materials: The following supporting information can be downloaded at: https://www.mdpi.com/article/10.3390/biomimetics7020051/s1, Table S1: Measurements during the drying process and average water content calculation.

Author Contributions: Conceptualization, H.V., L.G. and J.J.; Writing—Original Draft Preparation, H.V.; Writing—Review and Editing, H.V., L.G. and J.J.; Experiment—Conduction, J.J.; Visualization, L.G.; Supervision, M.S. All authors have read and agreed to the published version of the manuscript.

Funding: This work was funded by the Austrian Science Fund (FWF) project F77 (SFB "Advanced Computational Design").

Institutional Review Board Statement: Not applicable.

Informed Consent Statement: Not applicable.

Data Availability Statement: Not applicable.

Acknowledgments: We are most grateful for the technical support of the Institute of Technology and Testing of Building Materials, Graz University of Technology and Anita Klaus, Department for Industrial Microbiology, University of Belgrade.

Conflicts of Interest: The authors declare no conflict of interest.

References

1. Braungart, M.; McDonough, W.; Bollinger, A. Cradle-to-cradle design: Creating healthy emissions—A strategy for eco-effective product and system design. *J. Clean. Prod.* **2007**, *15*, 1337–1348. [CrossRef]
2. United Nations Environment Programme. *2020 Global Status Report for Buildings and Construction: Towards a Zero-Emission, Efficient and Resilient Buildings and Construction Sector*; United Nations Environment Programme: Nairobi, Kenya, 2020.
3. Is a World without Trash Possible? Available online: https://www.nationalgeographic.com/magazine/article/how-a-circular-economy-could-save-the-world-feature (accessed on 27 January 2022).
4. European Commission. *Communication from the Commission to the European Parliament, the Council, the European Economic and Social Committee and the Committee of the Regions—A New Circular Economy Action Plan—For a Cleaner and More Competitive Europe*; COM/2020/98 Final; The European Commission: Brussels, Belgium, 2020.
5. Jones, M.; Mautner, A.; Luenco, S.; Bismarck, A.; John, S. Engineered mycelium composite construction materials from fungal biorefineries: A critical review. *Mater. Des.* **2020**, *187*, 108397. [CrossRef]
6. Heisel, F.; Schlesier, K.; Lee, J.; Rippmann, M.; Saeidi, N.; Javadian, A.; Nugroho, A.R.; Hebel, D.; Block, P. Design of a load-bearing mycelium structure through informed structural engineering: The MycoTree at the 2017 Seoul Biennale of Architecture and Urbanism. In Proceedings of the World Congress on Sustainable Technologies (WCST-2017), Cambridge, UK, 11–14 December 2017.
7. The Growing Pavilion. 2019. Available online: https://thegrowingpavilion.com/biobased-materials/ (accessed on 13 January 2022).
8. Attias, N.; Danai, O.; Ezov, N.; Tarazi, E.; Grobman, Y.J. Developing novel applications of mycelium based bio-composite materials for design and architecture. In *Building with Bio-Based Materials: Best Practice and Performance Specification*; University of Zagreb, Faculty of Forestry: Zagreb, Croatia, 2017.
9. Attias, N.; Danai, O.; Abitbol, T.; Tarazi, E.; Ezov, N.; Pereman, I.; Grobman, Y.J. Mycelium bio-composites in industrial design and architecture: Comparative review and experimental analysis. *J. Clean. Prod.* **2020**, *246*, 119037. [CrossRef]
10. Appels, F.V.; Camere, S.; Montalti, M.; Karana, E.; Jansen, K.; Dijksterhuis, J.; Krijgsheld, P.; Wösten, H.A. Fabrication factors influencing mechanical, moisture- and water-related properties of mycelium-based composites. *Mater. Des.* **2019**, *161*, 64–71. [CrossRef]
11. Elsacker, E.; Vandelook, S.; Van Wylick, A.; Ruytinx, J.; De Laet, L.; Peeters, E. A comprehensive framework for the production of mycelium-based lignocellulosic composites. *Sci. Total Environ.* **2020**, *725*, 138431. [CrossRef] [PubMed]
12. Haneef, M.; Ceseracciu, L.; Canale, C.; Bayer, I.S.; Heredia-Guerrero, J.A.; Athanassiou, A. Advanced Materials from Fungal Mycelium: Fabrication and Tuning of Physical Properties. *Sci. Rep.* **2017**, *7*, 41292. [CrossRef] [PubMed]
13. Islam, M.R.; Tudryn, G.; Bucinell, R.; Schadler, L.; Picu, R.C. Morphology and mechanics of fungal mycelium. *Sci. Rep.* **2017**, *7*, 13070. [CrossRef] [PubMed]
14. Prosser, J. Growth kinetics of mycelial colonies and aggregates of ascomycetes. *Mycol. Res.* **1993**, *97*, 513–528. [CrossRef]
15. Kavanagh, K. *Fungi: Biology and Applications*, 2nd ed.; John Wiley & Sons: Hoboken, NJ, USA, 2005; p. 30.
16. Jones, M.; Huynh, T.; Dekiwadia, C.; Daver, F.; John, S. Mycelium Composites: A Review of Engineering Characteristics and Growth Kinetics. *J. Bionanoscience* **2017**, *11*, 241–257. [CrossRef]
17. Stamets, P. *Mycelium Running: How Mushrooms Can Help Save the World*, 1st ed.; Ten Speed Press: New York, NY, USA, 2005; pp. 279–280.
18. Kozłowski, R.M.; Mackiewicz-Talarczyk, M. *Handbook of Natural Fibres, Volume 1: Types, Properties and Factors Affecting Breeding and Cultivation*, 2nd ed.; Elsevier: Amsterdam, The Netherlands; Woodhead Publishing: Cambridge, UK, 2020; pp. 44–49.
19. *DIN EN 1015-18:2003-03*; Methods of Test for Mortar for Masonry—Part 18: Determination of Water Absorption Coefficient Due to Capillary Action of Hardened Mortar. European Committee for Standardization (CEN): Brussels, Belgium, 2002. [CrossRef]

20. *DIN EN 1015-11:2020-01*; Methods of Test for Mortar for Masonry—Part 11: Determination of Flexural and Compressive Strength of Hardened Mortar. European Committee for Standardization (CEN): Brussels, Belgium, 2019. [CrossRef]
21. Ghazvinian, A.; Farrokhsiar, P.; Vieira, F.; Pecchia, J.; Gursoy, B. Mycelium-Based Bio-Composites for Architecture: Assessing the Effects of Cultivation Factors on Compressive Strength. In Proceedings of the 37th eCAADe and 23rd SIGraDi Conference, Porto, Portugal, 11–13 September 2019; Volume 7, pp. 505–514.
22. Modal, H.; Ibrahim, M. *Fundamentals of Natural Fibres and Textiles*, 1st ed.; Elsevier: Duxford, UK, 2021; p. 602.
23. Sydor, M.; Bonenberg, A.; Doczekalska, B.; Cofta, G. Mycelium-Based Composites in Art, Architecture, and Interior Design: A Review. *Polymers* **2021**, *14*, 145. [CrossRef] [PubMed]
24. Dias, P.P.; Jayasinghe, L.B.; Waldmann, D. Investigation of Mycelium-Miscanthus composites as building insulation material. *Results Mater.* **2021**, *10*, 100189. [CrossRef]
25. Curling, S.F.; Clausen, C.A.; Winandy, J.E. Relationships between mechanical properties, weight loss and chemical composition of wood during incipient brown rot decay. *For. Prod. J.* **2002**, *52*, 34–39.
26. Van Wylick, A.; Elsacker, E.; Yap, L.L.; Peeters, E.; de Laet, L. Mycelium Composites and their Biodegradability: An Exploration on the Disintegration of Mycelium-Based Materials in Soil. In Proceedings of the 4th International Conference on Bio-Based Building Materials, Barcelona, Spain, 6 June 2021; Trans Tech Publications, Ltd.: Stafa-Zurich, Switzerland, 2022; Volume 1, pp. 652–659.

Article

Functional Grading of Mycelium Materials with Inorganic Particles: The Effect of Nanoclay on the Biological, Chemical and Mechanical Properties

Elise Elsacker [1,2,*,†], Lars De Laet [1] and Eveline Peeters [2,*]

[1] Architectural Engineering Research Group, Department of Architectural Engineering, Vrije Universiteit Brussel, Pleinlaan 2, B-1050 Brussels, Belgium; lars.de.laet@vub.be

[2] Research Group of Microbiology, Department of Bioengineering Sciences, Vrije Universiteit Brussel, Pleinlaan 2, B-1050 Brussels, Belgium

* Correspondence: elise.elsacker@ncl.ac.uk (E.E.); eveline.peeters@vub.be (E.P.); Tel.: +32-2-629-1906 (E.P.)

† Current address: Hub for Biotechnology in the Built Environment, Newcastle University, Devonshire Building, Newcastle upon Tyne NE1 7RU, UK.

Abstract: Biological materials that are created by growing mycelium-forming fungal microorganisms on natural fibers can form a solution to environmental pollution and scarcity of natural resources. Recent studies on the hybridization of mycelium materials with glass improved fire performance; however, the effect of inorganic particles on growth performance and mechanical properties was not previously investigated. Yet, due to the wide variety of reinforcement particles, mycelium nanocomposites can potentially be designed for specific functions and applications, such as fire resistance and mechanical improvement. The objectives of this paper are to first determine whether mycelium materials reinforced with montmorillonite nanoclay can be produced given its inorganic nature, and then to study the influence of these nanoparticles on material properties. Nanoclay–mycelium materials are evaluated in terms of morphological, chemical, and mechanical properties. The first steps are taken in unravelling challenges that exist in combining myco-fabrication with nanomaterials. Results indicate that nanoclay causes a decreased growth rate, although the clay particles are able to penetrate into the fibers' cell-wall structure. The FTIR study demonstrates that *T. versicolor* has more difficulty accessing and decaying the hemicellulose and lignin when the amount of nanoclay increases. Moreover, the addition of nanoclay results in low mechanical properties. While nanoclay enhances the properties of polymer composites, the hybridization with mycelium composites was not successful.

Keywords: mycelium-based composites; lignocellulosic fibers; natural fiber reinforcement; mechanical characteristics; manufacturing variables; nanoclay

Citation: Elsacker, E.; De Laet, L.; Peeters, E. Functional Grading of Mycelium Materials with Inorganic Particles: The Effect of Nanoclay on the Biological, Chemical and Mechanical Properties. *Biomimetics* **2022**, *7*, 57. https://doi.org/10.3390/biomimetics7020057

Academic Editors: Andrew Adamatzky, Han A.B. Wösten and Phil Ayres

Received: 23 March 2022
Accepted: 3 May 2022
Published: 5 May 2022

Publisher's Note: MDPI stays neutral with regard to jurisdictional claims in published maps and institutional affiliations.

1. Introduction

In recent years, the exciting characteristics of filamentous fungi did not go unnoticed in the context of biodegradable materials, providing a low-cost and environmentally sustainable solution compared to the production and life cycle of petroleum-based materials [1–7]. These composite materials are realized by growing the fungi into lignocellulosic fibers, thereby valorizing organic waste streams, and generating dense materials with a construction material application [8]. Typically, composites are composed of a matrix and a reinforcement. For mycelium materials, the matrix is mycelium, and the reinforcement is the natural fiber. Mycelium surrounds the fibers and maintains their relative position inside the material. The fibers, in turn, largely influence the mechanical and physical properties of the composite [1]. A key aspect of improving the mechanical and physical properties of mycelium materials is the implementation of organic or inorganic particles [9–11].

Due to the wide variety of reinforcement particles, mycelium nanocomposites can potentially be designed for specific functions and applications, such as fire resistance and mechanical improvement.

Recent studies of such hybridization using glass improved the fire performance of mycelium materials as a result of significantly higher silica (inflammable) concentrations and low combustible material content [11]. It takes almost six times as much time for mycelium materials incorporating 50 wt.% glass fines to flash over as it takes synthetic materials, such as extruded polystyrene insulation foam, and two times as much as particleboard [11]. This study suggests that mycelium materials are very economical and exhibit far better fire safety parameters than the traditional construction materials tested (extruded polystyrene foam and particleboard made from wood flakes and bonded with moisture-resistant synthetic resin) [11]. However, a major problem with incorporating high contents of inorganic matter in the substrate might be the biocompatibility with white-rot fungi. Minerals limit the growth of the hyphae over poor nutritional surfaces and can therefore influence the bond between mycelium and the lignocellulosic fibers that must hold the material together.

To maintain sufficient mycelial growth, glass fines that comprise primarily silica (SiO_2) and up to 30 wt.% organic surface matter were used in the particular research by Jones et al. [11]. Very limited research has further investigated the integration of additives in mycelium materials, with the exception of a study that indicates that the compressive strength of mycelium materials containing sand or gravel aggregates and wood chips increases up to 300% [12]. A patent makes note of the addition of components such as silica, clay and perlite to the fibers to retain moisture or enhance the viscosity of the substrate [13]. Thus far, no other study has investigated the fabrication, growth methods and mechanical properties of mycelium materials that incorporate inorganic (nano)particles.

Yet, nanotechnology and nanomaterials have great potential to improve the properties of different materials. Nanoparticles, like nanoclay, are widely used in various industries and areas of research, such as computing, adhesives, textiles, pharmaceutical and automotive [14–16]. For example, the reinforcement of particleboard and plywood panels with $nanoSiO_2$, $nanoAl_2O_3$, and nanoZnO was reported to significantly decrease formaldehyde emission [17,18]. During the past decades, rapid developments have occurred in the area of polymer/clay nanocomposites. Most early studies focused on synthetic polymer, such as polyamides [18], polyimides [19], methacrylates [20,21] or polystyrene [22]. Nanoclay also offers an improved dimensional stability in wood–plastic composites [23]. Moreover, for wood–plastic composites, it is reported that flexural strength, tensile strength and elongation and water absorption are improved by the addition of nanoclay; the most interesting properties are observed in specimens with 5% of nanoclay content [24]. This improvement is due to the formation of bonds between the hydroxyl groups of nanoclay and the wood flour components. The addition of clay nanoparticles to cotton-based polymer composites resulted in increased char yield, therefore rendering them flame retardant [25].

One of the most common nanoclay forms is montmorillonite with a particle thickness of 1 nm, crosswise 70 to 100 nm [26,27]. Montmorillonite clays have a layered structure, and each layer is constructed from tetrahedrally coordinated Si atoms fused onto an edge-shared octahedral plane of either $Al(OH)_3$ or $Mg(OH)_2$ [26]. The layers exhibit excellent mechanical properties parallel to the layer direction due to the nature of the bonding between these atoms [26]. The principle for using nanoclay is to separate not only clay aggregates, but also individual silicate layers in a polymer [26]. The choice for montmorillonite nanoparticles is mainly motivated by their wide availability and inexpensiveness [15,28]. The main advantage is that a minimal content (1–5 wt.%) of such additives can improve the reinforcement of the polymer matrix [15,29,30]. Moreover, several studies have shown that the resulting organic–inorganic hybrids possess tremendous improvement in tensile strength and modulus, gas permeability, heat distortion temperature, and flammability [31].

This paper focuses on the incorporation of nanoparticles and the development of intricate gradients in the material's arrangement. The goal of this work is to investigate the production of organic–inorganic hybrids and their material properties. The influence of nanoclay on the physical and mechanical properties of mycelium materials is studied. The hypothesis is that the hybridization of mycelium composites with nanoparticles improves mechanical properties as the nanoclay can reinforces the internal vessels of the lignocellulosic fibers. This is the first study undertaking a longitudinal analysis of nanoclay–mycelium hybrid materials by using different characterization methods, such as SEM, FTIR and mechanical testing.

The methodological approach focusses on the fabrication of nanoclay-coated fibers, inoculated with mycelium as binder (Figure 1). The experimental design was developed gradually and then combined in this work. The effect of nanoclay particles on the fabrication process and properties was mapped by analyzing the surface colonization rate, microscopic structure, chemical changes and mechanical properties.

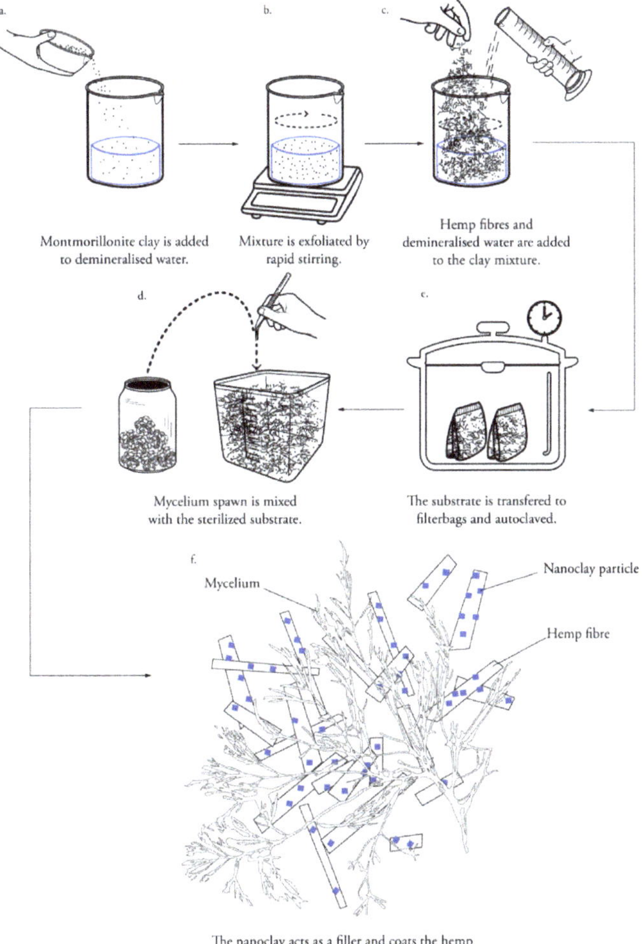

Figure 1. Schematic description of the fabrication process of nanoclay-reinforced mycelium composites.

2. Materials and Methods

2.1. Fungal Species

Trametes versicolor (M9912) spawn was purchased from Mycelia bvba (Veldeken 38A, 9850 Nevele, Belgium). The species were conserved on a grain mixture at 4 °C in a breathing Microsac 5 L bag (Sac O2 nv, Nevele, Belgium).

2.2. Materials

Studies were performed on the following fiber types: 5–25 mm hemp fibers (Aniserco S.A, Groot-Bijgaarden, Belgium). The superfine powder Ventoux montmorillonite clay was obtained from EMSPAC (Mons, France) in an untreated state.

2.3. Preparation of Nanoclay-Coated Substrate

Hemp nanocomposites containing 2.5 wt.% (weight percentage) montmorillonite clay as a filler material were prepared in batches of 820 g. The montmorillonite clay was exfoliated by rapid stirring for 30 min (ambient conditions) in 400 mL demineralized water, after which 25 wt.% hemp fibers and 72.5 wt.% ddH$_2$O were mixed in a bigger flask (Figure 1). The fiber/clay substrate was autoclaved at 121 °C for 20 min. The bags were left to cool down for 24 h. Other studies filtered and washed the material in acetonitrile and deionized water [25,32]. We chose not to follow this procedure in this study due to the toxicity of acetonitrile.

2.4. Particleboard Fabrication for Bending and Tensile Testing

In a laminar flow hood, 10 wt.% of mycelium spawn was mixed with the sterilized fibers (Figure 1). During the first growth phase, the substrate grew in bags with a depth-filtration system that allowed for airflow. The bags were stored in an incubation room at 26 °C and relative humidity of 60%. The mycelium homogeneously colonized the substrate in chunks. The bags were kneaded every day to stimulate the strengthening of the mycelium. After 5 days, the substrate was crumbled by hand to re-activate the mycelium. It was then distributed in Microbox containers (purchased at SacO2, Deinze, Belgium) with a depth-filtration system on top (185 × 185 × 78 mm). After 12 days, the substrate had taken the shape of the rectangular molds. Subsequently, the samples were removed from the molds and incubated again for 5 days to achieve a homogeneous colonization on the sides that were previously in contact with the mold.

The samples were compressed with an Instron 5900R test bench that had an oven built around it. A maximum force of 30 kN was applied at 2 kN/min. When a displacement of 50 mm was reached, the load was kept constant for 1 h at a temperature of 200 °C.

The samples were heat-pressed to an aimed thickness of ±15 mm. Heat-pressed samples were dried in a convection oven at a temperature of 70 °C for 10 h, until the weight stabilized, and all humidity evaporated. The particle boards were then stored at 21 °C and 65% RH for 3–4 weeks before testing.

Finally, the samples were cut with a thin blade saw, following the specimen dimensions set out by the several standards for mechanical tests (Figure 2). For the static bending tests, three specimens of 190 × 50 mm were cut from one particleboard. After loading, the non-damaged parts were reused for the tests on tensile strength perpendicular to the surface (internal bond) and cut to 50 × 50 mm. For tensile strength parallel to the surface, 5 specimens of 180 × 30 mm were cut from one particleboard.

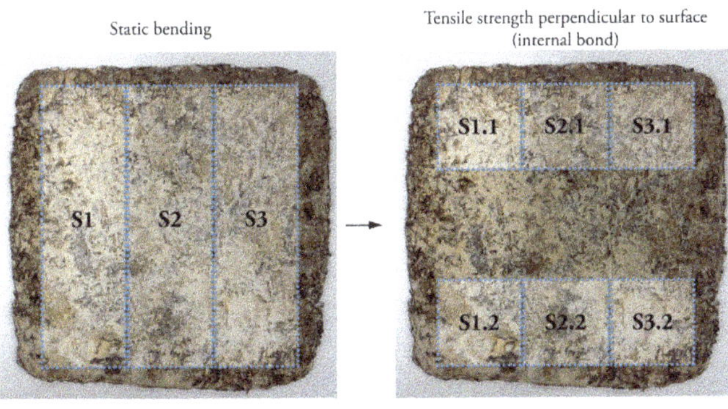

Figure 2. Example of the cutting pattern of specimens for the different mechanical tests.

2.5. Specimens for Compressive Testing

Compression tests were carried out with cylindrical specimens (h: 38 mm, d: 100 mm), whereby the quantity of nanoclay varied between 1 wt.%, 2.5 wt.%, 3 wt.% and 5 wt.%. The montmorillonite clay was exfoliated by rapid stirring for 30 min (ambient conditions) in demineralized water, after which hemp fibers and ddH$_2$O were mixed in a bigger flask. The fiber/clay substrate was autoclaved at 121 °C for 20 min. The bags were left to cool down for 24 h. Similarly, as described above, 10 wt.% of mycelium spawn was mixed with the sterilized substrate and incubated in bags for 5 days, after which it was packed in cylindrical Microbox containers (purchased at SacO2, Deinze, Belgium). After 12 days, the samples were removed from the molds and incubated again for 5 days to achieve a homogeneous colonization on the sides.

2.6. Substrate Inoculation for Growth Studies

The substrate for the growth studies was also prepared in the same way as described above. The samples were composed of varying nanoclay quantities (between 1 wt.%, 2.5 wt.%, 3 wt.% and 5 wt.%). Inoculation was performed by applying a 4–5 mm fragment of grain spawn in the middle of the petri dish containing 15 g of humid hemp fibers. Each experiment was conducted with triplicate dishes. The petri dishes were inverted and incubated at 26 °C in darkness for 9 days.

2.7. Surface Colonisation Rate Measurements and Analysis

Daily growth was monitored every 6 to 12 h, after an initial growth of 24–72 h, using a CanoScan 9000F Mark II (Canon), with a pixel density of 300 dpi.

Analysis of the mycelium extension rate was done by measuring the surface area covered by the mycelium. The scanned images were imported in Fiji [33], the contrast and brightness were adjusted and the outline of the mycelium was traced by using the freeform or circle tool. The pixel-to-surface measurement tool in ImageJ was employed to convert this pixel tracing in quantitative results.

2.8. Scanning Electron Microscopy

To evaluate the microstructural geometry of the mycelium, Scanning Electron Microscopy (SEM) characterization was carried out using a JEOL JSM-IT300 microscope in low vacuum, which allowed pressure in the chamber of up to 20 Pa, using a 5 kV accelerating voltage. The specimens were mounted on aluminum stubs with double-sided tape. Images were taken with the Backscattered Electron Detector (BED) or Secondary Electron Detector (SED) at zoom of $500\times$, $1000\times$ and $2000\times$. Samples were taken from the growth studies (see Section 2.6) grown for 22 days and air-dried for 5 days, at different locations in the Petri dish (5 replicates).

2.9. Fourier Transform Infrared Spectroscopy

Chemical characterization was performed for 1–2.5–3.5–5% nanoclay mycelium. For the Fourier Transform Infrared (FTIR) spectroscopy, all IR spectra were acquired on a Nicolet 6700 FT-IR spectrometer from Thermo Fischer Scientific (Waltham, MA, USA). The FTIR instrument is equipped with an IR source, DGTS KBr detector and KBr beam splitters and windows. FTIR spectra are recorded in single bounce Attenuated Total Refractance (ATR) mode using the Smart iTR accessory, equipped with a diamond plate ($42°$ angle of incidence). The spectra were recorded with automatic atmospheric correction for the background. All samples were measured at a spectral resolution of 4 cm^{-1}, with 64 scans per sample. For every species, three biological replicate specimens were analyzed, and five recordings of each sample were performed to ensure the reproducibility of obtained spectra. The processing of the spectra, such as calculating the average, the peak height and area, was done using Spectragryph v1.2.8 software. The spectra were cut between 4000 and 600 cm^{-1} bands, followed by the construction of a linear baseline. Finally, the peaks were normalized for peak maxima.

Samples were taken from the growth studies (see Section 2.6) grown for 22 days and air-dried for 5 days, at different locations in the Petri dish (5 replicates). These samples were placed in microcentrifuge tubes and then in a TissueLyser II from QIAGEN (Hilden, Germany) with a small metallic ball. The resulting fine powder was mounted on the FTIR.

2.10. Determination of Bending Behaviour

Since no standard exists for testing mycelium materials, three-point static flexural tests were performed according to specifications of norms that were expected to result in similar properties. The characteristics of mycelium materials are situated somewhere between foam and wood-based panels; therefore, we refer to the following standards: ISO 16978—*Wood-based panels—Determination of modulus of elasticity in bending and of bending strength* [34] and ISO 12344—*Thermal insulating products for building applications—Determination of bending behavior* [35]. According to ISO 16978, "the test pieces shall be rectangular, the width shall be 50 mm, the length shall be 20 times the nominal thickness plus 50 mm (minimum 150 mm)." On the other hand, ISO 12344 states that "specimens shall have a width of 150 mm and a length of 5 times the nominal thickness plus 50". For the mycelium samples, according to the first standard, the specimen dimensions should have been $350 \times 50 \times 15$ mm or $150 \times 50 \times 4$ mm. Following the second standard, the specimen dimensions should have been $125 \times 150 \times 15$ mm or $70 \times 150 \times 4$ mm. The test specimen dimensions were 170×50 mm. The distance between the supports was 150 mm for all tests. A loading speed of 2.5 mm/min was applied. The samples were tested using an Instron 5900R load bench with a load cell of 10 kN. The bending strength f_m of each test piece is calculated from the formula [34]:

$$f_\mathrm{m} = \frac{3F_\mathrm{max}l_1}{2bt^2} \ [\text{MPa}] \tag{1}$$

where F_{max} is the maximum load [N/mm^2], l_1 is the distance between the centers of the supports [mm], b is the width of the test piece [mm] and t is the thickness of the test piece [mm].

The modulus of elasticity E_m, is calculated from the formula [34]:

$$E_m = \frac{l_1^3 (F_2 - F_1)}{4\,bt^3 (a_2 - a_1)} \; [\text{MPa}] \tag{2}$$

where l_1, b and t are the dimensions as defined above, $F_2 - F_1$ is the linear portion of the load-deflection curve [N], F_1 is 10% and F_2 is 40% of the maximum load. The term $a_2 - a_1$ represents the increment of deflection at the mid-length of the test piece (corresponding to $F_2 - F_1$).

2.11. Determination of Tensile Behaviour Parallel to the Surface

Tensile strength parallel to the surface was measured according to ASTM 1037—*Standard Test Methods for Evaluating Properties of Wood-Base Fiber and Particle Panel Materials* [36]. The specimen dimensions were 170×30 mm. A loading speed of 1 mm/min was applied. Five samples of each treatment were tested using an Instron 5900R load bench with a load cell with a maximal capacity of 10 kN.

The specific strength and modulus were calculated using the following formulas:

$$T_\sigma = \frac{\sigma_u}{\rho} \; [\text{kN·m/kg}] \tag{3}$$

and

$$T_E = \frac{E}{\rho} \left[10^6 \; \text{m}^2 \; \text{s}^{-2} \right] \tag{4}$$

where T_σ is specific tensile strength (kN·m/kg or or MPa/(g/cm^3)), σ_u is ultimate tensile strength (MPa), ρ is density (g/cm^3), T_E is specific Young's modulus (10^6 m^2 s^{-2} or GPa/(g/cm^3)) and E is Young's modulus (GPa).

2.12. Determination of Tensile Behaviour Perpendicular to the Surface

This test was performed to determine the cohesion (internal bond) of the material. The test was performed according to *EN 319:1993 Particleboards And Fiberboards. Determination Of Tensile Strength Perpendicular To The Plane Of The Board* [37]. The specimens with dimension 50×50 mm were glued on aluminum loading blocks. After 24 h curing, the block was mounted into the grips. A loading speed of 0.5 mm/min was applied. The specimens were loaded at a uniform motion rate until failure occurred. Two samples of each treatment were tested using Instron 5900R load bench with a maximal capacity of 10 kN.

2.13. Determination of Compressive Behaviour

Compressive stiffness was determined following ISO 29469—*Thermal insulating products for building applications—Determination of compression behavior* [38] on an Instron 5900R load bench with a 100 kN capacity and a 10 kN load cell at ambient conditions (25 °C and ~50% RH). The 10 kN load cell was used in order to obtain the most accurate results, since low ultimate loads values were expected. The tests were displacement controlled with a rate of 5 mm/min. The contact surface was not perfect due to the rough surfaces of the samples. The test was stopped when a fixed strain was reached in the specimen, varying between 70% and 80%. The mechanical compressive stiffness or Young's modulus was obtained from the first slope of the stress–strain curve with the tangent modulus.

2.14. Statistical Analysis

The data were statistically analyzed in Microsoft Excel and graphed with GraphPad Prism (version 8.1.2). Data were checked for normality ($p \geq 0.05$) using a Kolmogorov–Smirnov test. An one-way analysis of variance (ANOVA) was used for normal data, and significant differences were considered at $p \leq 0.05$. The multiple comparisons test for normal data was generated based on Tukey's family error rate. For non-parametric data, the Kruskal–Wallis test was conducted, and significant differences were considered at $p \leq 0.05$. The Dunn's multiple comparison test was used for the non-parametric data.

3. Results

As this work deals with the fabrication of nanoclay-reinforced mycelium composites, it focused on the dispersion of inorganic nanoparticles in a hemp substrate as well as on mechanical performance. Particleboards were manufactured and mechanically tested to obtain first flexural modulus and strength, then elastic modulus and strength parallel and perpendicular to the surface.

3.1. Mycelium Growth and Surface Colonisation

The fungus used in this research is a white-rote species. Lignocellulose is the primary source of nutrient. Hence, the added minerals might limit the growth of the hyphae if the fibers are coated with clay particles. To assess the influence of the amount of nanoclay on the growth kinetics of mycelium, different proportions were mixed with hemp fibers. Remarkably, the results suggest that composites containing 5 wt.% of nanoclay grow 5% faster than composites with 1 wt.% of nanoclay during the first 7 days (Figure 3). The *T. versicolor* mycelium stays concentrated in the middle of the Petri dish during the first 7 days, and on the 10th day, only small, thin white branches start to explore the rest of the substrate. The graph gives an image of the growth rate, which shifted from 5 wt.% nanoclay having the highest rate after 7 days to 1 wt.% nanoclay having the highest rate after 22 days. After 10 and 22 days, composites with 1 wt.% fully colonized the Petri dish before the other compositions. The initial exploring hyphae slowly covered the whole substrate.

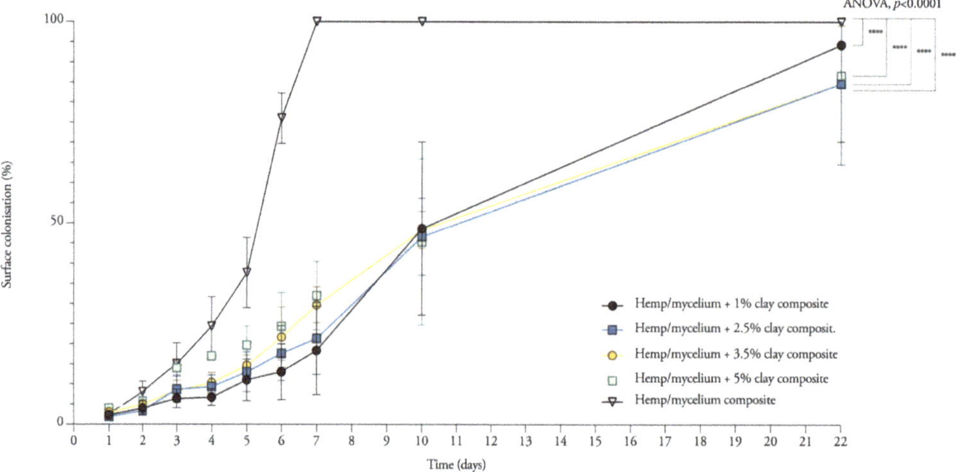

Figure 3. Radial mycelium extension rate measured as surface colonized (%) over 22 days for *T. versicolor* (M9912) on a hemp and 1–5% nanoclay substrate.

As expected, the nanoclay composites grew significantly slower than samples without nanoclay (two-way ANOVA, adjusted $p < 0.0001$). Samples containing nanoclay grew 24% to 29% slower than the control samples. The slower growth of mycelium makes the substrate more prone to contaminations. From the three replicates, at least one sample containing 3.5 wt.% and 5 wt.% was partially contaminated, blocking the hyphae from fully colonizing the rest of the substrate (Figure 4). Moreover, the hydrophobic nature of clay forms a barrier to the uptake of moisture and oxygen. In accordance with the present results, previous studies have demonstrated that the presence of nanoclay makes the wood–plastic composites less accessible to the fungus due to the reduction in oxygen content and moisture uptake, and nutrient shortage [39,40].

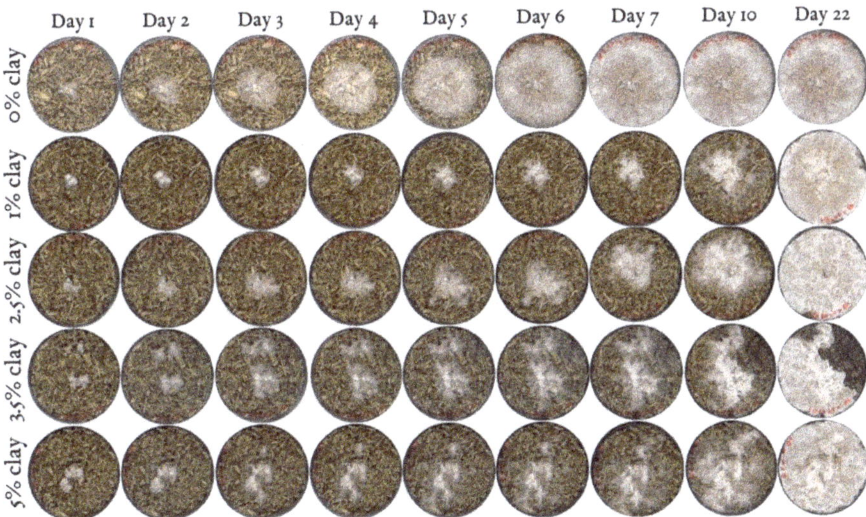

Figure 4. Colonization over 22 days for *T. versicolor* (M9912) on a hemp and 0–5% nanoclay substrate.

3.2. Morphological Analysis of Hyphae and Nanoclay

To further investigate the interactions between the organic and inorganic substrate with mycelium, different mixtures (1 wt.%, 2.5 wt.%, 3.5 wt.% and 5 wt.% nanoclay) were dried after having grown for 22 days and observed by SEM (Figure 5). The images show a distribution of nanoparticles around the hemp fibers. The amount of nanoclay (between 1 and 5%) does not seem to alter the ability of hyphae to grow in and around the fibers. The hyphae were able to penetrate the clay particles, as their growth tracks are marked in the clay (Figure 6). Interestingly, some nanoclay particles were observed around the hyphae. It is possible that the distribution of particles was altered while handling the samples, but the nanoparticles could also have been transported and moved during the growth of the hyphae. Fibers with 1 and 2% nanoclay were only slightly coated by the particles, while samples with 3.5 and 5% nanoclay were packed with particles that reached into the fibers' cell-wall structure. Still, the concertation of hyphae was higher in cavities and places where the fibers were not covered by nanoclay.

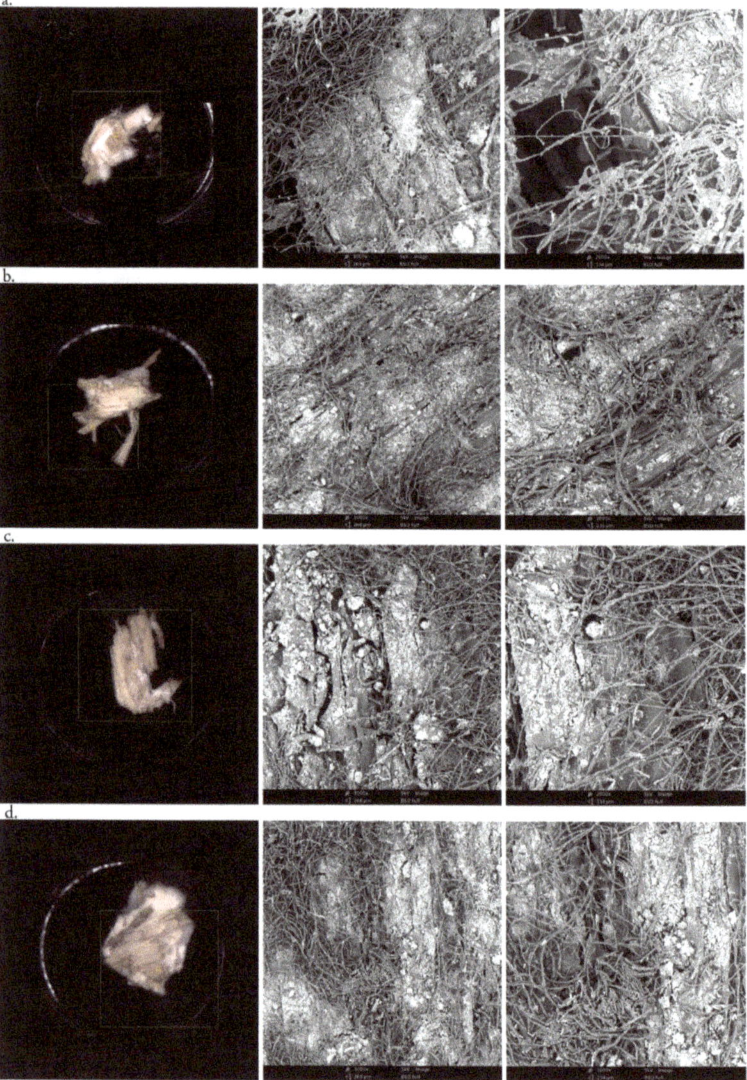

Figure 5. Morphological characterization by SEM images of *T. versicolor* (long threads) on a hemp (dark-gray, plate-like fibers) and nanoclay (light-gray particles) substrate after 7 days. (**a**) 1 wt.% nanoclay. (**b**) 2.5 wt.% nanoclay. (**c**) 3.5 wt.% nanoclay. (**d**) 5 wt.% nanoclay. Micrographs were taken at zoom 10×, zoom 1000× and zoom 2000×.

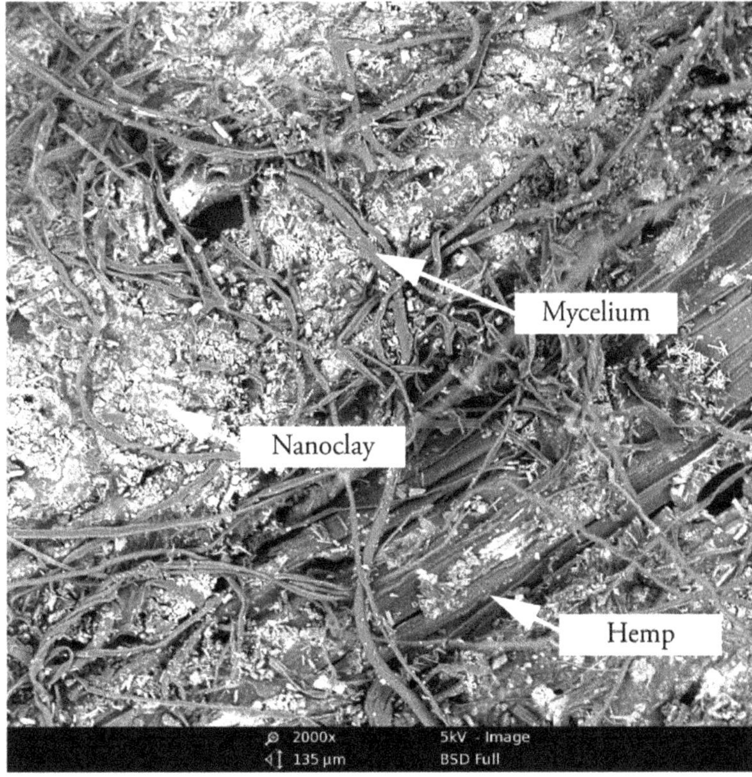

Figure 6. Micrograph of 2.5 wt.% nanoclay-mycelium composite showing the hyphae that grow inside the nanoclay, which is coated around the hemp fibers.

3.3. Spectral Response to Nanoclay Concentrations

Next, to understand the influence of different concentrations of nanoclay on the degradation capacity of *T. versicolor*, FTIR spectra were compared with undecayed hemp fibers and mycelium-hemp composites (Figure 7). For an exhaustive chemical analysis of the degradation between decayed and undecayed fibers, we refer to Elsacker et al. 2019 [1]. Here, we focus only on the impact of the nanoclay filler on the degradation capacity in the "fingerprint" region between 1800 and 600 cm^{-1}.

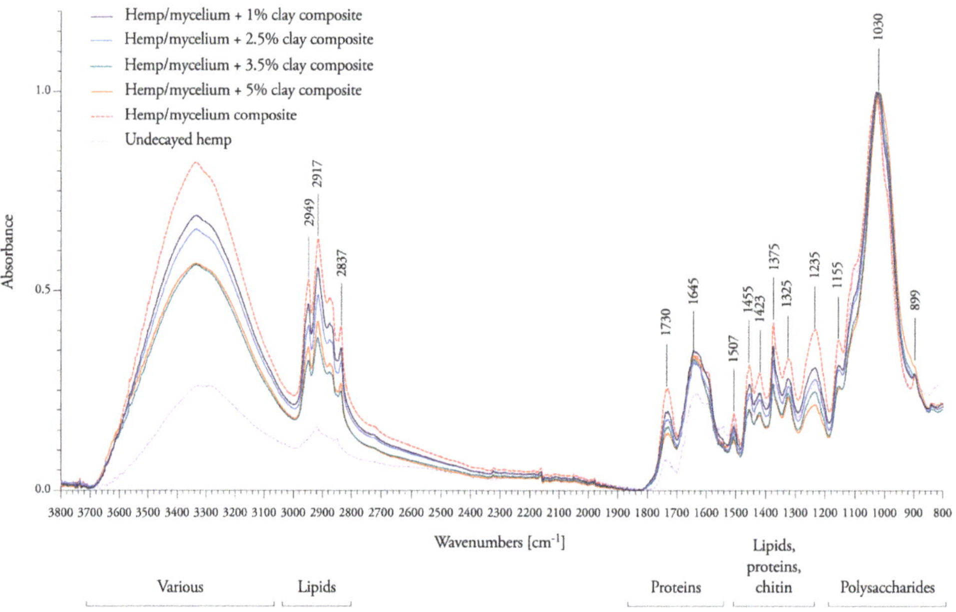

Figure 7. Mean FTIR spectra of 1–2.5–3.5–5% nanoclay–mycelium composites, mycelium composite without nanoclay (dotted line) on a hemp substrate after growth of 7 days, and undecayed hemp fibers (dashed line).

Mycelium-hemp composite samples showed the presence of bands at 3310 cm^{-1} for O-H stretching of bonded hydroxyl groups, 2949 and 2837 cm^{-1} for C-H stretching, 1731 cm^{-1} for C=O stretching, 1645 cm^{-1} for H-O-H deformation vibration of absorbed water and conjugated carbonyl groups, mainly originating from lignin, 1456 cm^{-1} for CH$_2$ deformation vibrations in lignin and xylan, 1423 cm^{-1} for aromatic skeletal vibrations combined with C-H in plane deformation and for +C-H deformation in lignin and carbohydrates, 1156 cm^{-1} for C-O-C vibration in cellulose and hemicelluloses, 1032 cm^{-1} for C=O stretching vibration in cellulose, hemicelluloses and lignin and 1000−800 cm^{-1} for C−H bending vibration in cellulose.

The intensities of all peaks of nanoclay composites decreased slightly between 1800 cm^{-1} and 1021 cm^{-1}, which indicates that *T. versicolor* has more difficulty accessing and decaying the hemicellulose and lignin when the amount of nanoclay increases. New bands were formed at 1005 cm^{-1}, revealing that cellulose is mostly resistant to degradation when nanoclay is present in the composite (subtracted peaks under the baseline in Figure 8). No changes in chemical composition were observed at 1645 cm^{-1} (H-O-H deformation vibration of absorbed water and conjugated carbonyl groups, mainly originating from lignin), or at 899 cm^{-1} (C−H bending vibration in cellulose). The percentage of nanoclay is visible on the spectral response as fibers containing 1% nanoclay are more degraded than the ones containing 5%. This suggests that nanoclay obstructs the nutritional access to the fibers.

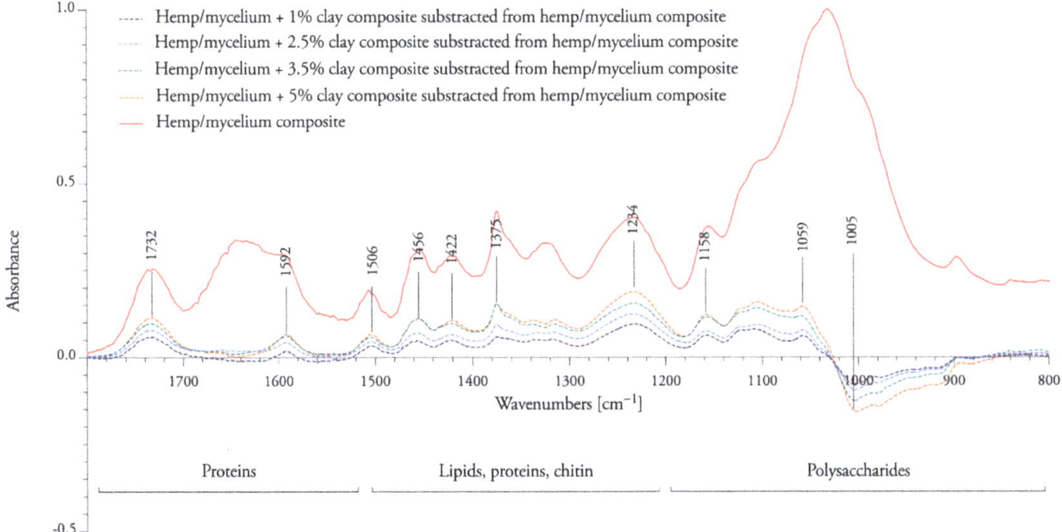

Figure 8. The subtracted peaks of nanoclay–mycelium composite (dotted line) from mycelium composite fibers without clay (solid line) below the baseline show an appearance of new bands due to degradation by the fungi, while the subtracted peaks above the baseline show the decrease of bands. The intensities are around zero when there is no change in the chemical composition of the fibers during the mycelium interaction.

3.4. Flexural Properties of Nanoclay-Mycelium Particleboards

Foremost, the goal of this study was to investigate the influence of the addition of nanoclay particles on the mechanical properties of mycelium composites. The particleboards were all visually different, showing variations in the growth pattern of the organism. These fungal morphological changes are associated with extracellular signaling and metabolic pathways [41].

The addition of nanoclay to the fibers did not significantly change the bending behavior of mycelium composites (Table 1). The control samples without nanoclay (1.46 MPa) are situated close to the NC-mycelium samples (1.47 MPa). The samples tested within this study displayed flexural properties that can be related to those of natural materials such as wood, cork and cancellous bone (Figure 9) [42]. It is not surprising that mycelium materials are efficient when bending, because natural materials such as wood, cork and bamboo also indicate high values for the flexure index [42]. Similarly, these natural cellular materials have low densities due to the high volume of voids.

Table 1. Overview of the material properties in three-point bending of nanoclay–mycelium composites. The standard deviation is performed with triplicate specimens (mean ± one standard deviation).

Label	Dry Density [kg/m³]	Flexural Strength [MPa]	Flexural Modulus [GPa]
NC-mycelium hemp composite	426.98 ± 9.19	1.47 ± 0.02	0.19 ± 0.002
Mycelium hemp composite (control)	488.89 ± 41.09	1.46 ± 0.48	0.22 ± 0.06

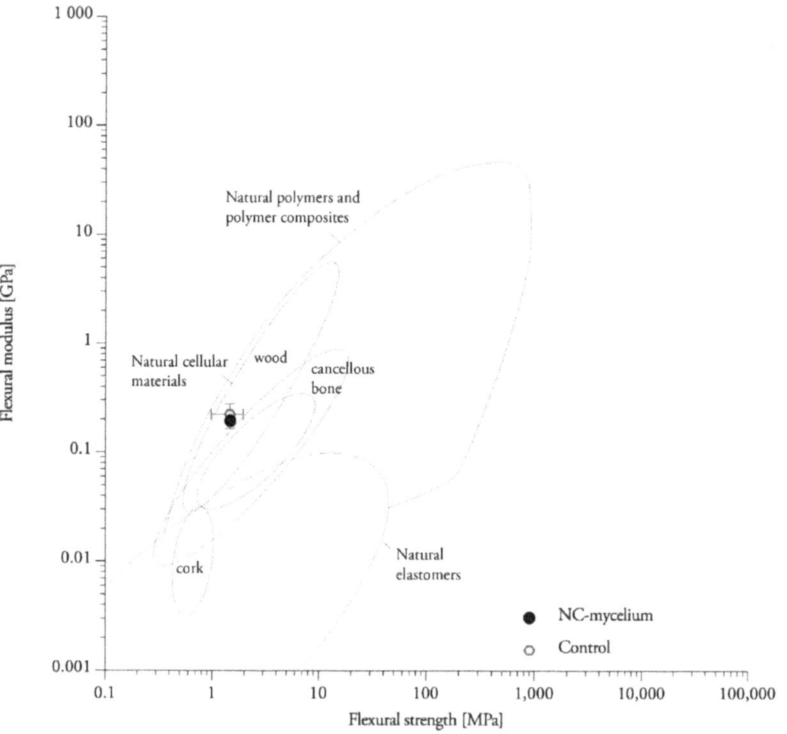

Figure 9. A material property, plotting Young's modulus against strength for nanoclay–mycelium composites on an Ashby chart for natural materials.

3.5. Tensile Properties Parallel to the Surface

The composites show ductile behavior under tension, with the failure occurring due to the breakage of the mycelium binding between fibers (Figure 10). The values of tensile strength and elastic modulus of NC-mycelium are lower (0.62 MPa) than those of the control samples without nanoclay particles (1.14 MPa) (Table 2). Although pneumatic grips with a cardboard lining were used to avoid failure before testing, the stress concentrations and failure developed for some replicates in the clamping region. Other studies show that analysis of specimens that failed at the grip–specimen interface versus those that failed at mid-substance showed no significant difference in their tensile properties [43]. However, these data must be interpreted with caution, and further studies are recommended.

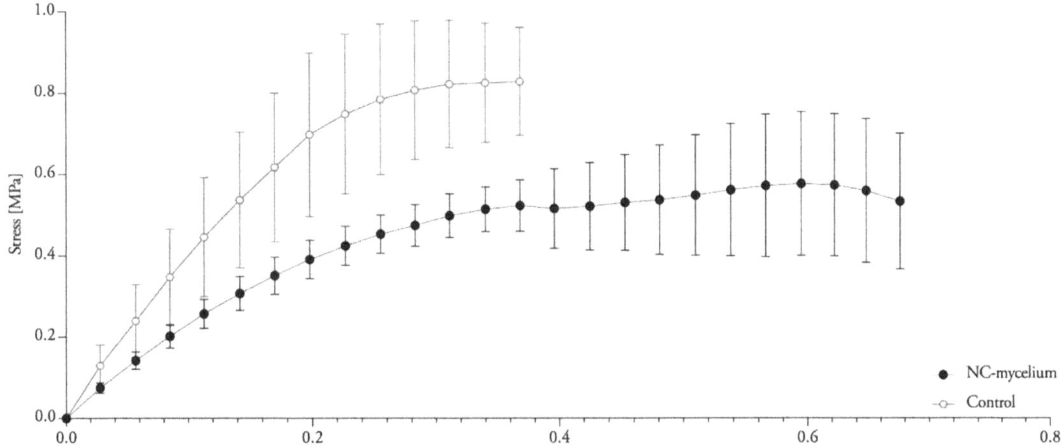

Figure 10. Stress–strain curves of tensile tests of nanoclay–mycelium composites.

Table 2. Overview of the material properties in tension of nanoclay–mycelium composites. The standard deviation is performed with triplicate specimens (mean ± one standard deviation).

Label	Dry Density [kg/m³]	Ultimate Tensile Strength [MPa]	Specific Tensile Strength [kN·m/kg]	Elastic Modulus [GPa]	Specific Modulus [10^6 m² s⁻²]
NC-mycelium hemp composite	654.55 ± 24.39	0.62 ± 0.10	0.95 ± 0.17	0.27 ± 0.04	0.42 ± 0.08
Mycelium hemp composite (control)	980.67 ± 84.77	1.14 ± 0.13	1.15 ± 0.07	0.59 ± 0.15	0.61 ± 0.17

3.6. Tensile Properties Perpendicular to the Surface

Samples were submitted to tension perpendicular to the surface in order to quantify the adhesion of the fibers (Figure 11). The results (Table 3) indicate higher internal bonding for nanoclay-based mycelium composites (0.023 MPa) compared to the control sample (0.007 MPa). These results contrast with the previous tests and might be related to the orientation of the specimen during testing. The samples were cut from a particleboard and teared apart in the opposite direction from the tensile tests described in the previous section. If the materials were grown directly in the geometry of the specimen, the mycelium layer, which grows at the surface of the materials, would probably impact the internal bond. Additionally, the composites do not meet the requirements for lightweight particleboards (0.1 MPa) as established by EN 622-3 (Figure 12).

Figure 11. Detail of specimens and fixtures before and after loading for tension tests perpendicular to the surface (internal bond).

Table 3. Overview of the material properties in internal bond of nanoclay–mycelium composites. The standard deviation is performed with 9 specimens (mean ± one standard deviation).

Label	Dry Density [kg/m^3]	Ultimate Strength [MPa]	Specific Strength [kN·m/kg]	Elastic Modulus [GPa]	Specific Modulus [10^6 m^2 s^{-2}]
NC-mycelium	340.85 ± 2.09	0.017 ± 0.006	0.049 ± 0.017	0.003 ± 0.000	0.009 ± 0.001
Control	492.32 ± 45.40	0.007 ± 0.003	0.01 ± 0.006	0.005 ± 0.002	0.01 ± 0.0007

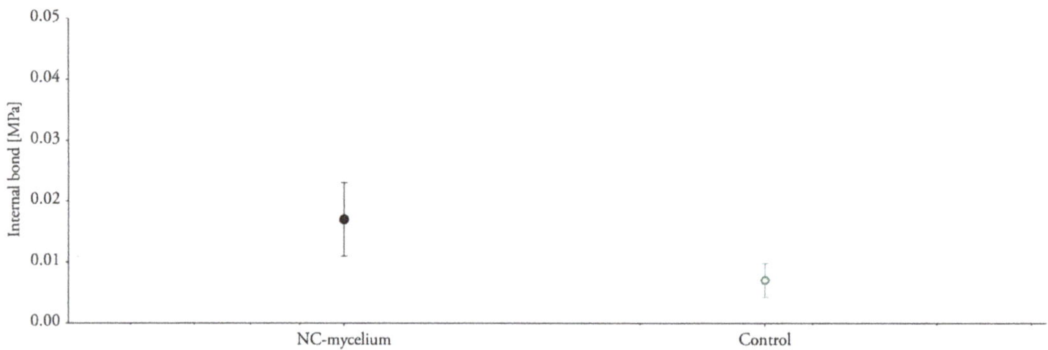

Figure 12. Internal bond strength of nanoclay–mycelium composite. Labels with different letters indicate a statistical difference ($p < 0.05$) among the specimens.

3.7. Compressive Properties

Although nanoclay was expected to influence the compressive strength, it did not significantly (Table 4). The samples showed progressive load-deflection behavior (Figure 13). Mechanical compressive stiffness was obtained from the slope (first distinct straight portion) of the stress–strain curve with the tangent modulus. The compressive stiffness ranged between 0.45 and 0.54 MPa for nanoclay-reinforced composites, compared with 0.34 Mpa for mycelium composites without nanoclay (Figure 14). The values of the compressive Young's modulus range in the same order as previously reported compressive properties for mycelium composites [1]. Yet, we expected to see a more pronounced fluctuation in the results with the addition of different percentages of inorganic particles.

Table 4. Overview of material properties in compression of mycelium composites grown with 1 to 5% nanoclay. The standard deviation is performed with three specimens.

Label	Dry Density [kg/m³]	Compressive Strength [Mpa]	Corresponding Strain [%]	Young's Modulus [Mpa]
1% NC-mycelium	180.14 ± 0.01	0.12354	42.85	0.52 ± 0.02
2.5% NC-mycelium	183.16 ± 0.08	0.12356	42.52	0.54 ± 0.02
3.5% NC-mycelium	176.82 ± 0.02	0.11932	42.41	0.45 ± 0.01
5% NC-mycelium	175.31 ± 0.02	0.12804	43.00	0.45 ± 0.01
Control	172.34 ± 0.04	0.10715	43.50	0.34 ± 0.01

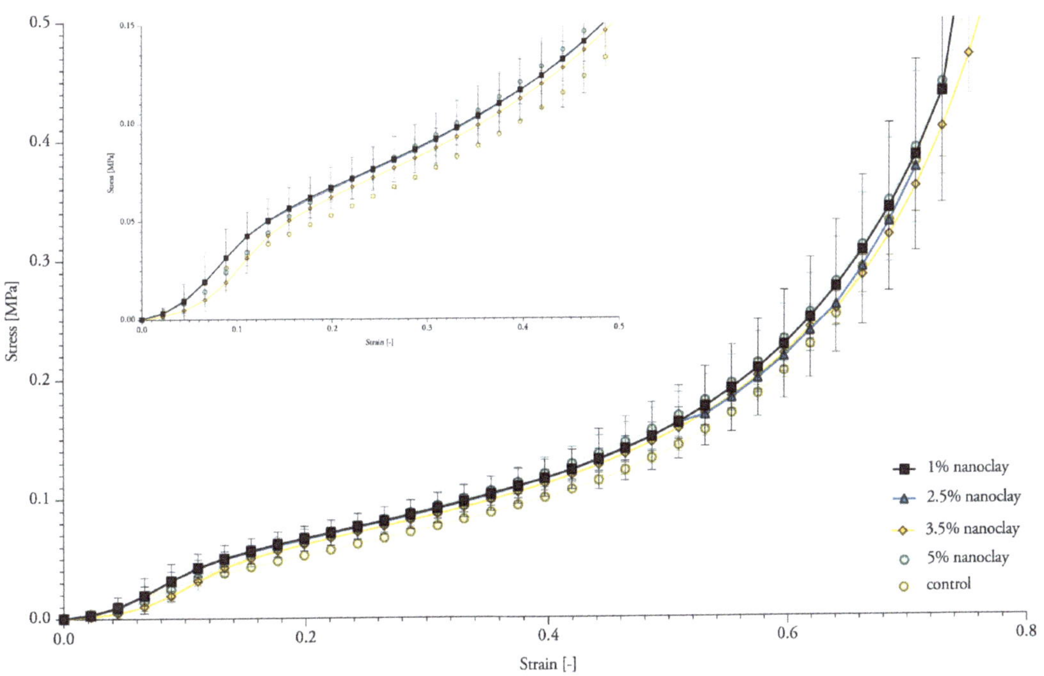

Figure 13. Typical stress–strain curve of compression for mycelium composites grown with 1 to 5% nanoclay, showing a zoom on the first straight portion of the curve. The standard deviation is performed with 3 specimens (mean ± one standard deviation).

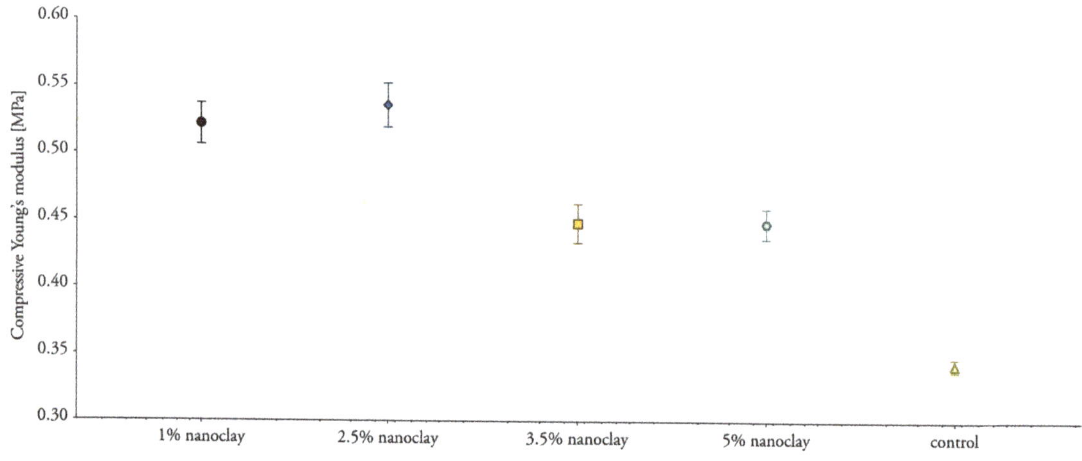

Figure 14. Compressive Young's modulus of mycelium composites during uniaxial compression with 1 to 5% nanoclay. The standard deviation is performed with 3 specimens (mean ± one standard deviation).

4. Discussion

This section discusses the mechanical properties of mycelium materials and nanoclay–mycelium materials and gives an overview of values obtained in other investigations.

4.1. Nanoclay-Mycelium Materials Meet the Requirements of Softboards for the Flexure Index

Nanoclay–mycelium materials meet the requirements of bending for softboards as defined in EN 622-4 (2019) (type SB.LS) in dry, humid, and exterior conditions and load-bearing use. The findings combined extend some of the existing data on the flexural properties for mycelium materials obtained in other studies (Figure 15). The flexural strength of heat-pressed rapeseed and cotton mycelium materials ranges between 0.62 and 0.87 MPa. The flexural modulus of the same samples ranges between 0.03 and 0.07 GPa [3]. However, researchers were able to achieve higher flexural modulus (1.4–4.6 MPa) with mycelium–cotton stalk composites under 160, 180 and 200 °C pressing temperatures [44]. Heat-pressing was done with a pressure of 3.5–4.0 MPa to a density of 600 kg/m^3 [44], while the density of the previous research was only 350 kg/m^3 [3] and average density of samples in this research was 363.18 kg/m^3. Non-pressed cotton plant materials were reported to have flexural strengths in the range of 0.007–0.026 MPa (These data must be compared with caution, because it was normalized by the authors to a standard polystyrene density) [45]. One other study on non-pressed mycelium materials made from crop residues and coated with edible films (carrageenan, chitosan and xanthan gum) reports a flexural strength of 0.01 MPa [46]. The flexural modulus ranges between 66.14 and 71.77 MPa for cotton and hemp mycelium materials [47]. A higher porosity in the discussed studies' material strongly contributes to lower flexural strength, while a high density achieved by heat-pressing can further enhance the material properties.

The amount of existing literature on the mechanical properties of polymer-based montmorillonite nanoclay composites points to the great attention paid to this in academic and industrial sectors. For example, the flexural strength of wood flour–PLA composites with 0.5% sodium-montmorillonite clay was 13% larger than the value for the control sample without nanoclay [48]. In this thesis, flexural strength increased by only 1% for nanoclay-infused materials.

4.2. Heat-Pressing Is a Major Factor in Increasing Tensile Properties

Tensile strength (σ) of pure *Ganoderma lucidum* and *Pleurotus ostreatus* mycelium biofilms ranges between 0.7 and 1.1 MPa [5]. For pure wild types of *S. commune* biofilms, tensile strength ranges between 5.1 and 9.5 MPa, while for modified strains (deletion of the hydrophobin gene sc3), tensile strength increases to 15.6–40.4 MPa [49]. Reported tensile properties of composite mycelium materials vary for different types of feedstocks (Figure 16), and are for example higher for sawdust substrates (σ 0.05 MPa) than for straw substrates (σ 0.01–0.04 MPa) [3]. Heat-pressing was reported to be the major factor in increasing tensile strength and modulus (σ 0.13–0.24 MPa, E 35–97 MPa) in comparison to cold-pressing (σ 0.03 MPa, E 6–9 MPa) or non-pressing (σ 0.01–0.05 MPa, E 2–13 MPa) [3]. In comparison, the developed samples and conducted experiments in this research resulted on average in tensile strength that was six times higher (σ 1.14 MPa, E 585.98 MPa). A possible explanation for the divergence of these findings with the literature may the use of different production and testing methods, as there is no adequate standard for the production and testing of mycelium materials.

Tensile strength of wood flour–PLA composites slightly improves by 4% with the addition of montmorillonite clay [48]. According to the authors, the higher strength of the composites is caused by the intercalated structure of montmorillonite in the wood flour cell wall [48]. In addition, incorporation of nanoclay reduces the polarity of natural fiber and benefited the interface between the fiber and the PLA matrix [50]. The intercalated montmorillonite in polymer matrix improves the capability to transfer load from the polymer to the strong montmorillonite [51].

4.3. Mycelium Composites Have a Low Internal Bond

Other studies found an internal bond strength between 0.05 and 0.18 MPa, with mycelium–cotton stalk composites under 160, 180 and 200 °C pressing temperatures [44]. In a follow-up study, the same authors improved the internal bond to 0.34 MPa by immersing the materials in water until an uptake of 30% before heat-pressing. In comparison with that study, the internal bonding results presented in this work are very low: 0.023 MPa for nanoclay-based mycelium composites and 0.007 MPa for mycelium composites without nanoclay. Nonetheless, no other satisfying results were achieved by other researchers, as for example Sun et al. note that the composites were too weak to be measured in the internal bond strength test [52]. The values for the internal bond strength of the wood- and resin-based boards can significantly increase with the addition of 1–5% nanoclay [53,54]. The authors observed that the increase in brittleness due to the addition of nanoclay inclusions was balanced by the increase in ductility due to the higher press temperature. Previous studies reported that the maximum internal bond value (1.99 MPa) was obtained for particleboard with 2% nanoclay loading, which accounted for an improvement of 18.03% compared to the reference board [55]. The improved behavior can be explained by the percolation theory, which is applied to the networking capability of the clay nano-platelets [56].

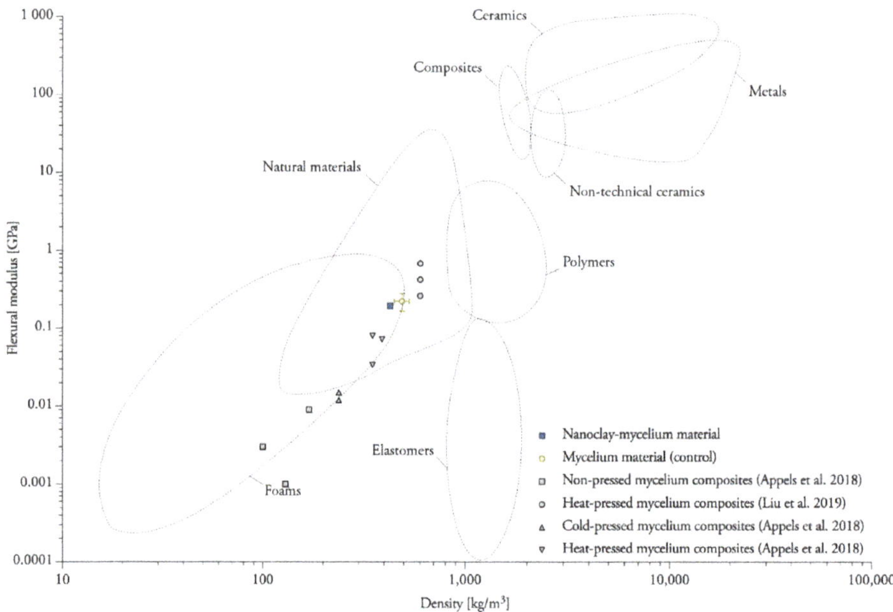

Figure 15. Overview of data on the flexural modulus for mycelium materials obtained in this and other studies. Data compiled from [3,44]. Plotted on Ashby chart for engineering materials [57].

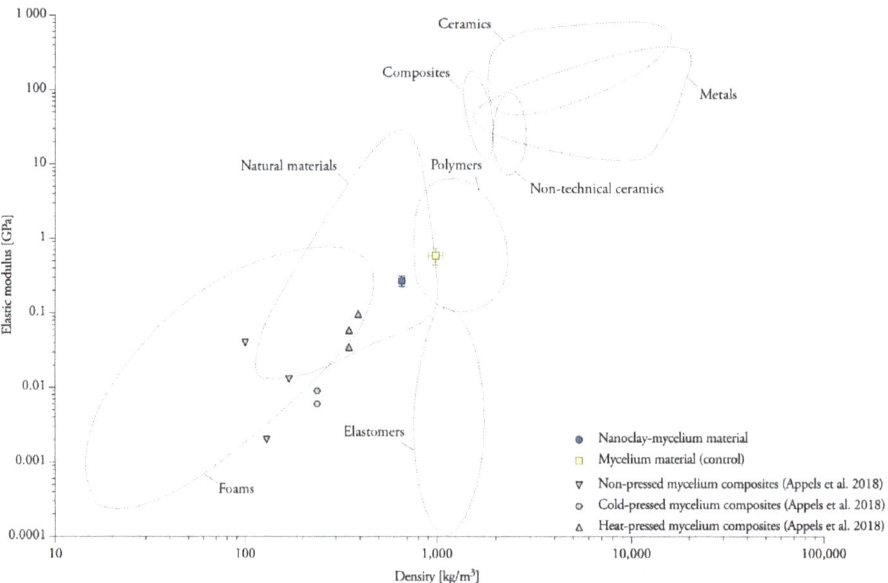

Figure 16. Overview of data about the elastic modulus for mycelium materials obtained in this and other studies. Data compiled from [3]. Plotted on Ashby chart for engineering materials [57].

5. Conclusions

The effects of hybridization of mycelium composites with nanoclay were evaluated in terms of morphological, chemical, and mechanical properties. Nanoclay causes a decreased growth rate, which appears to be up to 29% slower than that of control samples. Moreover, the slower growth of mycelium makes the substrate prone to contaminations. The distribution of clay particles was visualized through SEM imagery. The particles can penetrate into the fibers' cell-wall structure, and some nanoclay particles were observed around the hyphae. The FTIR study demonstrated that *T. versicolor* has more difficulty accessing and decaying the hemicellulose and lignin when the amount of nanoclay increases. It seems that cellulose is more difficult to degrade when nanoclay is present in the composite. The ability of the fungus to degrade the nanoclay-coated fibers decreases with the increase of nanoclay.

The addition of nanoclay to the fibers does not significantly affect the bending behavior of mycelium composites. Tensile strength and elastic modulus of nanoclay–mycelium materials are lower than those of the control samples without nanoclay particles. While nanoclay enhances the properties of polymer composites, the hybridization with mycelium composites was not successful. On the other hand, strength values are higher than most of the existing data about flexural and tensile properties for mycelium materials obtained in other studies. The results indicate a slightly higher but negligible internal bonding and compressive stiffness for nanoclay-based mycelium composites.

The challenges of the hybridization of mycelium materials with inorganic reinforcements have become clearer. In general, this work provides a useful method for combining renewable organic sources, such as natural fibers and fungal mycelium, with nanoparticles. There is still a long way to go to understand all biological and engineering mechanisms needed to achieve specific structural properties using nano-mycelium materials.

Author Contributions: E.E. conceived the study, designed the experiments, analyzed the tests, carried out the tests. E.E., L.D.L. and E.P. interpreted results and edited the manuscript. All authors have read and agreed to the published version of the manuscript.

Funding: This research was funded by Vrije Universiteit Brussel and by Research Foundation Flanders (FWO-Vlaanderen), grant number 1S36417N.

Institutional Review Board Statement: Not applicable.

Informed Consent Statement: Not applicable.

Data Availability Statement: Data is available upon request.

Acknowledgments: This article is based on Chapter 5 of the first author's Ph.D. thesis [2]. The authors would like to thank Svetlana Verbruggen and Frans Boulpaep of the MEMC research group for their support with the mechanical tests and the set-up of the instruments.

Conflicts of Interest: The authors declare no conflict of interest.

References

1. Elsacker, E.; Vandelook, S.; Brancart, J.; Peeters, E.; De Laet, L. Mechanical, Physical and Chemical Characterisation of Mycelium-Based Composites with Different Types of Lignocellulosic Substrates. *PLoS ONE* **2019**, *14*, e0213954. [CrossRef] [PubMed]
2. Elsacker, E. Mycelium Matters—An Interdisciplinary Exploration of the Fabrication and Properties of Mycelium-Based Materials. Ph.D. Thesis, VUBPRESS, Vrije Universiteit Brussel, Brussles, Belgium, 2021.
3. Appels; Camere, S.; Montalti, M.; Karana, E.; Jansen, K.M.B.; Dijksterhuis, J.; Krijgsheld, P.; Wösten, H.A.B. Fabrication Factors Influencing Mechanical, Moisture- and Water-Related Properties of Mycelium-Based Composites. *Mater. Des.* **2018**, *161*, 64–71. [CrossRef]
4. Islam, M.R.; Tudryn, G.; Bucinell, R.; Schadler, L.; Picu, R.C. Morphology and Mechanics of Fungal Mycelium. *Sci. Rep.* **2017**, *7*, 13070. [CrossRef]
5. Haneef, M.; Ceseracciu, L.; Canale, C.; Bayer, I.S.; Heredia-Guerrero, J.A.; Athanassiou, A. Advanced Materials from Fungal Mycelium: Fabrication and Tuning of Physical Properties. *Sci. Rep.* **2017**, *7*, 41292. [CrossRef]
6. Jones, M.; Mautner, A.; Luenco, S.; Bismarck, A.; John, S. Engineered Mycelium Composite Construction Materials from Fungal Biorefineries: A Critical Review. *Mater. Des.* **2020**, *187*, 108397. [CrossRef]
7. Attias, N.; Danai, O.; Abitbol, T.; Tarazi, E.; Ezov, N.; Pereman, I.; Grobman, Y.J. Mycelium Bio-Composites in Industrial Design and Architecture: Comparative Review and Experimental Analysis. *J. Clean. Prod.* **2020**, *246*, 119037. [CrossRef]
8. Elsacker, E.; Vandelook, S.; Van Wylick, A.; Ruytinx, J.; De Laet, L.; Peeters, E. A Comprehensive Framework for the Production of Mycelium-Based Lignocellulosic Composites. *Sci. Total Environ.* **2020**, *725*, 138431. [CrossRef]

9. Elsacker, E.; Vandelook, S.; Damsin, B.; Van Wylick, A.; Peeters, E.; De Laet, L. Mechanical Characteristics of Bacterial Cellulose-Reinforced Mycelium Composite Materials. *Fungal Biol. Biotechnol.* **2021**, *8*, 18. [CrossRef]

10. Attias, N.; Reid, M.; Mijowska, S.C.; Dobryden, I.; Isaksson, M.; Pokroy, B.; Grobman, Y.J.; Abitbol, T. Biofabrication of Nanocellulose–Mycelium Hybrid Materials. *Adv. Sustain. Syst.* **2020**, *5*, 2000196. [CrossRef]

11. Jones, M.; Bhat, T.; Huynh, T.; Kandare, E.; Yuen, R.; Wang, C.H.; John, S. Waste-Derived Low-Cost Mycelium Composite Construction Materials with Improved Fire Safety. *Fire Mater.* **2018**, *42*, 816–825. [CrossRef]

12. Moser, F.J.; LWormit, A.; Reimer, J.J.; Jacobs, G.; Trautz, M.; Hillringhaus, F. Fungal Mycelium as a Building Material; Fachgruppe Biologie, Lehrstuhl für Botanik und Institut für Biologie I (Botanik), Lehrstuhl und Institut für Allgemeine Konstruktionstechnik des Maschinenbaus, Lehrstuhl für Tragkonstruktionen. In Proceedings of the Annual Symposium of the International Association for Shell and Spatial Structures (IASS 2017), Hamburg, Germany, 25–28 September 2017.

13. Ross, P. Method for Producing Fungus Structures. U.S. Patent 9,410,116, 9 August 2016.

14. Abend, S.; Lagaly, G. Sol-Gel Transitions of Sodium Montmorillonite Dispersions. *Appl. Clay Sci.* **1999**, *16*, 201–227. [CrossRef]

15. Bensadoun, F.; Kchit, N.; Billotte, C.; Bickerton, S.; Trochu, F.; Ruiz, E. A Study of Nanoclay Reinforcement of Biocomposites Made by Liquid Composite Molding. *Int. J. Polym. Sci.* **2011**, *2011*, 964193. [CrossRef]

16. Lee, W.-F.; Fu, Y.-T. Effect of Montmorillonite on the Swelling Behavior and Drug-Release Behavior of Nanocomposite Hydrogels. *J. Appl. Polym. Sci.* **2003**, *89*, 3652–3660. [CrossRef]

17. Kristak, L.; Antov, P.; Bekhta, P.; Lubis, M.A.R.; Iswanto, A.H.; Reh, R.; Sedliacik, J.; Savov, V.; Taghiyari, H.R.; Papadopoulos, A.N.; et al. Recent Progress in Ultra-Low Formaldehyde Emitting Adhesive Systems and Formaldehyde Scavengers in Wood-Based Panels: A Review. *Wood Mater. Sci. Eng.* **2022**, 1–20. [CrossRef]

18. Yano, K.; Usuki, A.; Okada, A.; Kurauchi, T.; Kamigaito, O. Synthesis and Properties of Polyimide-Clay Hybrid. *J. Polym. Sci. Part A Polym. Chem.* **1993**, *31*, 2493–2498. [CrossRef]

19. Yano, K.; Usuki, A.; Okada, A. Synthesis and Properties of Polyimide-Clay Hybrid Films. *J. Polym. Sci. Part A Polym. Chem.* **1997**, *35*, 2289–2294. [CrossRef]

20. Lee, D.C.; Jang, L.W. Preparation and Characterization of PMMA-Clay Hybrid Composite by Emulsion Polymerization. *J. Appl. Polym. Sci.* **1996**, *61*, 1117–1122. [CrossRef]

21. Okamoto, M.; Morita, S.; Taguchi, H.; Kim, Y.H.; Kotaka, T.; Tateyama, H. Synthesis and Structure of Smectic Clay/Poly(Methyl Methacrylate) and Clay/Polystyrene Nanocomposites via in Situ Intercalative Polymerization. *Polymer* **2000**, *41*, 3887–3890. [CrossRef]

22. Fu, X.; Qutubuddin, S. Synthesis of Polystyrene-Clay Nanocomposites. *Mater. Lett.* **2000**, *42*, 12–15. [CrossRef]

23. DePolo, W.S.; Baird, D.G. Particulate Reinforced PC/PBT Composites. II. Effect of Nano-Clay Particles on Dimensional Stability and Structure-Property Relationships. *Polym. Compos.* **2009**, *30*, 200–213. [CrossRef]

24. Rangavar, H.; Taghiyari, H.R.; Oromiehie, A.; Gholipour, T.; Safarpour, A. Effects of Nanoclay on Physical and Mechanical Properties of Wood-Plastic Composites. *Wood Mater. Sci. Eng.* **2017**, *12*, 211–219. [CrossRef]

25. Delhom, C.D.; White-Ghoorahoo, L.A.; Pang, S.S. Development and Characterization of Cellulose/Clay Nanocomposites. *Compos. Part B Eng.* **2010**, *41*, 475–481. [CrossRef]

26. Gao, F. Clay/Polymer Composites: The Story. *Mater. Today* **2004**, *7*, 50–55. [CrossRef]

27. Sinha Ray, S.; Okamoto, M. Polymer/Layered Silicate Nanocomposites: A Review from Preparation to Processing. *Prog. Polym. Sci.* **2003**, *28*, 1539–1641. [CrossRef]

28. Utracki, L.A.; Sepehr, M.; Boccaleri, E. Synthetic, Layered Nanoparticles for Polymeric Nanocomposites (PNCs). *Polym. Adv. Technol.* **2007**, *18*, 1–37. [CrossRef]

29. Dean, D.; Obore, A.M.; Richmond, S.; Nyairo, E. Multiscale Fiber-Reinforced Nanocomposites: Synthesis, Processing and Properties. *Compos. Sci. Technol.* **2006**, *66*, 2135–2142. [CrossRef]

30. Laoutid, F.; Bonnaud, L.; Alexandre, M.; Lopez-Cuesta, J.-M.; Dubois, P. New Prospects in Flame Retardant Polymer Materials: From Fundamentals to Nanocomposites. *Mater. Sci. Eng. R Rep.* **2009**, *63*, 100–125. [CrossRef]

31. Bee, S.-L.; Abdullah, M.A.A.; Bee, S.-T.; Sin, L.T.; Rahmat, A.R. Polymer Nanocomposites Based on Silylated-Montmorillonite: A Review. *Prog. Polym. Sci.* **2018**, *85*, 57–82. [CrossRef]

32. White, L.A. Preparation and Thermal Analysis of Cotton-Clay Nanocomposites. *J. Appl. Polym. Sci.* **2003**, *92*, 2125–2131. [CrossRef]

33. Schindelin, J.; Arganda-Carreras, I.; Frise, E.; Kaynig, V.; Longair, M.; Pietzsch, T.; Preibisch, S.; Rueden, C.; Saalfeld, S.; Schmid, B.; et al. Fiji: An Open-Source Platform for Biological-Image Analysis. *Nat. Methods* **2012**, *9*, 676–682. [CrossRef]

34. *ISO 16978*; Wood-Based Panels—Determination of Modulus of Elasticity in Bending and of Bending Strength. BSI: London, UK, 2003.

35. *ISO 12344*; Thermal Insulating Products for Building Applications—Determination of Bending Behaviour. BSI: London, UK, 2010.

36. *ASTM D1037*; Test Methods for Evaluating Properties of Wood-Base Fiber and Particle Panel Materials. ASTM International: West Conshohocken, PA, USA, 2020.

37. *EN 319*; Particleboards and Fibreboards. Determination of Tensile Strength Perpendicular to the Plane of the Board. BSI: London, UK, 1993.

38. *ISO 29469*; Thermal Insulating Products for Building Applications—Determination of Compression Behaviour. BSI: London, UK, 2008.

39. Bari, E.; Taghiyari, H.R.; Schmidt, O.; Ghorbani, A.; Aghababaei, H. Effects of Nano-Clay on Biological Resistance of Wood-Plastic Composite against Five Wood-Deteriorating Fungi. *Maderas Cienc. Tecnol.* **2015**, *17*, 205–212. [CrossRef]
40. Kord, B.; Jari, E.; Najafi, A.; Tazakorrezaie, V. Effect of Nanoclay on the Decay Resistance and Physicomechanical Properties of Natural Fiber-Reinforced Plastic Composites against White-Rot Fungi (*Trametes Versicolor*). *J. Thermoplast. Compos. Mater.* **2014**, *27*, 1085–1096. [CrossRef]
41. Luo, F.; Zhong, Z.; Liu, L.; Igarashi, Y.; Xie, D.; Li, N. Metabolomic Differential Analysis of Interspecific Interactions among White Rot Fungi Trametes Versicolor, Dichomitus Squalens and Pleurotus Ostreatus. *Sci. Rep.* **2017**, *7*, 5265. [CrossRef]
42. Wegst, U.G.K.; Ashby, M.F. The Mechanical Efficiency of Natural Materials. *Philos. Mag.* **2004**, *84*, 2167–2186. [CrossRef]
43. Ng, B.H.; Chou, S.M.; Krishna, V. The Influence of Gripping Techniques on the Tensile Properties of Tendons. *Proc. Inst. Mech. Eng. H* **2005**, *219*, 349–354. [CrossRef]
44. Liu, R.; Long, L.; Sheng, Y.; Xu, J.; Qiu, H.; Li, X.; Wang, Y.; Wu, H. Preparation of a Kind of Novel Sustainable Mycelium/Cotton Stalk Composites and Effects of Pressing Temperature on the Properties. *Ind. Crops Prod.* **2019**, *141*, 111732. [CrossRef]
45. Holt, G.A.; Mcintyre, G.; Flagg, D.; Bayer, E.; Wanjura, J.D.; Pelletier, M.G. Fungal Mycelium and Cotton Plant Materials in the Manufacture of Biodegradable Molded Packaging Material: Evaluation Study of Select Blends of Cotton Byproducts. *J. Biobased Mater. Bioenergy* **2012**, *6*, 431–439. [CrossRef]
46. López Nava, J.A.; Méndez González, J.; Ruelas Chacón, X.; Nájera Luna, J.A. Assessment of Edible Fungi and Films Bio-Based Material Simulating Expanded Polystyrene. *Mater. Manuf. Processes* **2016**, *31*, 1085–1090. [CrossRef]
47. Ziegler, A.R.; Bajwa, S.G.; Holt, G.A.; McIntyre, G.; Bajwa, D.S. Evaluation of Physico-Mechanical Properties of Mycelium Reinforced Green Biocomposites Made from Cellulosic Fibers. *Appl. Eng. Agric.* **2016**, *32*, 931–938. [CrossRef]
48. Liu, R.; Chen, Y.; Cao, J. Effects of Modifier Type on Properties of in Situ Organo-Montmorillonite Modified Wood Flour/Poly(Lactic Acid) Composites. *ACS Appl. Mater. Interfaces* **2016**, *8*, 161–168. [CrossRef]
49. Appels; Dijksterhuis, J.; Lukasiewicz, C.E.; Jansen, K.M.B.; Wösten, H.A.B.; Krijgsheld, P. Hydrophobin Gene Deletion and Environmental Growth Conditions Impact Mechanical Properties of Mycelium by Affecting the Density of the Material. *Sci. Rep.* **2018**, *8*, 4704. [CrossRef] [PubMed]
50. Liu, R.; Liu, M.; Hu, S.; Huang, A.; Ma, E. Comparison of Six WPCs Made of Organo-Montmorillonite-Modified Fibers of Four Trees, Moso Bamboo and Wheat Straw and Poly(Lactic Acid) (PLA). *Holzforschung* **2018**, *72*, 735–744. [CrossRef]
51. Gurunathan, T.; Mohanty, S.; Nayak, S.K. Effect of Reactive Organoclay on Physicochemical Properties of Vegetable Oil-Based Waterborne Polyurethane Nanocomposites. *RSC Adv.* **2015**, *5*, 11524–11533. [CrossRef]
52. Sun, W.; Tajvidi, M.; Hunt, C.G.; McIntyre, G.; Gardner, D.J. Fully Bio-Based Hybrid Composites Made of Wood, Fungal Mycelium and Cellulose Nanofibrils. *Sci. Rep.* **2019**, *9*, 3766. [CrossRef] [PubMed]
53. Ashori, A.; Nourbakhsh, A. Effects of Nanoclay as a Reinforcement Filler on the Physical and Mechanical Properties of Wood-Based Composite. *J. Compos. Mater.* **2009**, *43*, 1869–1875. [CrossRef]
54. Salari, A.; Tabarsa, T.; Khazaeian, A.; Saraeian, A. Effect of Nanoclay on Some Applied Properties of Oriented Strand Board (OSB) Made from Underutilized Low Quality Paulownia (Paulownia Fortunei) Wood. *J. Wood Sci.* **2012**, *58*, 513–524. [CrossRef]
55. Ismita, N.; Lokesh, C. Effects of Different Nanoclay Loadings on the Physical and Mechanical Properties of Melia Composita Particle Board. *Bois. Trop.* **2018**, *334*, 7. [CrossRef]
56. Lei, H.; Du, G.; Pizzi, A.; Celzard, A. Influence of Nanoclay on Urea-Formaldehyde Resins for Wood Adhesives and Its Model. *J. Appl. Polym. Sci.* **2008**, *109*, 2442–2451. [CrossRef]
57. Ashby, M. Designing Architectured Materials. *Scr. Mater.* **2013**, *68*, 4–7. [CrossRef]

 *biomimetics*

Article

Mycelium-Based Composite Graded Materials: Assessing the Effects of Time and Substrate Mixture on Mechanical Properties

Ali Ghazvinian * and Benay Gürsoy

Department of Architecture, Penn State University, University Park, PA 16802, USA; bug61@psu.edu
* Correspondence: axg1370@psu.edu; Tel.: +1-(618)-425-5170

Abstract: Mycelium-based composites (MBC) are biodegradable, lightweight, and regenerative materials. Mycelium is the vegetative root of fungi through which they decompose organic matter. The proper treatment of the decomposition process results in MBC. MBC have been used in different industries to substitute common materials to address several challenges such as limited resources and large landfill waste after the lifecycle. One of the industries which started using this material is the architecture, engineering, and construction (AEC) industry. Therefore, scholars have made several efforts to introduce this material to the building industry. The cultivation process of MBC includes multiple parameters that affect the material properties of the outcome. In this paper, as a part of a larger research on defining a framework to use MBC as a structural material in the building industry, we defined different grades of MBC to address various functions. Furthermore, we tested the role of substrate mixture and the cultivation time on the mechanical behavior of the material. Our tests show a direct relationship between the density of the substrate and the mechanical strength. At the same time, there is a reverse relation between the cultivation time and the material mechanical performance.

Keywords: mycelium-based composites; compressive structures; compressive strength; digital image correlation; masonry

Citation: Ghazvinian, A.; Gürsoy, B. Mycelium-Based Composite Graded Materials: Assessing the Effects of Time and Substrate Mixture on Mechanical Properties. *Biomimetics* **2022**, *7*, 48. https://doi.org/10.3390/biomimetics7020048

Academic Editors: Andrew Adamatzky, Phil Ayres and Han A.B. Wösten

Received: 15 March 2022
Accepted: 16 April 2022
Published: 19 April 2022

Publisher's Note: MDPI stays neutral with regard to jurisdictional claims in published maps and institutional affiliations.

1. Introduction

The architecture, engineering, and construction (AEC) industry consumes half of the mineral resources and contributes the most to landfill waste [1]. Therefore, there is a need for alternative construction materials and greener energy resources to reduce the AEC industry's global greenhouse gas emissions and landfill waste. A circular approach must replace the current linear approach of extract-produce-use-dump. This approach emphasizes reuse, remanufacturing, refurbishment, repair, and upgrading of materials and utilization of solar, wind, biomass, and waste-derived energy during the product's life cycle. [2]. One possible path to conform with circular economies is through the use of bio-based materials as alternatives to conventional materials in construction [3,4]. These materials can be produced using waste as one of their initial ingredients and can become reusable, recyclable, or compostable at the end of their lifecycles.

Due to the rapid population growth worldwide, the global demand for food and agricultural wastes and byproducts has increased [5]. Besides, this growing population needs affordable habitat. Since the traditional ways of dumping agricultural waste into landfills or burning them impact global warming [6], converting agricultural waste into building components seems an optimal solution for these problems. Various bio-based materials are studied in this context [7]. The development of these materials can be costly, time-consuming, and inefficient due to the problematic methods of processing and functionalization [8], although they have a multitude of advantages. One of these materials is mycelium-based composites (MBC). MBC is manufactured using a low-energy and natural process that sequesters carbon and uses waste as the input. There are several applications of mycelium-based matter and fungal biotechnologies [9]. This research focuses on using MBC as load-bearing masonry components in construction.

2. Background

Mycelium is the vegetative root of fungi. Mycelium has a long, branching, and filamentous structure called hyphae that secrete enzymes to break down the biopolymers into simpler bodies of digestible carbon-based nutrients. The outcome of this process is an organic colony of hyphae. The organic matter bounds with this hyphal structure and forms fungal skin during this process. When this process is ceased through drying or heating, the incomplete process results in MBC. This composite material is made of the substrate as the filler and the hyphal mycelium as the binder. Without the hyphal binder, the substrate works as an inconsistent mass of particles and shows negligible mechanical performance. This bio-based composite can be shaped to produce panels, bricks, or various objects [10]. The properties of MBC depend on various cultivation parameters, such as the fungal species, substrates, growth conditions, processing of material, and additives [5,11]. This dependence on the controllable parameters enables MBC to meet specific application requirements [5]. Among these applications are acoustic insulation [12,13], thermal insulation [14–17], packaging [18–21], fire retardants [22–24], and structural building components [11,25–35].

Scholars are making various efforts to make MBC meet the performance requirements of the AEC industry. One approach is to enhance the material properties of MBC by investigating the cultivation parameters. Another approach is to develop novel design and fabrication techniques around the specific material properties of MBC via geometry and form optimizations [32]. This paper focuses on the former approach.

2.1. The Cultivation Process of MBC

The cultivation process of MBC has three major phases: inoculation, growth, and ceasing. Figure 1 illustrates the cultivation process of MBC.

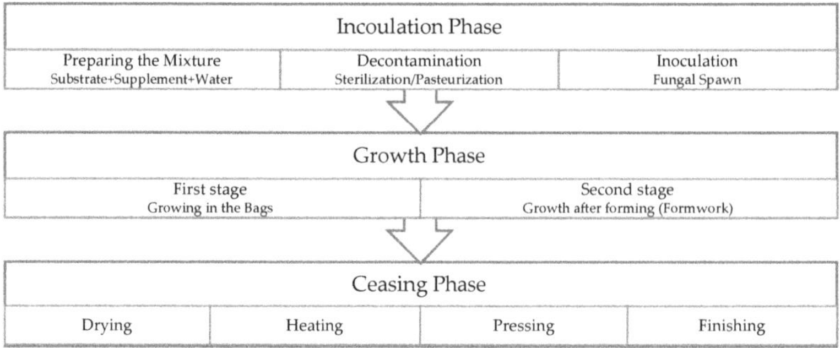

Figure 1. MBC cultivation process.

The substrate mixture is first prepared with organic matter as the filler, optional supplements that provide additional carbon and nitrogen, and water in the inoculation phase. This mixture is then pasteurized or sterilized to eliminate any other living organism that can pose a threat to mycelium growth. The substrate mixture is typically placed within autoclavable bags and heated in autoclave machines for less than an hour at 121 °C. Alternative methods such as the use of herbal remedies [15] and heating in lower temperatures for longer durations are also available. Once the sterilization is complete, the substrate mixture is cooled down to room temperature, and the fungal spawns are added. The ingredients of the substrate mixture and the fungal species used in inoculation affect the material properties of MBC [36,37]. The second phase of the MBC cultivation process, the growth, starts in the autoclavable bags. Depending on the application, growth can continue within formworks [38,39]. Some studies explored the extrusion of the substrate mixtures through additive manufacturing [40–48]. Some of the parameters that affect the resulting material properties in this phase are the duration of growth, environmental conditions such as temperature and relative humidity, and CO_2 concentration [49,50]. The third phase, ceasing, also influences the resulting MBC. In this phase, mycelial growth is stopped by either drying or heating the colonized substrate mixture. Researchers have also explored experimental processes such as rubbing herbal oils and hot or cold pressing the composite. The method of ceasing and parameters associated with the chosen method (for example, pressing temperature) [51] can alter the material properties of MBC.

2.2. Role of Cultivation Parameters on the Mechanical Behavior of MBC

Various scholars have studied the mechanical behavior of MBC, specifically compressive strength. The consensus is that the mechanical behavior of MBC is comparable to that of synthetic foams, with room for enhancement [10]. MBC made of low-weight substrate mixtures have similar compressive strength to polystyrene foams and are weaker than polyurethane and phenolic formaldehyde resins [5].

Among the various research that explore the role of cultivation parameters on the outcome, Holt et al. (2012) [20] studied the substrate mixtures of six different cotton plant biomass. Yang et al. (2017) [7] experimented with the degree of compaction of substrates within formworks. They also tested the role of the duration of cultivation (two and six weeks) on the outcome. Their results show that the densely packed samples have higher compressive strength and elastic moduli. In comparison, the longer duration of cultivation results in better compressive strength and lower elastic moduli. Islam et al. (2018) [52] defined three sizes: small (from 0.4 to 0.9 mm), medium (from 0.9 to 1.7 mm), and large (from 1.7 to 6.7 mm) fillers (such as sawdust), and a mixture of these three to study the effects of filler size on compressive strength. They reported that the mechanical behavior of MBC is not affected by the filler size. On the contrary, the experiments by Elsacker et al. (2019) [53] show that fiber size is more influential on the mechanical strength than the type of fibrous substrate used. Except for the dust material that yielded poor growth and mechanical properties, the more chopped material resulted in better strength. In conformation with Yang et al. (2017) [7], their experiments showed that densely packed substrates had better mechanical properties than loose fibers.

Attias et al. (2019) [54] experimented with three different spawns and two growing protocols. They used *Colorius*, *Trametes*, and *Ganoderma* species and cultivated them with a 7-day difference in incubation time to establish their final experiments on the suitability of these conditions. They continued their study in Attias et al. (2020) [25] and reported better mechanical behavior for the samples cultivated with *Ganoderma* species. They also reported a reverse relation between the mycelium colonization and compressive strength, suggesting that shorter incubation periods restrict the organic matter digestion and preserve the mechanical characteristics of the substrate mixture. On the other hand, longer incubation times change the material content of the digestible substrate more and weaken the produced MBC. Bruscato et al. (2019) [55] utilized three different species of *Pycnoporus sanguineus*, *Pleurotus albidus*, and *Lentinus velutinus* for cultivation with sawdust and wheat bran. They found *L. velutinus* to be resulting in weaker composites because of the way mycelium colonizes during the growth. They suggest that the colonization of this species is more accentuated around the interstices of the mixture fillers than the overall agglomerate, which is different from the other two species, and that this is the reason for lower mechanical strength.

Appels et al. (2019) [50] Studied the role of different species, substrates, and pressing conditions on material behavior. They tested the bending capacity of MBC in their studies, with *T. multicolor* and *P. ostreatus* growing on rapeseed straw and beech sawdust. They also used three different conditions for ceasing: non-pressed, cold-pressed, and heat-pressed. Their most important result is the direct relation of mechanical strength and elastic moduli with the pressing, mainly through hot-pressing. They reported that heat-pressing shifts MBC performance from foam-like to wood-like. They also explained that colonization of mycelium occurs better at the outer parts of the substrates than the cores, emphasizing the importance of forcing air through the center of the substrate. Additional research on the optimal temperatures for hot-pressing the substrates reveals that lower temperatures result in weaker materials, and higher temperatures may burn the materials [51].

Ghazvinian et al. (2019) [28] studied the role of supplements on two different substrates with *P. ostreatus*. Their results show slightly stronger materials with 7% wheat bran in the inoculation phase. There is also a considerable difference between MBC cultivated with oak sawdust and wheat straw. Ongpeng et al. (2020) [33] utilized clay, rice bran, and sawdust mixed with different waste materials to make MBC bricks and tested them to compare with masonry minimum limits. They also used the compressed substrates without mycelium to study the role of mycelium as the binding agent in these bricks. For the clay samples, the mycelium content was not modifying the characteristics, while for the other samples, mycelium bound the substrates, which resulted in stronger materials. Besides, all the mycelium-based bricks passed the minimum compressive strength for masonry bricks. One other important aspect studied by Zimele et al. (2020) [27] is the biodegradability of this material after use. Compared to hemp magnesium oxychloride concrete and cemented wood wool panel, two other bio-based materials, MBC showed quadruple biodegradability. This biodegradability is a testimony of the circularity of MBC when used in the AEC industry. An LCA analysis of MBC bricks on the lab and industrial scale shows reductions in most impact categories. Biodegradability might reduce the AEC industry's environmental footprint if conventional building materials can be substituted with MBC [56].

In a more recent study, Elsacker et al. (2021) [57] investigated the addition of other organisms, such as bacterial cellulose to *T. versicolor* inoculated on hemp-based substrates to make particleboards. They found the enhancing role of bacterial cellulose in improving internal bonding.

3. Materials and Methods

This paper studied the effects of three different MBC cultivation parameters on compressive strength. The first parameter studied is the substrate mixture type. We created various mixtures by combining particle-based (i.e., sawdust) and fibrous (i.e., straw) materials. The other two parameters we studied are related to the duration of cultivation. The entire and partial cultivation time in bags and formworks has been investigated.

As mentioned before, there are three primary phases in the cultivation process of MBC. Figure 2 illustrates the materials and methods employed in these phases as part of this research.

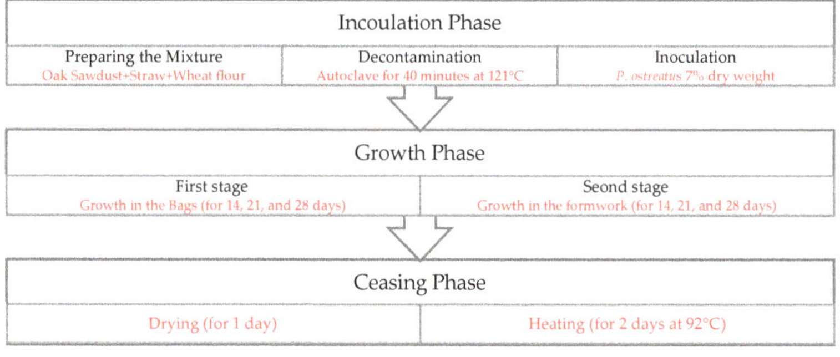

Figure 2. Details of MBC cultivation.

3.1. Substrate Mixtures Preparation

Five different substrate mixtures are created for the experiment based on the sawdust to straw ratios. Oakwood pellets (Atlanta, GA, USA) and wheat straws (chopped, 3 cm long) are the base materials used for the mixtures. The various mixtures have straw and sawdust ratios of 1 to 1, 2, 3, 7, and one with sawdust only, as shown in Table 1. In addition, unbleached whole wheat flour (Bentonville, AR, USA) has been added to the mixtures by

7% of the dry weight to enhance the growth process. The water content of the mixtures has been controlled to stay between 65% to 70%, following the best practices for *Pleurotus ostreatus* mushroom cultivation.

Table 1. Characteristics of different mixtures.

Mixture	Sawdust Ratio	Straw Ratio	Water Content	Wheat Flour Content	*Fungal* sp. Content
A	1	0	65–70%	7% DW	7% DW
B	1	7	65–70%	7% DW	7% DW
C	1	3	65–70%	7% DW	7% DW
D	1	2	65–70%	7% DW	7% DW
E	1	1	65–70%	7% DW	7% DW

DW = dry weight of the mixture.

The five substrate mixtures have been hand-mixed for 120 s to distribute the ingredients and water evenly. They were then moved to autoclavable bags (Impresa Mushroom Growing Bags) for sterilization and test paper bags for humidity check. Bags were sterilized in the autoclave machine at 121 °C temperature for 40 min and then removed from the autoclave machine and left to cool down (overnight, at room temperature).

3.2. Cultivation of Materials

The fungal spawns of *Pleurotus ostreatus* were purchased from local suppliers (Lambert Spawn, Coatesville, PA, USA). Oyster mushroom spawn has been used because 1) it is widely available locally, and 2) satisfactory results have been obtained with similar genera according to the literature [8]. Sterilized substrates were inoculated with the spawns by 7% of the dry weight in a sterilized environment. The bags were then placed in a growth room with environmental control. The temperature was set to 21 °C, and the relative humidity to 95%. The room was kept dark to help with the growth process.

To study the effects of different durations of growth on the mechanical behavior of MBC, three different durations of growth for 5, 6, and 7 weeks were selected, regarding the best results from the literature [7]. The growth process of MBC has been divided into two phases: within bags and formworks. To compare the partial growth of MBC in bags and formworks, the substrate mixtures have been placed in formworks at different times. These different substrate mixtures and timeframes made 35 different treatments. The treatments are coded as X_{ij}: X indicates the substrate mixture used for the cultivation, and i and j indicate the cultivation time (weeks) in bags and formworks. The details about the total and partial duration of the treatments are shown in Table A1.

3.3. Preparing Samples for Mechanical Test

The formworks for this experiment were made of cardboard covered with plastic tapes to avoid the hyphae feeding on the cardboard and decrease the cardboard's humidity level by capillary action. All formworks were cubes of 5 cm in length. The materials have been moved to the formworks after the first growth phase in the bags. We filled the formworks in three steps and hand-pressed the materials at each step. To control the conditions of similar samples for later experiments, the amount of material added in each step to the formwork was controlled by weight. For example, while all the formworks that were filled for samples A_{ij} weighed 120 g in total, each formwork was filled in three stages, adding 40 g of the material at each stage. Finally, the formworks were placed in the same room for the second phase of the growth process.

Following the growth phase, the samples were unmolded and placed in the oven for 48 h at 92 °C. After this heating process, almost all samples lost about two-thirds of their weight, showing that they were thoroughly dried and ceased the growth process. The cubic samples were then moved to the lab for the mechanical tests.

3.4. Mechanical Tests

The mechanical tests were performed on an MTS machine (Figure 3). ASTM C109 standard procedure requires testing three samples of each treatment. Therefore, for each treatment, we have created three samples. Each sample has been compressed to 80% of its initial height with a 0.05 mm per second rate to study its behavior under compression. We considered material strength as the stress in which material collapsed (the peak stress in stress-strain diagrams when a peak occurs) or the stress at the 10% strain, whichever comes first. This paper reports the average result of each sample group when the difference is less than 8.7%.

Figure 3. Mechanical tests were performed on MBC sample C_{33} with a 0.05 mm per second rate with the MTS machine (Figures are taken at 15-s intervals).

For the treatments X_{24}, X_{33}, and X_{42}, the Correlated Solutions VIC-3D Digital Image Correlation (DIC) system was also used to enable a more detailed study of MBC behavior under compression (Figure 4). In this system, a setup with two (or more) cameras captures images from the samples while they are reacting to an external force or stimulus. The samples are prepared with speckle points, lights, or other readable signs. After the experiment is conducted, the images from the cameras are correlated to present the alterations of the samples throughout the process. The DIC shows the exact deformations of the sample and enables studying the quantitative and qualitative mechanical behavior.

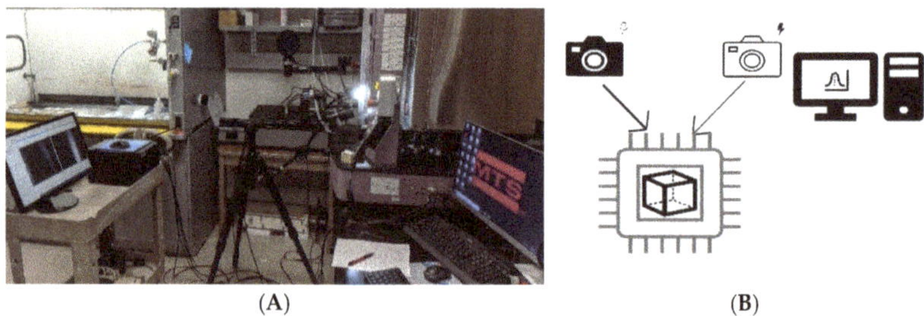

(A) (B)

Figure 4. (**A**) The Digital Image Correlation (DIC) system and (**B**) its scheme. DIC includes two cameras that capture the alterations of the sample while under compression and correlate the images.

4. Results and Discussion

4.1. The Effects of Sawdust to Straw Ratio in Substrate Mixtures on Compressive Strength

Five substrate mixtures (A, B, C, D, and E) with different ratios of sawdust and straw have been prepared (Table 1). Each mixture has been subjected to seven different differential growth times, resulting in 35 treatments. Table 2 shows the results of the mechanical tests for all 35 treatments. The treatments with only sawdust content (A) showed the best mechanical behavior. This result is in line with the literature [50,57]. While the treatments with 1 to 1 sawdust to the straw ratio (E) showed the weakest mechanical behavior and lowest compressive strength, the other three substrate mixtures with sawdust to straw ratios of 1 to 2 (D), 1 to 3 (C), and 1 to 7 (B) exhibited negligible differences.

Table 2. Compressive strength of different treatments (kPa).

		Treatments						
		X_{23}	X_{24}	X_{32}	X_{33}	X_{34}	X_{42}	X_{43}
Substrate Mixtures	A	498	416	330	360	303	325	288
	B	187	168	107	143	97	69	71
	C	177	159	118	103	82	65	62
	D	192	142	62	95	74	58	73
	E	116	135	34	68	75	39	58

The test results show that stronger substrates with more lignin content, such as sawdust, result in MBC with better compressive strengths, while weaker substrates, such as straw, result in MBC with weaker compressive strengths. There is a direct correlation between the density of the substrate, the density of MBC, and the mechanical strength of the material. However, the difference in the mechanical strength of MBC cultivated with substrates that include both straw and sawdust is negligible.

4.2. The Effects of Total Growth Time on Compressive Strength

Substrate mixtures have been grown for five (X_{23}, X_{32}), six (X_{24}, X_{33}, X_{42}), and seven (X_{34}, X_{43}) weeks. According to the literature [25], the longer growth time causes more organic substrate degradation, which means less substrate and more hyphal structures. Since most of the compressive strength of MBC is from the substrates, longer growth times result in less compressive strength. The results from our mechanical tests are also in line with the literature [7,25]. Figure 5 shows the average compressive strength for each substrate mixture (A, B, C, D, and E) grown for five, six, and seven weeks. For each substrate mixture, the compressive strength decreases by increasing the total cultivation time.

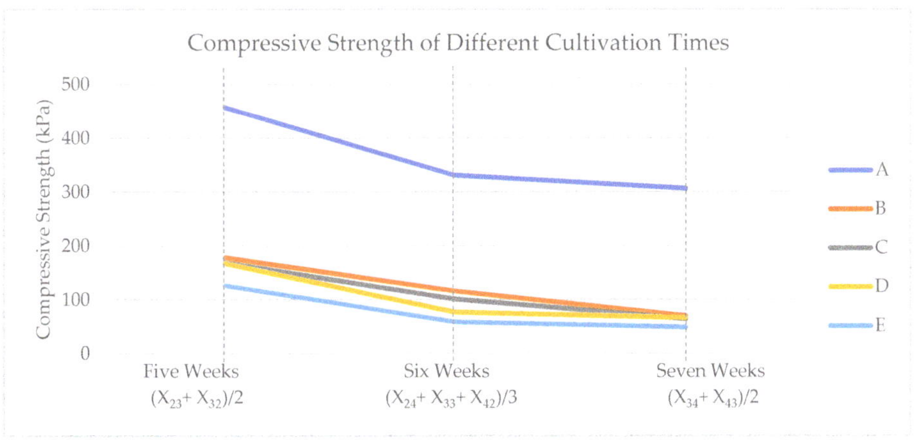

Figure 5. Compressive strength (kPa) of treatments with different growth times.

4.3. The Effects of Varied Bag/Formwork Growth Times on Compressive Strength

Treatments with the exact total growth times have been grown for different time frames within bags and formworks. We tested this feature to find the optimal duration of growth in each phase of MBC cultivation. For each substrate mixture (A, B, C, D, and E), we have two sets of treatments cultivated for five weeks (X_{23} and X_{32}), three sets for six weeks (X_{24}, X_{33}, and X_{42}), and two for seven weeks (X_{34} and X_{43}). Our tests show that more extended cultivation in formworks yields better mechanical performance than longer cultivation times within bags for all substrate mixtures and cultivation sets. Figure 6 shows the average compressive strength of all the substrate mixtures cultivated for six weeks (X_{24}, X_{33}, and X_{42}) regarding the cultivation time. The samples grown in molds for a more extended time show better mechanical performance.

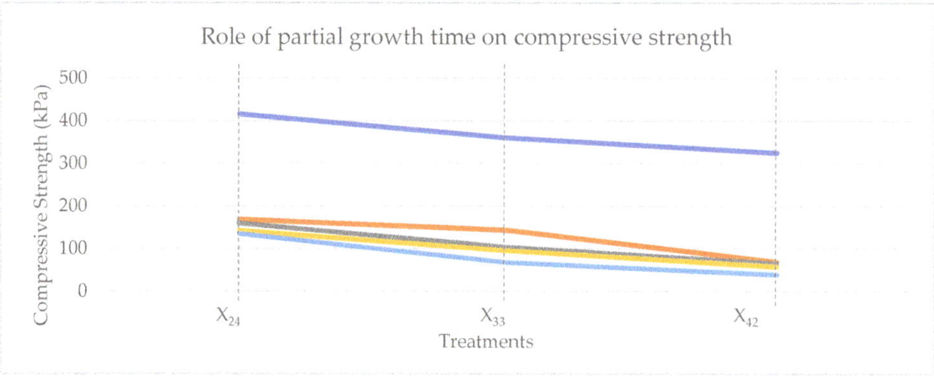

Figure 6. Compressive strength (kPa) of treatments with the six-week cultivation time and different partial growth times.

The duration of cultivation has an inverse relation to the compressive strength of the material. We cultivated mixtures for 5, 6, and 7 weeks. The material cultivated for five weeks showed higher compressive strength in all the cases. This result is in accordance with the published literature [7]. For all the samples, the extended time of growth in formworks compared to the growth in bags yielded stronger compressive strengths.

4.4. Digital Image Correlation (DIC)

For the cubic samples of treatments X_{24}, X_{33}, and X_{42}, we used the DIC setup with two 5 MP (2448 × 2048 pixels by 50 fps) and two Schneider Xenoplan 1.9/35 mm compact series lenses. This setup lets us capture the behavior of cubic samples under compression. The cubic samples were speckled prior to the test, and the speckles' movement during the test was monitored. This monitoring allowed us to have a more detailed quantitative study of the material's mechanical behavior and qualitatively study its behavior. The detailed movement data of speckles enables access to different displacement and strain amounts happening throughout the loading process. Mapping these on the loading timeline enables us to have more precise results. First, the samples' mechanical strength and elastic moduli were calculated using the DIC system data (reported in Table A1). These results verified the results from the extensometer attached to the MTS machine with less than a 5% difference. The system also allowed us to calculate other engineering characteristics of the material, such as the principal strains, shear moduli, and the Poison ratio. Besides that, the images taken from the system and the correlation between images reveal how the material behaves under compression.

One of the results from the tests and the study of the images show that treatments with substrate mixtures with more sawdust content (Sample A) behave with toughness and show a peak in their stress/strain diagrams. In comparison, treatments with more straw content in their substrate mixtures (Sample E) behave with hardness and reach the fracture point without showing plastic behaviors. Figures 7 and 8 show the stress/strain diagrams of cubic samples of treatments A_{33} and $E_{33\,and}$ some images of their behavior under compression.

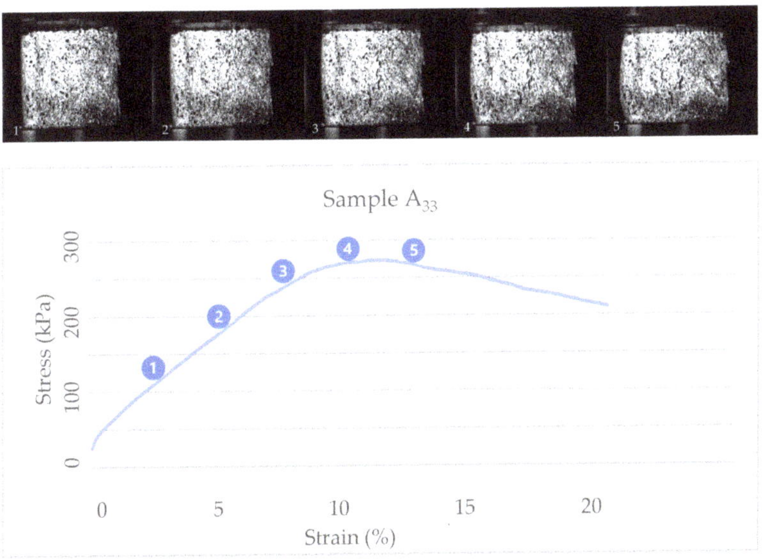

Figure 7. Stress-strain diagram and actual images of cubic sample A_{33}.

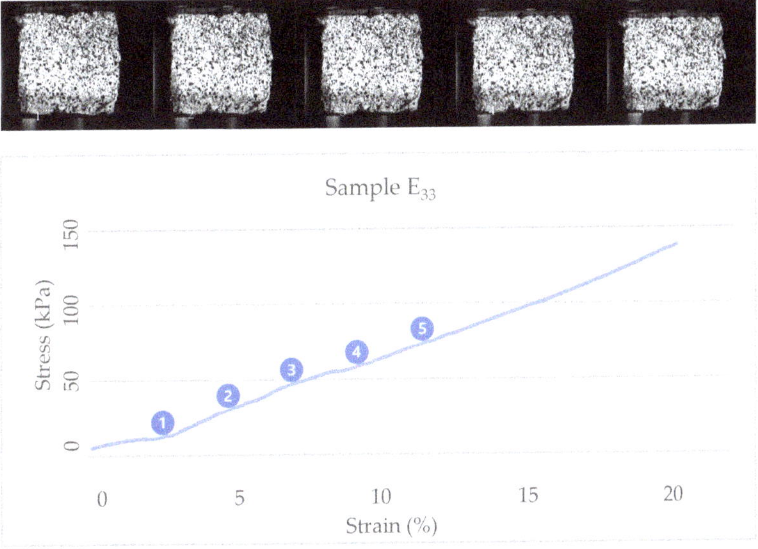

Figure 8. Stress-strain diagram and actual images of cubic sample E_{33}.

The images show that the cubic sample of the A_{33} treatment develops a crack in the center and deforms before reaching the peak point (near its 10% strain). In comparison, the sample of the E_{33} treatment does not show large cracks and reaches the maximum strain without fracture. Most treatments with substrate mixtures with more straw content have shown this behavior.

Figures 9 and 10 show the correlation of images from cubic samples E24 and E42 and their stress/strain diagrams. The other samples of substrate mixture E show the same behavior. Studying the correlated images and the internal strains of the cubic samples also show that the material works in compression with more tendency to use its toughness.

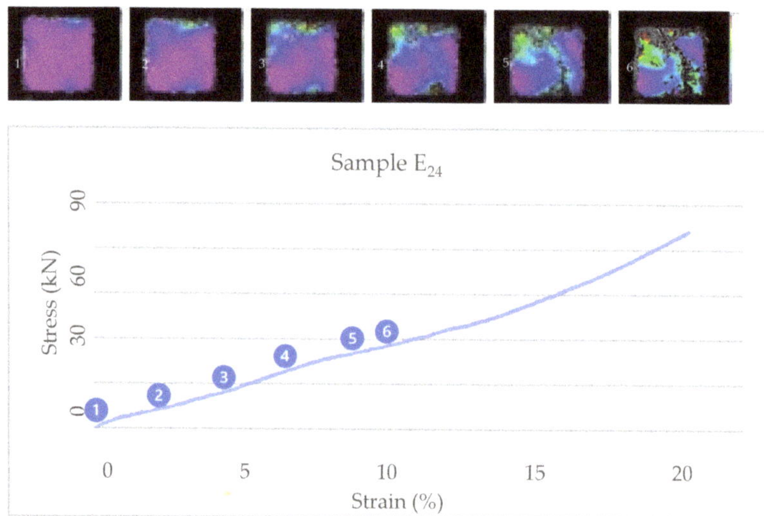

Figure 9. Stress-strain diagram and correlated images of cubic sample E_{24}.

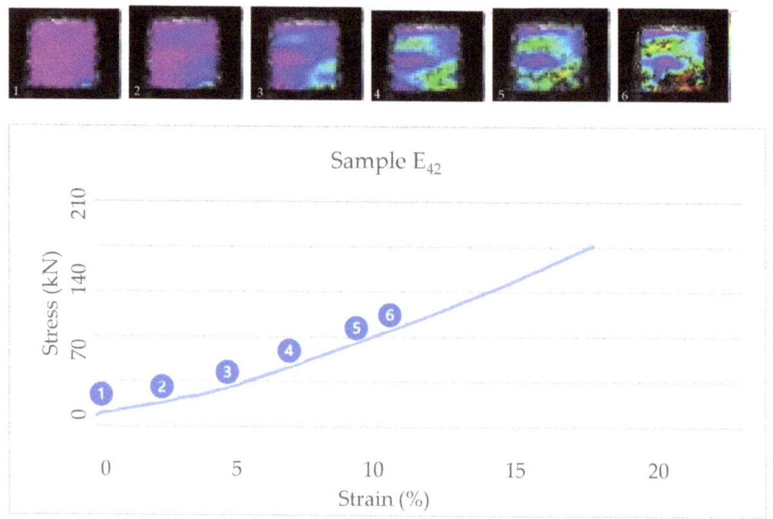

Figure 10. Stress-strain diagram and correlated images of sample E_{42}.

The results from the DIC system showed that substrate mixtures with more sawdust content tend to use their hardness, while straw-based substrate mixtures tend to behave with toughness. As tough materials are more resistant to fracturing and are not easily breakable, they seem to be better options for the compressive structural systems working through form. While, for functions that need materials that bear the load by their strength, materials with hardness tendencies are preferable.

5. Conclusions

The sustainable aspects of MBC make them suitable alternatives for their non-sustainable counterparts in several industries. From foam-like materials in the packaging industry to panels in the building industry, MBC cover many functions with different performance parameters. As studied in this paper, the AEC industry can also benefit more from the mechanical strength of MBC. These composites offer lightweight, graded, and biodegradable alternatives to conventional building materials and can help address the environmental problems caused by the AEC industry.

This research used agricultural waste (sawdust and straw) to cultivate MBC using locally available fungal species. We presented experiments in which we prepared treatments with five different substrate mixtures of varying sawdust to straw ratios. We tested the effects of the total duration of growth on the compressive strength of MBC cultivated with these treatments. We also tested the effects of varying the duration of growth in bags and the duration of growth in formworks on compressive strength. Our mechanical tests showed the possibility of cultivating a gradient of compressive materials. The results are also verified with a Digital Image Correlation (DIC) system, which also enabled the extraction of other material characteristics for future use in structural form-finding and the study of the qualitative behavior of the samples under compression.

MBC material is lightweight, its dead load is negligible, and it bears the load through the form. So, material grades with tougher properties can enable the designing and building of compressive structural forms. Besides, the harder grades can be used for the functions that bear the light loads through the strength of materials. In the following stages of this research, our goal is to develop computational form-finding methods to design and fabricate compressive structures with MBC that employ the results of our mechanical tests as the main inputs in optimizing the structural forms.

Author Contributions: Conceptualization, A.G., and B.G.; methodology, A.G.; software, A.G.; validation, A.G.; formal analysis, A.G.; investigation, A.G.; resources, A.G.; data curation, A.G.; writing—original draft preparation, A.G.; writing—review and editing, B.G.; visualization, A.G., and B.G.; supervision, B.G.; project administration, B.G.; funding acquisition, A.G., and B.G. All authors have read and agreed to the published version of the manuscript.

Funding: This research has been partially funded by Penn State Institute of Energy and Environment Seed Grant, and The H. Campbell and Eleanor R. Stuckeman Fund for Collaborative Design Research Fund.

Institutional Review Board Statement: Not Applicable.

Informed Consent Statement: Not Applicable.

Data Availability Statement: Data might be requested by contacting the corresponding author.

Acknowledgments: The authors appreciate the contributions of John A. Pecchia, Fabricio Vieira, and the Mushroom Research Center at Penn State University staff for helping with the cultivation process, and Beth Last, and the Material Characterization Lab at Penn State staff for their support with the mechanical tests.

Conflicts of Interest: The authors declare no conflict of interest.

Appendix A

The table below shows the characteristics of treatments used for the experiments and their mechanical strength and elastic moduli regarding the DIC results.

Table A1. Characteristics of Treatments and their Mechanical Properties.

Treatment	Substrate Mixture	Growth in Bags (Days)	Growth in Formwork (Days)	Total Growth Time (Days)	Mechanical Strength (kPa)	Elastic Moduli (MPa)
A_{23}	A: Sawdust only	14	21	35	491	6.1
A_{24}	65–70% Water	14	28	42	418	3.4
A_{32}	Content	21	14	35	335	3.3
A_{33}	7% DW Wheat	21	21	42	355	3.5
A_{34}	Flour	21	28	49	300	3.4
A_{42}	7% DW *Fungal* sp.	28	14	42	321	3.4
A_{43}		28	21	49	292	3.0
B_{23}	B: Straw to	14	21	35	190	1.8
B_{24}	Sawdust = 7/1	14	28	42	170	1.6
B_{32}	65–70% Water	21	14	35	110	1.1
B_{33}	Content	21	21	42	140	0.9
B_{34}	7% DW Wheat	21	28	49	98	0.9
B_{42}	Flour	28	14	42	70	0.8
B_{43}	7% DW *Fungal* sp.	28	21	49	68	0.7
C_{23}	C: Straw to	14	21	35	181	1.6
C_{24}	Sawdust = 3/1	14	28	42	161	1.4
C_{32}	65–70% Water	21	14	35	121	1.7
C_{33}	Content	21	21	42	101	0.9
C_{34}	7% DW Wheat	21	28	49	81	0.8
C_{42}	Flour	28	14	42	66	0.8
C_{43}	7% DW *Fungal* sp.	28	21	49	61	0.7
D_{23}	D: Straw to	14	21	35	193	0.6
D_{24}	Sawdust = 2/1	14	28	42	142	2.0
D_{32}	65–70% Water	21	14	35	63	1.5
D_{33}	Content	21	21	42	96	0.5
D_{34}	7% DW Wheat	21	28	49	75	0.8
D_{42}	Flour	28	14	42	59	0.7
D_{43}	7% DW *Fungal* sp.	28	21	49	72	0.6
E_{23}	E: Straw to	14	21	35	120	0.6
E_{24}	Sawdust = 1/1	14	28	42	130	1.3
E_{32}	65–70% Water	21	14	35	33	1.1
E_{33}	Content	21	21	42	69	0.5
E_{34}	7% DW Wheat	21	28	49	74	0.7
E_{42}	Flour	28	14	42	40	0.7
E_{43}	7% D.W. *Fungal* sp.	28	21	49	60	0.6

References

1. Ajayi, S.O.; Oyedele, L.O.; Akinade, O.; Bilal, M.; Owolabi, H.A.; Alaka, H.A.; Kadiri, K.O. Reducing waste to landfill: A need for cultural change in the UK construction industry. *J. Build. Eng.* **2015**, *5*, 185–193. [CrossRef]
2. Korhonen, J.; Honkasalo, A.; Seppälä, J. Circular Economy: The Concept and its Limitations. *Ecol. Econ.* **2018**, *143*, 37–46. [CrossRef]
3. Xing, Y.; Jones, P.; Donnison, I. Characterisation of Nature-Based Solutions for the Built Environment. *Sustainability* **2017**, *9*, 149. [CrossRef]
4. Karana, E.; Blauwhoff, D.; Hultink, E.J.; Camere, S. When the material grows: A case study on designing (with) mycelium-based materials. *Int. J. Des.* **2018**, *12*, 119–136. Available online: www.ijdesign.org (accessed on 15 January 2022).
5. Jones, M.; Mautner, A.; Luenco, S.; Bismarck, A.; John, S. Engineered mycelium composite construction materials from fungal biorefineries: A critical review. *Mater. Des.* **2019**, *187*, 108397. [CrossRef]
6. Yevich, R.; Logan, J.A. An assessment of biofuel use and burning of agricultural waste in the developing world. *Glob. Biogeochem. Cycles* **2003**, *17*. [CrossRef]
7. Yang, Z.; Zhang, F.; Still, B.; White, M.; Amstislavski, P. Physical and Mechanical Properties of Fungal Mycelium-Based Biofoam. *J. Mater. Civ. Eng.* **2017**, *29*, 04017030. [CrossRef]
8. Haneef, M.; Ceseracciu, L.; Canale, C.; Bayer, I.S.; Heredia-Guerrero, J.A.; Athanassiou, A. Advanced Materials From Fungal Mycelium: Fabrication and Tuning of Physical Properties. *Sci. Rep.* **2017**, *7*, 41292. [CrossRef]

9.   Meyer, V.; Basenko, E.Y.; Benz, J.P.; Braus, G.H.; Caddick, M.X.; Csukai, M.; De Vries, R.P.; Endy, D.; Frisvad, J.C.; Gunde-Cimerman, N.; et al. Growing a circular economy with fungal biotechnology: A white paper. *Fungal Biol. Biotechnol.* **2020**, *7*, 5. [CrossRef]

10.   Girometta, C.; Picco, A.M.; Baiguera, R.M.; Dondi, D.; Babbini, S.; Cartabia, M.; Pellegrini, M.; Savino, E. Physico-Mechanical and Thermodynamic Properties of Mycelium-Based Biocomposites: A Review. *Sustainability* **2019**, *11*, 281. [CrossRef]

11.   Elsacker, E.; Vandelook, S.; Van Wylick, A.; Ruytinx, J.; De Laet, L.; Peeters, E. A comprehensive framework for the production of mycelium-based lignocellulosic composites. *Sci. Total Environ.* **2020**, *725*, 138431. [CrossRef] [PubMed]

12.   Pelletier, M.; Holt, G.; Wanjura, J.; Bayer, E.; McIntyre, G. An evaluation study of mycelium based acoustic absorbers grown on agricultural by-product substrates. *Ind. Crop. Prod.* **2013**, *51*, 480–485. [CrossRef]

13.   Pelletier, M.; Holt, G.; Wanjura, J.; Lara, A.; Tapia-Carillo, A.; McIntyre, G.; Bayer, E. An evaluation study of pressure-compressed acoustic absorbers grown on agricultural by-products. *Ind. Crop. Prod.* **2017**, *95*, 342–347. [CrossRef]

14.   Wimmers, G.; Klick, J.; Tackaberry, L.; Zwiesigk, C.; Egger, K.; Massicotte, H. Fundamental studies for designing insulation panels from wood shavings and filamentous fungi. *BioResources* **2019**, *14*, 5506–5520. [CrossRef]

15.   Dias, P.P.; Jayasinghe, L.B.; Waldmann, D. Investigation of Mycelium-Miscanthus composites as building insulation material. *Results Mater.* **2021**, *10*, 100189. [CrossRef]

16.   Schritt, H.; Vidi, S.; Pleissner, D. Spent mushroom substrate and sawdust to produce mycelium-based thermal insulation composites. *J. Clean. Prod.* **2021**, *313*, 127910. [CrossRef]

17.   Gauvin, F.; Tsao, V.; Vette, J.; Brouwers, H.J.H. Physical Properties and Hygrothermal Behavior of Mycelium-Based Composites as Foam-Like Wall Insulation Material. *Constr. Technol. Archit.* **2022**, *1*, 643–651. [CrossRef]

18.   Abhijith, R.; Ashok, A.; Rejeesh, C. Sustainable packaging applications from mycelium to substitute polystyrene: A review. *Mater. Today Proc.* **2018**, *5*, 2139–2145. [CrossRef]

19.   Sivaprasad, S.; Byju, S.K.; Prajith, C.; Shaju, J.; Rejeesh, C. Development of a novel mycelium bio-composite material to substitute for polystyrene in packaging applications. *Mater. Today Proc.* **2021**, *47*, 5038–5044. [CrossRef]

20.   Holt, G.A.; McIntyre, G.; Flagg, D.; Bayer, E.; Wanjura, J.D.; Pelletier, M.G. Fungal Mycelium and Cotton Plant Materials in the Manufacture of Biodegradable Molded Packaging Material: Evaluation Study of Select Blends of Cotton Byproducts. *J. Biobased Mater. Bioenergy* **2012**, *6*, 431–439. [CrossRef]

21.   Jose, J.; Uvais, K.N.; Sreenadh, T.S.; Deepak, A.V.; Rejeesh, C.R. Investigations into the Development of a Mycelium Biocomposite to Substitute Polystyrene in Packaging Applications. *Arab. J. Sci. Eng.* **2021**, *46*, 2975–2984. [CrossRef]

22.   Jones, M.; Bhat, T.; Huynh, T.; Kandare, E.; Yuen, K.K.R.; Wang, C.-H.; John, S. Waste-derived low-cost mycelium composite construction materials with improved fire safety. *Fire Mater.* **2018**, *42*, 816–825. [CrossRef]

23.   Jones, M.; Bhat, T.; Kandare, E.; Thomas, A.; Joseph, P.; Dekiwadia, C.; Yuen, K.K.R.; John, S.; Ma, J.; Wang, C.-H. Thermal Degradation and Fire Properties of Fungal Mycelium and Mycelium—Biomass Composite Materials. *Sci. Rep.* **2018**, *8*, 17583. [CrossRef] [PubMed]

24.   Jones, M.; Bhat, T.; Wang, C.H.; Moinuddin, K.; John, S. Thermal degradation and fire reaction properties of mycelium composites. In Proceedings of the 21st International Conference on Composite Materials, Xi'an, China, 20–25 August 2017. Available online: https://www.researchgate.net/publication/319065199 (accessed on 15 January 2022).

25.   Attias, N.; Danai, O.; Abitbol, T.; Tarazi, E.; Ezov, N.; Pereman, I.; Grobman, Y.J. Mycelium bio-composites in industrial design and architecture: Comparative review and experimental analysis. *J. Clean. Prod.* **2019**, *246*, 119037. [CrossRef]

26.   Javadian, A.; le Ferrand, H.; Hebel, D.E.; Saeidi, N. Application of Mycelium-Bound Composite Materials in Construction Industry: A Short Review. *SOJ Mater. Sci. Eng.* **2020**, *7*, 1–9. Available online: www.symbiosisonlinepublishing.com (accessed on 15 January 2022).

27.   Zimele, Z.; Irbe, I.; Grinins, J.; Bikovens, O.; Verovkins, A.; Bajare, D. Novel Mycelium-Based Biocomposites (MBB) as Building Materials. *J. Renew. Mater.* **2020**, *8*, 1067–1076. [CrossRef]

28.   Ghazvinian, A.; Farrokhsiar, P.; Vieira, F.; Pecchia, J.; Gursoy, B. Mycelium-Based Bio-Composites for Architecture: Assessing the Effects of Cultivation Factors on Compressive Strength. In Proceedings of the 37th eCAADe/23rd SIGraDi Conference, Porto, Portugal, 11–13 September 2019; pp. 505–514.

29.   Escaleira, R.M.; Campos, M.J.; Alves, M.L. Mycelium-Based Composites: A New Approach to Sustainable Materials. In *Sustainability and Automation in Smart Constructions*; Springer: Cham, Switzerland, 2020; pp. 261–266. [CrossRef]

30.   Yang, L.; Park, D.; Qin, Z. Material Function of Mycelium-Based Bio-Composite: A Review. *Front. Mater.* **2021**, *8*, 737377. [CrossRef]

31.   Manan, S.; Ullah, M.W.; Ul-Islam, M.; Atta, O.M.; Yang, G. Synthesis and applications of fungal mycelium-based advanced functional materials. *J. Bioresour. Bioprod.* **2021**, *6*, 1–10. [CrossRef]

32.   Ghazvinian, A. A Sustainable Alternative to Architectural Materials: Mycelium-Based Bio-Composites. In Proceedings of the Divergence in Architectural Research, Atlanta, GA, USA, 15 February 2021; Georgia Institute of Technology: Atlanta, GA, USA, 2021; pp. 159–167.

33.   Ongpeng, J.M.C.; Inciong, E.; Sendo, V.; Soliman, C.; Siggaoat, A. Using Waste in Producing Bio-Composite Mycelium Bricks. *Appl. Sci.* **2020**, *10*, 5303. [CrossRef]

34.   Gou, L.; Li, S.; Yin, J.; Li, T.; Liu, X. Morphological and physico-mechanical properties of mycelium biocomposites with natural reinforcement particles. *Constr. Build. Mater.* **2021**, *304*, 124656. [CrossRef]

35. Ghazvinian, A.; Gursoy, B. Basics of Building with Mycelium-Based Bio-Composites: A Review of Built Projects and Related Material Research. *J. Green Build.* **2022**, *17*, 37–69. [CrossRef]
36. Jones, M.; Huynh, T.; Dekiwadia, C.; Daver, F.; John, S. Mycelium Composites: A Review of Engineering Characteristics and Growth Kinetics. *J. Bionanosci.* **2017**, *11*, 241–257. [CrossRef]
37. Nashiruddin, N.I.; Chua, K.S.; Mansor, A.F.; Rahman, R.A.; Lai, J.C.; Azelee, N.I.W.; El Enshasy, H. Effect of growth factors on the production of mycelium-based biofoam. *Clean Technol. Environ. Policy* **2021**, *24*, 351–361. [CrossRef]
38. Heisel, F.; Lee, J.; Schlesier, K.; Rippmann, M.; Saeidi, N.; Javadian, A.; Nugroho, A.R.; Van Mele, T.; Block, P.; Hebel, D.E. Design, Cultivation and Application of Load-Bearing Mycelium Components: The MycoTree at the 2017 Seoul Biennale of Architecture and Urbanism. *Int. J. Sustain. Energy Dev.* **2017**, *6*, 296–303. [CrossRef]
39. Dessi-Olive, J. Monolithic Mycelium: Growing Vault Structures. In Proceedings of the 18th International Conference on Non-Conventional Materials and Technologies (NOCMAT), Nairobi, Kenya, 24–26 July 2019. Available online: https://www.academia.edu/39909593/Monolithic_Mycelium_Growing_Vault_Structures?auto=download&campaign=weekly_digest (accessed on 15 January 2022).
40. Bhardwaj, A.; Rahman, A.M.; Wei, X.; Pei, Z.; Truong, D.; Lucht, M.; Zou, N. 3D Printing of Biomass–Fungi Composite Material: Effects of Mixture Composition on Print Quality. *J. Manuf. Mater. Process.* **2021**, *5*, 112. [CrossRef]
41. Bhardwaj, A.; Vasselli, J.; Lucht, M.; Pei, Z.; Shaw, B.; Grasley, Z.; Wei, X.; Zou, N. 3D Printing of Biomass-Fungi Composite Material: A Preliminary Study. *Manuf. Lett.* **2020**, *24*, 96–99. [CrossRef]
42. Modanloo, B.; Ghazvinian, A.; Matini, M.; Andaroodi, E. Tilted Arch; Implementation of Additive Manufacturing and Bio-Welding of Mycelium-Based Composites. *Biomimetics* **2021**, *6*, 68. [CrossRef] [PubMed]
43. Chen, H.; Abdullayev, A.; Bekheet, M.F.; Schmidt, B.; Regler, I.; Pohl, C.; Vakifahmetoglu, C.; Czasny, M.; Kamm, P.H.; Meyer, V.; et al. Extrusion-based additive manufacturing of fungal-based composite materials using the tinder fungus Fomes fomentarius. *Fungal Biol. Biotechnol.* **2021**, *8*, 21. [CrossRef]
44. Soh, E.; Chew, Z.Y.; Saeidi, N.; Javadian, A.; Hebel, D.; Le Ferrand, H. Development of an extrudable paste to build mycelium-bound composites. *Mater. Des.* **2020**, *195*, 109058. [CrossRef]
45. Goidea, A.; Floudas, D.; Andréen, D. Pulp Faction: 3d printed material assemblies through microbial biotransformation. *Fabricate* **2020**, *2020*, 42–49.
46. Alima, N.; Snooks, R.; McCormack, J. Bio Scaffolds: The orchestration of biological growth through robotic intervention. *Int. J. Intell. Robot. Appl.* **2022**, 1–8. [CrossRef]
47. Cheng, A.; Lim, S.; Thomsen, M.R. Multi-Material Fabrication for Biodegradable Structures—Enabling the printing of porous mycelium composite structures. In Proceedings of the 39th eCAADe Conference, Novi Sad, Serbia, 8–10 September 2021; pp. 85–94.
48. Goidea, A.; Floudas, D.; Andréen, D. Transcalar Design: An Approach to Biodesign in the Built Environment. *Infrastructures* **2022**, *7*, 50. [CrossRef]
49. Appels, F.V.W.; Dijksterhuis, J.; Lukasiewicz, C.E.; Jansen, K.; Wösten, H.A.B.; Krijgsheld, P. Hydrophobin gene deletion and environmental growth conditions impact mechanical properties of mycelium by affecting the density of the material. *Sci. Rep.* **2018**, *8*, 4703. [CrossRef]
50. Appels, F.V.; Camere, S.; Montalti, M.; Karana, E.; Jansen, K.; Dijksterhuis, J.; Krijgsheld, P.; Wösten, H.A. Fabrication factors influencing mechanical, moisture- and water-related properties of mycelium-based composites. *Mater. Des.* **2018**, *161*, 64–71. [CrossRef]
51. Liu, R.; Long, L.; Sheng, Y.; Xu, J.; Qiu, H.; Li, X.; Wang, Y.; Wu, H. Preparation of a kind of novel sustainable mycelium/cotton stalk composites and effects of pressing temperature on the properties. *Ind. Crop. Prod.* **2019**, *141*, 111732. [CrossRef]
52. Islam, M.R.; Tudryn, G.; Bucinell, R.; Schadler, L.; Picu, R.C. Mechanical behavior of mycelium-based particulate composites. *J. Mater. Sci.* **2018**, *53*, 16371–16382. [CrossRef]
53. Elsacker, E.; Vandelook, S.; Brancart, J.; Peeters, E.; De Laet, L. Mechanical, physical and chemical characterisation of mycelium-based composites with different types of lignocellulosic substrates. *PLoS ONE* **2019**, *14*, e0213954. [CrossRef] [PubMed]
54. Attias, N.; Danai, O.; Tarazi, E.; Pereman, I.; Grobman, Y.J. Implementing bio-design tools to develop mycelium-based products. *Des. J.* **2019**, *22*, 1647–1657. [CrossRef]
55. Bruscato, C.; Malvessi, E.; Brandalise, R.N.; Camassola, M. High performance of macrofungi in the production of mycelium-based biofoams using sawdust—Sustainable technology for waste reduction. *J. Clean. Prod.* **2019**, *234*, 225–232. [CrossRef]
56. Stelzer, L.; Hoberg, F.; Bach, V.; Schmidt, B.; Pfeiffer, S.; Meyer, V.; Finkbeiner, M. Life Cycle Assessment of Fungal-Based Composite Bricks. *Sustainability* **2021**, *13*, 11573. [CrossRef]
57. Elsacker, E.; Vandelook, S.; Damsin, B.; Van Wylick, A.; Peeters, E.; De Laet, L. Mechanical characteristics of bacterial cellulose-reinforced mycelium composite materials. *Fungal Biol. Biotechnol.* **2021**, *8*, 18. [CrossRef] [PubMed]

 *biomimetics*

Article

Tilted Arch; Implementation of Additive Manufacturing and Bio-Welding of Mycelium-Based Composites

Behzad Modanloo [1], Ali Ghazvinian [2,*], Mohammadreza Matini [3] and Elham Andaroodi [1]

[1] Faculty of Architecture, University College of Fine Arts, University of Tehran, Tehran 1415564583, Iran; behzad.modanloo@ut.ac.ir (B.M.); andaroodi@ut.ac.ir (E.A.)
[2] Department of Architecture, Penn State University, University Park, PA 16802, USA
[3] Faculty of Architecture and Urban Planning, University of Art, Tehran 1136813518, Iran; m.matini@art.ac.ir
* Correspondence: axg1370@psu.edu; Tel.: +1-6184255170

Abstract: Bio-based materials have found their way to the design and fabrication in the architectural context in recent years. Fungi-based materials, especially mycelium-based composites, are a group of these materials of growing interest among scholars due to their light weight, compostable and regenerative features. However, after about a decade of introducing this material to the architectural community, the proper ways of design and fabrication with this material are still under investigation. In this paper, we tried to integrate the material properties of mycelium-based composites with computational design and digital fabrication methods to offer a promising method of construction. Regarding different characteristics of the material, we found additive manufacturing parallel to bio-welding is an appropriate fabrication method. To show the feasibility of the proposed method, we manufactured a small-scale prototype, a tilted arch, made of extruded biomass bound with bio-welding. The project is described in the paper.

Keywords: mycelium-based composites; additive manufacturing; bio-based materials; circular construction; digital fabrication

Citation: Modanloo, B.; Ghazvinian, A.; Matini, M.; Andaroodi, E. Tilted Arch; Implementation of Additive Manufacturing and Bio-Welding of Mycelium-Based Composites. *Biomimetics* **2021**, *6*, 68. https://doi.org/10.3390/biomimetics6040068

Academic Editors: Andrew Adamatzky, Phil Ayres and Han A.B. Wösten

Received: 1 November 2021
Accepted: 29 November 2021
Published: 30 November 2021

Publisher's Note: MDPI stays neutral with regard to jurisdictional claims in published maps and institutional affiliations.

1. Introduction

The United Nations predicts that by 2050, two-third of the world's population will live in urban areas. Population growth in urban areas increases the demand for habitat construction. In addition, natural resources are dwindling, which has resulted in a search for sustainable and renewable alternatives for existing materials. One of the proposed solutions for these challenges is to work with bio-based materials. In addition to well-known bio-based materials such as bioplastics, materials made of bacteria, algae, and fungi have been increasingly interesting for design and fabrication. These alternative materials have led to the emergence of new design methods at the intersection of design, materials science, biology, arts, and crafts, which fundamentally changes the designer's role from a passive receiver to an active material maker [1]. In the last decade, architects and designers have begun working with biologists and materials scientists to discover how to design and build using biomaterials [2].

In this regard, mycelium-based composites (MBCs) offer sustainable and biodegradable options for a wide range of design and fabrication processes, including architectural applications. The industrial potential of fungi in areas ranging from food production to medical biotechnology has been studied for a long time [3]. However, little research has been done in the field of architecture, engineering, and construction (AEC) [4]. Therefore, detailed research is needed to implement and build these biomaterials in the AEC industry. Furthermore, although MBCs offer many advantages for lightweight, sustainable materials [5], there are still challenges for using these materials, especially for large-scale production [6].

This paper presents the early stages of an interdisciplinary research project exploring the applications of mycelium-based composites in architecture as a sustainable, renewable, and biodegradable alternative material. This comprehensive research aims to introduce the best practices of using mycelium-based composites in architecture. In this paper, we tried to examine additive manufacturing as the forming technique for this material. In the first phase, a systematic material study was conducted to evaluate the different effects of substrate mixtures (sawdust, paper, sawdust + paper; with and without additives) on the growth and compressive strength of mycelium-based composite blocks. Following the material study, a recursive computational design and digital fabrication process have been done to construct a series of prototypes to show how the material works. The final prototype of this stage of our study, a tilted arch introduced as a proof of concept, is described in the last section.

2. Background

Fungi are a group of heterotrophic organisms that unlike plants, cannot produce their food through photosynthesis. Therefore, to survive, they feed on organic compounds as parasites or saprophytes. The cell walls of fungi are made of a substance called chitin. Chitin is a large and complex polysaccharide made from modified glucose chains. This substance has the primary role in the material characteristics of the fungi-based matter [7]. Mycelium is the mass fibrous and branched vegetative root of fungi, made of hyphae. Through the mycelium, a fungus absorbs nutrients from its environment. Hyphae first secrete enzymes into food sources that break down biological polymers of organic substrates into smaller units, such as monomers. These monomers are then adsorbed to the mycelium. Through this process, the mycelial branches bind the organic matter together and make a lightweight, foam-like material called mycelium-based composites [8].

Many factors affect mycelial growth, and by changing these factors, different MBC are obtained [9]. These factors include fungi used for inoculation, type of substrate, environmental conditions during growth, and formation and storage techniques [10]. The fungi and substrates used for MBC cultivation can change the characteristics of the resulting material by modifying the chemical and biological formation [7]. The environmental conditions, such as nutrients, temperature, relative humidity, pH, and aeration, can also affect the outcome of MBC cultivation [11]. The curing and post-processing of MBC can also influence the material characteristics of MBC. For example, different ways of pressing MBC via hot or cold pressing systems result in different material grades to increase material density and decrease porosity [12]. Therefore, scholars try to improve this material's mechanical and durability behavior by tweaking each of these influencing parameters [13].

Various techniques can be used to form this material and fabricate architectural elements with MBCs [14]. One of the unique features of this material is its ability to grow in molds, allowing designers to directly grow mycelium in the final desired body shape. Processing techniques such as laser cutting and hot and cold compression can also be used to achieve the shape and structure required for grown materials [15]. In addition, the outer layer of MBC, called fungal skin [8], increases the compressive strength of the material and its water repellency. Thus, the only technique used to build large-scale architectural prototypes is to form mycelium composites into molds. However, there are also attempts on smaller scales to produce MBC elements by additive [16,17] and subtractive [6] manufacturing and fabric molds [18]. Early efforts to use MBC focused more on designing blocks and panels for walls and ceilings. As a result, a limited number of architectural projects were designed and constructed using these materials.

The first architectural structure designed and built using MBC was Mycotectural Alpha fabricated by Phil Ross in 2009 as part of an art performance at the Kunsthalle Dusseldorf [19]. The Hy-Fi Tower is the largest MBC structure based on discrete elements. The tower was designed and built by David Benjamin of The Living Studios in 2014 in collaboration with Arup and Ecovative as part of the MoMA Young Architects Program.

The entire 13-meter-tall structure was built using about 10,000 mycelium-based blocks (Figure 1).

Figure 1. The Hy-fi Tower (Image credit: Iwan Baan).

BEETLES3.3 and Yassin-Areddia designed and built a temporary shell-like pavilion of wood and mycelium, called Shell Mycelium. They intended to create temporary structures without waste. Shell Mycelium was the first attempt to fabricate non-discrete structures. The triangular wooden frame of the pavilion is filled with mycelium and covered by coconut husk. As mycelium bounds the organic substrate within the growth, the result was a shell-like structure made of MBC. Mycotree is a research pavilion designed and built by the Block Research Group at ETH Zurich and the Karlsruhe Institute for Technology for the 2017 Seoul Architecture and Urban Planning Biennale (Figure 2). The structure was designed using the 3D graphic statics method [20], which allows the design of components that bear loads only under pressure.

Figure 2. The Mycotree Pavilion (Image credit: Carlina Teteris).

El Monolito Micelio is the first large-scale effort of fabricating monolithic structures with MBC. For the design and fabrication of this fungi-based vault, a large formwork and a complex internal falsework have been designed, and about a ton of fresh mycelium has been cultivated [21].

Studying the outcome of these efforts and prototypes to build with MBC shows that working with this material in smaller components is more promising, as the uncertainties of coping with bio-based materials are more controllable on smaller scales. Besides, as the material's growth depends on the presence of air, and the outer surfaces are better grown than the core of the components, working with larger pieces requires some advanced solutions and equipment to insert and distribute the air [21].

The other solution to form the MBC material in desired shapes and provide enough air to grow is the use of additive manufacturing. Compared to molding the MBC, additive manufacturing can improve fungal growth conditions, allowing faster growth and complete coverage [17]. Therefore, this method can lead to better material performance and more efficient production. In addition, it is possible to produce a complex and customized form in this process beyond what can be achieved through molding.

The first scholarly published paper about 3D printing of MBC was [22]. After the initial growth process, they opened and mixed the biomass material with additives to make it extrudable. Next, they added water and psyllium husk powder to the mixture. This powder prevents the separation of the solid phase from the liquid. In the final stage, they printed the material with a 2040 wasp printer, in the dimensions of 10 cm by 10 cm by 2 cm. Then they put the printed piece in a sterile package away from direct sunlight to grow the second stage for five days. In the final stage, the pieces were placed in a heater for four hours at 95 °C to stop the fungal growth. The results showed that, like in formworks, the fungi could not grow in all printed parts, and most of the growth took place in the exterior parts. The detailed quality differences of various material mixtures have been covered in [23]. In another research, scholars used shredded bamboo fibers and chitosan as the primary materials of MBC [24]. Bamboo fibers provide the nutrients needed for mycelial growth. The fungus used in this study is *Ganoderma lucidum*, a common fungus with fast growth and white roots. Chitosan is a biopolymer derived from chitin, the main constituent of crustaceans' shells extracted from food waste. When chitosan is dissolved in a mildly acidic liquid medium, it forms a gel that acts as a physical stabilizer to increase the material's efficiency and make extrusion possible.

One of the prototypes of 3D-printed MBC is the Pulp-Faction project [17]. This project demonstrates the challenges and potentials of additive fabrication of MBC (Figure 3). The authors used wood, pulp, and kaolin clay in combination with water to make printable pulp. Wood and paper pulp formed the bulk of the material and served to feed the fungi. The composite also contains materials that put solid and liquid components together in a cohesive manner. This study used two types of fungi, *Byssomerulius corium*, and *Gloeophyllum*, both wood decomposers. The inoculation took place in two stages, the first stage before printing and the next stage after printing. When the growth reached the final stage, the printed parts were dried to stop the decomposition process. Next, they printed the composite using a Vormvrij Lutum v4 printer. The results showed that the final volume shrinks by 30%. A set of the aluminum grid was used as a base for printing and coating to minimize the resulting distortion. The grid secured the position of the first and last layers, limiting the vertical contraction.

In another prototype, mixed clay and mycelium have been used to prepare a novel form of printable material called Mycera [25]. They mixed these two constituents with various proportions to enable different outcome materials. They fabricated a bar-node prototype in which the bar elements are more rigid composites, while the nodes are made of mixtures that can bind better (Figure 4).

Figure 3. The Pulp Faction Prototype (Image credit: Ana Goidea).

Figure 4. The Bar-node Structure Fabricated with Mycera (Image credit: Julian Jauk, Hana Vasatko, Lukas Gosch).

Another line of research concentrates on using additive manufacturing of soil and fungi for mycoremediation [16]. This study tried to find the best practice of 3D printing soil-based structures and using mycelium as a remediating agent for weakened soil environments. Studying the outcome of these efforts and prototypes to print MBC shows that taking advantage of additive manufacturing for working with the MBC material enables many possibilities to enhance the forming and growth process. 3D printing is one of the best fabrication methods to increase manufacturing speed and create free forms to enable a higher strength due to geometry with a certain amount of MBC material. First, however, it is crucial to find the proper ways of printing this material due to its material characteristics.

3. Methodology

In order to find the best practices of using MBC for architecture, we took advantage of using Material-driven design (MDD) methodology. MDD, which aims to design by considering the limitations and affordances of the materials [1]. The MDD addresses the challenges of using this uncertain bio-based material and supports the design process when a particular material is the starting point in the design process. Relying on the concept of material experience, the MDD emphasizes that the designer's journey of material properties and experiments leads to a broader view of material applications. The MDD method offers four main steps: (1) understanding the material, (2) creating a material experience landscape, (3) material manifestation experience patterns, and (4) product design [1].

While working with novel materials and modifying the material properties, the desired forms and geometries and the techniques to achieve these forms are also necessary [2]. These interdependent factors make us use a recursive framework with material, geometry, and technique in which each stage informs the other two stages for the subsequent iterations (Figure 5). The other important factor in this recursive study method is starting with smaller scales and scaling up the prototypes for the geometry and technique stages [26]. It means that we must consider a bottom-up part of study for enhancing our material characteristics, parallel to a top-down method for enabling our geometries and techniques with the material of choice. Therefore, we mapped the MDD on our framework, which matches the first two steps of MDD with the bottom-up part of our framework and the next two steps with the top-down stage.

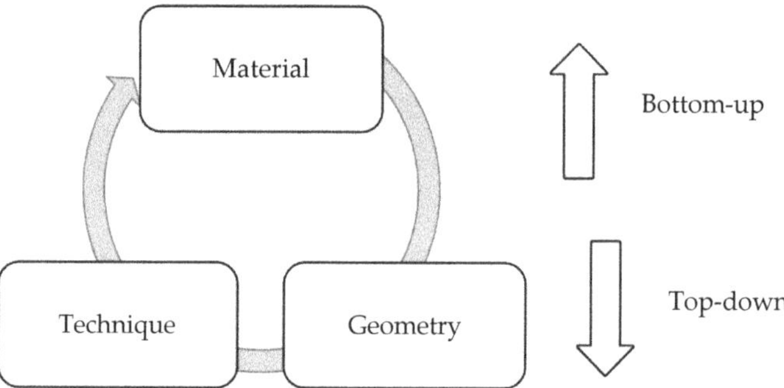

Figure 5. The recursive framework defined to work with novel materials/techniques (adapted from [2]).

3.1. Understanding the Material and Material Exploration

For the first two stages of MDD, we studied the literature about MBC and started an experiment with the material. The first critical decision in MBC production is the mixture of the substrate. The substrate can be composed of any organic matter if the fungus can decompose it. The substrates that are selected to produce these materials can be divided into three general categories:

- The wood waste consists of sawdust produced in carpentry—municipal waste like paper, cardboard, egg combs, and paper cups.
- Agricultural waste usually refers to the severed stems of post-harvest plants, often incinerated and producing CO_2 in the air. Substances such as wheat straw, rice, barley, sugarcane pulp bagasse, oilseed pulp, and sesame pulp fall into this category.

Initially, to measure the growth rate of the fungus in different substrates, several organic substrates such as shredded cardboard (smaller than 5 mm), shredded paper (smaller than 5 mm), sawdust (1 to 3 mm), and chopped straw (1 to 3 mm) were selected for inoculation testing.

The other influencing parameter is the fungal species to cultivate the MBC. According to the literature, among the fungi that grow well on wood-based products, *Pleurotus* and *Ganoderma* show tremendous growth compared to other fungi [7]. Therefore, due to the abundance and fast growth of the former, we chose *Pleurotus ostreatus* (oyster mushroom), with 10% of the dry weight content, for our experiments. We bought the oyster mushroom spawns grown on wheat grains from a local provider in Tehran, Iran. As the optimized relative humidity for the cultivation of *P. ostreatus* is around 65–70%, we mixed our soaked substrates with distilled water to maintain this relative humidity.

Fungi are heterotrophic organisms that need foreign nutrients to grow. Carbon and nitrogen are the two primary nutrients needed for fungi growth. When the environment cannot provide the fungi with nutrients, additives might help [9]. We added wheat bran, with 7% of dry weight content, to help the growth process.

The MBC cultivation process requires proper sterilization to achieve good results and prevent contamination by other organisms. Sterilization is needed for both the substrate and the growth environment. For this purpose, all equipment was disinfected by soaking in 70% ethanol (Kimia Alcohol Zanjan, Zanjan, Iran) for 30 min. For the substrate sterilization, we heated the treatments in autoclave bags for 45 min at 121 °C. After inoculation, the mycelium must be maintained under controlled light, temperature, and humidity conditions to ensure sustained growth. Optimal temperature and humidity conditions vary significantly depending on the type of fungus used. However, most species grow around 25–35 °C [7].

For the initial experiments on the substrates, we cultivated the mixtures for 14 days in autoclave bags and then moved to our sample formworks for the second growth phase. Next, the samples were grown for another 14 days, then unmolded, dried for 24 h, and heated at 90 °C for 6 h.

The criterion for choosing the best growth was the visual test regarding the amount of mycelium covering the substrate on the surface of cultivated samples. After comparing the surfaces, a cross-section was cut from each sample to observe the growth of mycelium. Observations have shown that mycelium often grows in places that are in contact with oxygen. However, as predicted, shallow mycelium growth was seen inside the samples. Therefore, as the third stage of the MDD method, we decided to increase the surface contact with oxygen via 3D printing instead of molding.

In order to enable extrusion of MBC material, gelling agents are needed. These additives, when dissolved in a liquid phase as a colloidal mixture, form a homogeneous paste. Most of them are organic hydrocolloids or inorganic hydrophilic substances. Examples of these substances are tragacanth, pectin, starch, carbomer, sodium alginate, and gelatin [27]. In the literature related to the additive manufacturing of MBC, psyllium husk and chitosan have been used as gelling agents [22,23,27,28].

The gelling agents used in this study are entirely herbal and not only do not interfere with the growth of the fungus but also act as an additive to help the growth process. We experimented with several types of additives, such as Persian gum, Arabic gum, guar gum, and psyllium husk, with different proportions during the material studies. We tried to extrude straight lines of MBC that are smooth, well-grown, and with the minimum shrinkage after drying.

Regarding the results of the cultivation and gelling agent tests, we selected two substrates and two gelling agents for printing: paper as an example of wood products and sawdust as a wood waste substrate. We chose to proceed with Arabic gum and guar gum, as they showed more consistent extruded lines and they are less expensive than other options. The most crucial characteristic of these gelling agents is that they hydrate rapidly in cold water to produce highly viscous solutions. Guar gum, exceptionally when fully hydrated, produces a homogeneous viscous colloidal mixture that is a thixotropic rheological system. Like other gums, the viscosity of guar gum depends on time, temperature, concentration, pH, and the stirring protocol [29]. Figure 6 shows parts of the initial material tests.

As the gelling agents adsorb water to make the paste homogeneous, the amount of water added to the substrate increases compared to the treatments formed with formworks. Table 1 shows the characteristics of the treatment mixtures for the second set of tests. In the second set of tests, we experimented with the mechanical behavior of different material treatments.

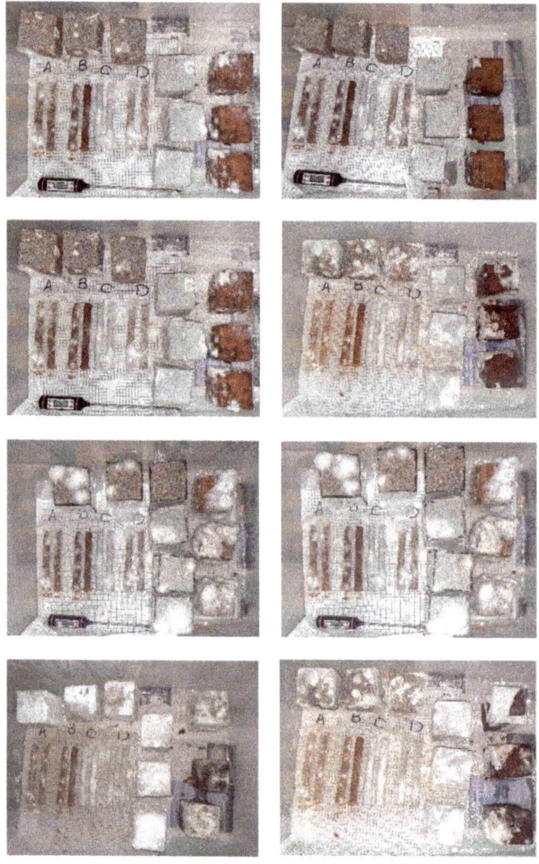

Figure 6. Initial material tests for examining treatments' cultivation and gelling agents' performance (Image credit: Authors).

Table 1. Mixture Properties for the First set of Experiments.

Treatment	Substrate	Gelling Agent	Water Content (g)
A	Sawdust (1 to 3 mm)	Guar Gum (10 g)	500
B	Sawdust (1 to 3 mm)	Arabic Gum (10 g)	500
C	Shredded Paper (>5 mm)	Guar Gum (30 g)	500
D	Shredded Paper (>5 mm)	Arabic Gum (30 g)	500

To test the mechanical strength of these material treatments, we used a UTM machine for loading (Figure 7). First, we formed three $5 \times 5 \times 5$ cm^3 cubes from each treatment and loaded them at the rate of 3 mm/min. Then, examining the behavior of each sample, we drew stress-strain diagrams to obtain the yield stress and the Young modulus of each mixture.

Figure 7. The Compressive Strength Test (Image credit: Authors).

The results of these tests showed that the paper substrate could be a suitable option for 3D printing of MBC due to its better mechanical behavior. Table 2 depicts the results of these tests. The results related to the gelling agents showed that for Arabic gum, more content is required than guar gum, while they behave the same in printability. Also, the samples cultivated with guar gum showed plastic behavior while Arabic gum showed the brittle fracture. Thus, we used paper as a substrate for the following stages of the research, with guar gum as the gelling agent. We used the same time frame for the cultivation for this stage, with 14 days in bags and 14 days in the sample formworks, followed by drying for one day and heating for 6 h at 90 °C.

Table 2. Material Characteristics of Tested Mixtures.

Treatment	Yield Stress (KPa)	Young Modulus (MPa)	Behavior
A	171.44	19.1	Plastic
B	167.68	18.2	Brittle
C	524.14	41.4	Plastic
D	536.27	78.3	Brittle

3.2. Material Manifestation

Due to the lack of access to the market extruders, and the need for having an extruder that can be decontaminated for bio-based materials, we fabricated an extruder from scratch (Figure 8).

Figure 8. The Fabricated Extruder (Image credit: Authors).

After fabricating the extruder, we defined an experimental printing session to inquire about the best extrusion scenario. We printed several cylinders with 30 cm diameter with 5 mm to 9 mm nozzles (Table 3). We observed that the paste comes out of the smaller nozzles inconsistently and irregularly. This result showed that the low nozzle diameters do not work with the paste because oyster mushroom spawn grows on wheat and form larger particles. The best scenario happened with the 9-mm nozzle (Figure 9).

Table 3. The Extrusion Properties of Different Nozzles.

Nozzle Size (mm)	5	6	7	8	9
Step (mm)	4	4	6	6	9
Pressure (bar)	2.2	2.1	1.8	1.7	1.7
Speed (mm/s)	100	100	100	100	100
Fresh Thickness (mm)	10	12	14	16	18
Dried Thickness (mm)	9	10	12	14	16

Figure 9. The initial extrusion experiment (Image credit: Authors).

After the extrusion, the printed materials become cylindrical due to the shape of the nozzle. We tested several samples of 100 × 9 mm to inquire about the deformation after printing. When dried, the parts were measured, and no change in their dimensions was found, although they became lighter and more porous. A slight increase in cross-section was observed in layered printing due to weight accumulation in the lower layers.

3.3. Product Design

In order to take advantage of using the MBC material and additive manufacturing together, various criteria must be met. First, as the material works in compression and has little tensile strength [2,30], forms that can bear the load in a compressive manner are favorable. Second, from another perspective, additive manufacturing enables a high surface-to-volume ratio, so it is vital to design a form that can offer this feature. Finally, the limiting parameter for us, with the equipment that we fabricated, was the size of the printable object. Therefore, we started by printing diverse lattice bricks (Figure 10).

Figure 10. The extruded lattice brick (Image credit: Authors).

The results showed us a negligible difference between the mechanical strength of the lattice bricks and molded regular bricks of the same size. Thus, we decided to examine the catenary as the simplest compression-based form to better use the material and technology. We generated catenary forms within the computational environment to find the proper form to print an arch. We intended to fabricate a vault with the repetition of continuously printed arches. After extrusion and material cultivation, the arches will be tilted 90 degrees and become a vertical vault (Figure 11).

Figure 11. The concept design of the extruded vault (Image credit: Authors).

In order to extrude the generated catenary form, it must consist of one continuous line. For this purpose, we defined a polyline that ends where it starts and then moves to the upper layer. After that, the layers are repeated as many as desired. Then, using a PRC plugin, the path is converted to a G-Code for the robotic arm connected to the extruder and printed (Figure 12).

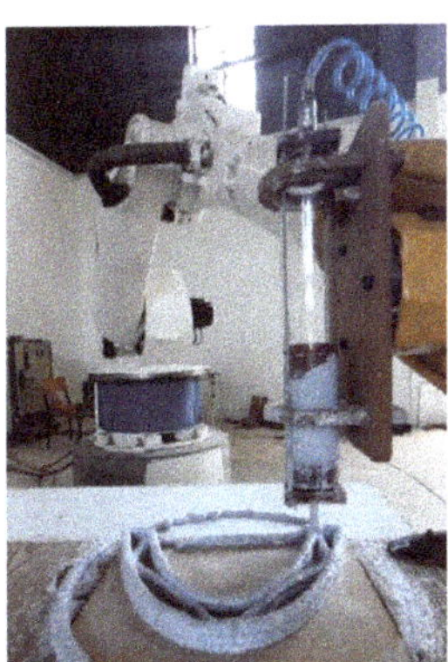

Figure 12. The printing session of one of the arches (Image credit: Authors).

We printed six similar arches with the material grown for two weeks in the bags and left them in the chamber to grow for five weeks (Figure 13).

Figure 13. The printed arches growing in the chamber (Image credit: Authors).

After this period, regarding the ability of the living material, we put the arches on top of each other to enable the bio-welding. This feature of the material lets us bind the material without adding external binders or connections. After seven days, the arches were well-bound together and formed a vault (Figure 14).

Figure 14. The extruded vault made of MBC material bound by bio-welding (Image credit: Authors).

4. Result and Discussion

The primary purpose of this study was to integrate the MBC material and the design computation and digital fabrication to enable the proper use of this bio-based material in the architectural context. Regarding the material characteristics of the MBC, utilizing this lightweight and quasi-weak material in a compressive manner was intended. Fabricating a catenary form that enables bearing the loads only in compression addresses this challenge. Unlike other catenary arches made with the MBC material formed with pipe-like formworks, our work is 3D printed. This fabrication method let us first fabricate with little waste due to eliminating formworks commonly made of inorganic materials such as plastics. Second, regarding the flexibility of forms that can be extruded, in comparison with molding, we could access a form with voids and divisions that make the outcome more mechanically strong, with less material and weight. Finally, as discussed in the paper, the extrusion also helped with the growth of the MBC material, as the surface-to-volume ratio is higher than the traditional molding process.

In addition to the mentioned outcome, in this project, we could take advantage of the bio-welding of the living mycelium for the first time (Figure 15). The concept of bio-welding enables the binding of the material without using external binding agents. This feature can minimize the need for dry or wet joints and mortars in the architectural context.

Figure 15. The Bio-welded Arches (Image credit: Authors).

5. Conclusions

As studied in this paper, mycelium-based composites are potential materials to substitute conventional matter in the architectural context. They offer light weight, regenerative, and compostable material features that could help address some challenges of temporary structures in the future. In this study, we aimed to inquire about the proper ways of fabrication with this material regarding its deficiencies in mechanical behavior and its compliance with additive manufacturing. We used shredded paper, oyster mushroom, wheat bran, and guar gum to prepare a treatment mixture that can be extruded and offers sufficient mechanical strength. The results showed that we could fabricate compression-only forms with this material by generating intelligent forms and geometries and using the bio-welding ability. The future challenges related to this research are scaling up the process of additive manufacturing of the MBC material and generating other innovative geometries for incremental printing of the bio-based material.

Author Contributions: Conceptualization, B.M.; methodology, B.M. and A.G.; software, B.M.; validation, B.M.; formal analysis, B.M.; investigation, B.M. and A.G.; writing—original draft preparation, A.G. and B.M.; writing—review and editing, A.G.; visualization, B.M.; supervision, M.M. and E.A.; project administration, B.M. All authors have read and agreed to the published version of the manuscript.

Funding: This research received no external funding.

Institutional Review Board Statement: Not Applicable.

Informed Consent Statement: Not Applicable.

Data Availability Statement: Data is available upon request.

Conflicts of Interest: The authors declare no conflict of interest.

References

1. Karana, E.; Blauwhoff, D.; Hultink, E.J.; Camere, S. When the material grows: A case study on designing (with) mycelium-based materials. *Int. J. Des.* **2018**, *12*, 119–136.
2. A Sustainable Alternative to Architectural Materials: Mycelium-based Bio-Composites. In Proceedings of the ConCave Ph.D. Symposium: Divergence in Architectural Research, Georgia Institute of Technology, Atlanta, GA, USA, February 2021; pp. 159–167. [CrossRef]
3. Meyer, V.; Basenko, E.Y.; Benz, J.P.; Braus, G.H.; Caddick, M.X.; Csukai, M.; De Vries, R.P.; Endy, D.; Frisvad, J.C.; Gunde-Cimerman, N.; et al. Growing a circular economy with fungal biotechnology: A white paper. *Fungal Biol. Biotechnol.* **2020**, *7*, 5. [CrossRef]
4. Jones, M.; Mautner, A.; Luenco, S.; Bismarck, A.; John, S. Engineered mycelium composite construction materials from fungal biorefineries: A critical review. *Mater. Des.* **2020**, *187*, 108397. [CrossRef]
5. Stelzer, L.; Hoberg, F.; Bach, V.; Schmidt, B.; Pfeiffer, S.; Meyer, V.; Finkbeiner, M. Life Cycle Assessment of Fungal-Based Composite Bricks. *Sustainability* **2021**, *13*, 11573. [CrossRef]
6. Elsacker, E.; Søndergaard, A.; Van Wylick, A.; Peeters, E.; De Laet, L. Growing living and multifunctional mycelium composites for large-scale formwork applications using robotic abrasive wire-cutting. *Constr. Build. Mater.* **2021**, *283*, 122732. [CrossRef]
7. Jones, M.; Huynh, T.; Dekiwadia, C.; Daver, F.; John, S. Mycelium composites: A review of engineering characteristics and growth kinetics. *J. Bionanosci.* **2017**, *11*, 241–257. [CrossRef]
8. Appels, F.V.; Camere, S.; Montalti, M.; Karana, E.; Jansen, K.; Dijksterhuis, J.; Krijgsheld, P.; Wösten, H.A. Fabrication factors influencing mechanical, moisture- and water-related properties of mycelium-based composites. *Mater. Des.* **2019**, *161*, 64–71. [CrossRef]
9. Ghazvinian, A.; Farrokhsiar, P.; Vieira, F.; Pecchia, J.; Gursoy, B. Mycelium-Based Bio-Composites for Architecture: Assessing the Effects of Cultivation Factors on Compressive Strength. In Proceedings of the eCAADe and SIGraDi Conference, University of Porto, Porto, Portugal, 11–13 September 2019; pp. 505–514. Available online: http://papers.cumincad.org/cgi-bin/works/paper/ecaadesigradi2019_465 (accessed on 5 November 2021).
10. Elsacker, E.; Vandelook, S.; Van Wylick, A.; Ruytinx, J.; De Laet, L.; Peeters, E. A comprehensive framework for the production of mycelium-based lignocellulosic composites. *Sci. Total Environ.* **2020**, *725*, 138431. [CrossRef] [PubMed]
11. Rafiee, K.; Kaur, G.; Brar, S.K. Fungal biocomposites: How process engineering affects composition and properties? *Bioresour. Technol. Rep.* **2021**, *14*, 100692. [CrossRef]
12. Manan, S.; Ullah, M.W.; Ul-Islam, M.; Atta, O.M.; Yang, G. Synthesis and applications of fungal mycelium-based advanced functional materials. *J. Bioresour. Bioprod.* **2021**, *6*, 1–10. [CrossRef]

13. Attias, N.; Danai, O.; Abitbol, T.; Tarazi, E.; Ezov, N.; Pereman, I.; Grobman, Y.J. Mycelium bio-composites in industrial design and architecture: Comparative review and experimental analysis. *J. Clean. Prod.* **2020**, *246*, 119037. [CrossRef]
14. Almpani-Lekka, D.; Pfeiffer, S.; Schmidts, C.; Seo, S. A review on architecture with fungal biomaterials: The desired and the feasible. *Fungal Biol. Biotechnol.* **2021**, *8*, 17. [CrossRef] [PubMed]
15. Grünewald, J.; Parlevliet, P.; Altstädt, V. Manufacturing of thermoplastic composite sandwich structures. *J. Thermoplast. Compos. Mater.* **2016**, *30*, 437–464. [CrossRef]
16. Colmo, C.; Ayres, P. 3d Printed Bio-hybrid Structures Investigating the architectural potentials of Mycoremediation. In Proceedings of the eCAADe Conference, Berlin, Germany, 16–18 September 2020; pp. 573–582. Available online: http://papers.cumincad.org/cgi-bin/works/BrowseTree=series:acadia/Show?ecaade2020_299 (accessed on 5 November 2021).
17. Goidea, A.; Floudas, D.; Andréen, D. Pulp Faction: 3d printed material assemblies through microbial biotransformation. *Fabricate* **2020**, *2020*, 42–49.
18. Elbasdi, G.; Alaçam, S. An investigation on growth behaviour of mycelium in a fabric formwork. In Proceedings of the Parametricism Vs. Materialism: Evolution of Digital Technologies for Development [8th ASCAAD Conference], London, UK, 7–8 November 2016; pp. 65–74. Available online: http://papers.cumincad.org/cgi-bin/works/paper/ascaad2016_009 (accessed on 5 November 2021).
19. McGaw, J. Dark Matter. *Arch. Theory Rev.* **2018**, *22*, 120–139. [CrossRef]
20. Heisel, F.; Lee, J.; Schlesier, K.; Rippmann, M.; Saeidi, N.; Javadian, A.; Nugroho, A.R.; Van Mele, T.; Block, P.; Hebel, D.E. Design, Cultivation and Application of Load-Bearing Mycelium Components: The MycoTree at the 2017 Seoul Biennale of Architecture and Urbanism. *Int. J. Sustain. Energy Dev.* **2017**, *6*, 296–303. [CrossRef]
21. Dessi-Olive, J. Monolithic Mycelium: Growing Vault Structures. In Proceedings of the 18th International Conference on Non-Conventional Materials and Technologies (NOCMAT), Nairobi, Kenya, 24–26 July 2019. Available online: https://www.academia.edu/39909593/Monolithic_Mycelium_Growing_Vault_Structures?auto=download&campaign=weekly_digest (accessed on 28 April 2020).
22. Bhardwaj, A.; Vasselli, J.; Lucht, M.; Pei, Z.; Shaw, B.; Grasley, Z.; Wei, X.; Zou, N. 3D Printing of Biomass-Fungi Composite Material: A Preliminary Study. *Manuf. Lett.* **2020**, *24*, 96–99. [CrossRef]
23. Bhardwaj, A.; Rahman, A.M.; Wei, X.; Pei, Z.; Truong, D.; Lucht, M.; Zou, N. 3D Printing of Biomass–Fungi Composite Material: Effects of Mixture Composition on Print Quality. *J. Manuf. Mater. Process.* **2021**, *5*, 112. [CrossRef]
24. Soh, E.; Chew, Z.Y.; Saeidi, N.; Javadian, A.; Hebel, D.; Le Ferrand, H. Development of an extrudable paste to build mycelium-bound composites. *Mater. Des.* **2020**, *195*, 109058. [CrossRef]
25. Jauk, J.; Vašatko, H.; Gosch, L.; Christian, I.; Klaus, A.; Stavric, M. DIGITAL FABRICATION OF GROWTH Combining digital manufacturing of clay with natural growth of mycelium. In Proceedings of the 26th CAADRIA Conference, The Chinese University of Hong Kong, Hong Kong, China, 29 March–1 April 2021; pp. 753–762. Available online: http://papers.cumincad.org/cgi-bin/works/paper/caadria2021_282 (accessed on 5 November 2021).
26. Oghazian, F.; Vazquez, E. A Multi-Scale Workflow for Designing with New Materials in Architecture: Case Studies across Materials and Scales Case Studies across Materials and Scales. In Proceedings of the 26th CAADRIA Conference, The Chinese University of Hong Kong, Hong Kong, China, 29 March–1 April 2021; pp. 533–542. Available online: http://papers.cumincad.org/cgi-bin/works/paper/caadria2021_213 (accessed on 5 November 2021).
27. Bhatia, S.; Bera, T. Somatic Embryogenesis and Organogenesis. In *Modern Applications of Plant Biotechnology in Pharmaceutical Sciences*; Elsevier Inc.: Amsterdam, The Netherlands, 2015; pp. 209–230. [CrossRef]
28. Silverman, J.; Cao, H.; Cobb, K. Development of Mushroom Mycelium Composites for Footwear Products. *Cloth. Text. Res. J.* **2020**, *38*, 119–133. [CrossRef]
29. Mudgil, D.; Barak, S.; Khatkar, B.S. Guar gum: Processing, properties and food applications—A Review. *J. Food Sci. Technol.* **2014**, *51*, 409–418. [CrossRef] [PubMed]
30. Zimele, Z.; Irbe, I.; Grinins, J.; Bikovens, O.; Verovkins, A.; Bajare, D. Novel mycelium-based biocomposites (Mbb) as building materials. *J. Renew. Mater.* **2020**, *8*, 1067–1076. [CrossRef]

Article

Wood-Veneer-Reinforced Mycelium Composites for Sustainable Building Components

Eda Özdemir [1,*], Nazanin Saeidi [2], Alireza Javadian [2,*], Andrea Rossi [1], Nadja Nolte [1], Shibo Ren [3,*], Albert Dwan [3], Ivan Acosta [3], Dirk E. Hebel [2], Jan Wurm [3,4] and Philipp Eversmann [1]

[1] Department of Experimental and Digital Design and Construction, University of Kassel, 34127 Kassel, Germany; rossi@asl.uni-kassel.de (A.R.); nadjanolte@uni-kassel.de (N.N.); eversmann@asl.uni-kassel.de (P.E.)

[2] Chair of Sustainable Construction, Karlsruhe Institute of Technology, 76131 Karlsruhe, Germany; nazanin.saeidi@kit.edu (N.S.); dirk.hebel@kit.edu (D.E.H.)

[3] Arup Deutschland GmbH, 10623 Berlin, Germany; albert.dwan@arup.com (A.D.); ivan.acosta@arup.com (I.A.); jan.wurm@kuleuven.be (J.W.)

[4] Design and Engineering of Construction and Architecture, KU Leuven, 1030 Brussels, Belgium

* Correspondence: eda.ozdemir@asl.uni-kassel.de (E.Ö.); alireza.javadian@kit.edu (A.J.); shibo.ren@arup.com (S.R.)

Abstract: The demand for building materials has been constantly increasing, which leads to excessive energy consumption for their provision. The looming environmental consequences have triggered the search for sustainable alternatives. Mycelium, as a rapidly renewable, low-carbon natural material that can withstand compressive forces and has inherent acoustic and fire-resistance properties, could be a potential solution to this problem. However, due to its low tensile, flexural and shear strength, mycelium is not currently widely used commercially in the construction industry. Therefore, this research focuses on improving the structural performance of mycelium composites for interior use through custom robotic additive manufacturing processes that integrate continuous wood fibers into the mycelial matrix as reinforcement. This creates a novel, 100% bio-based, wood-veneer-reinforced mycelium composite. As base materials, *Ganoderma lucidum* and hemp hurds for mycelium growth and maple veneer for reinforcement were pre-selected for this study. Compression, pull-out, and three-point bending tests comparing the unreinforced samples to the veneer-reinforced samples were performed, revealing improvements on the bending resistance of the reinforced samples. Additionally, the tensile strength of the reinforcement joints was examined and proved to be stronger than the material itself. The paper presents preliminary experiment results showing the effect of veneer reinforcements on increasing bending resistance, discusses the potential benefits of combining wood veneer and mycelium's distinct material properties, and highlights methods for the design and production of architectural components.

Keywords: mycelium; bio-composites; bio-fabrication; digital fabrication; additive manufacturing; ultrasonic welding; wood printing; circular construction; robotic fabrication; reinforced composites

Citation: Özdemir, E.; Saeidi, N.; Javadian, A.; Rossi, A.; Nolte, N.; Ren, S.; Dwan, A.; Acosta, I.; Hebel, D.E.; Wurm, J.; et al. Wood-Veneer-Reinforced Mycelium Composites for Sustainable Building Components. *Biomimetics* **2022**, 7, 39. https://doi.org/10.3390/biomimetics7020039

Academic Editors: Andrew Adamatzky, Han A.B. Wösten and Phil Ayres

Received: 17 March 2022
Accepted: 29 March 2022
Published: 31 March 2022

1. Introduction

In the past decades, the construction industry has been challenged by the rapidly increasing population and the proportional demand in housing and construction material supply [1]. Concurrently, the excessive energy used, the pollution and the waste generated to produce traditional building materials, such as steel, cement and plastics, impose severe environmental challenges [2]. The majority of greenhouse gas (GHG) emissions results from the processing of materials that are commonly used in the construction industry [3]. The diminution of natural resources and the growing recognition of climate change have been encouraging researchers and companies to seek sustainable alternatives to the currently used materials [4]. The 4R concept of Reduce, Reuse, Recycle and Recover has been increasingly becoming more prevalent to reduce waste and promote circular economy models within industries.

Growing biological materials using plant-based waste from industries can be a potential solution [5]. Among these, the development of bio-based composite materials from mycelium has been introduced recently and could potentially transform the construction sector. Indeed, mycelium-based bio-composites could support the transition towards the utilization of the available organic waste resources by binding them through the mycelium network, further facilitating the development of sustainable and circular alternatives to energy- and resource-intensive construction materials and building products.

Mycelium has been proven to deliver a range of properties significant to construction, from good acoustic to mechanical properties, including compressive strength, while being a renewable and low-carbon alternative material with relatively good fire-resistance properties [6]. However, one of the major limitations for its application within the construction industry is caused by its low resistance to tension and bending [7]. On the other hand, wood has been known for centuries for its high structural performance. It is an inherently tension-resistant material due to its fiber arrangement [8]. Therefore, this research aims at combining the advantages of each material: exploiting the intrinsic properties of mycelium and wood veneer, and exploring the development of novel, 100% bio-based mycelium-wood veneer composites with improved mechanical properties.

We explore two methods for increasing material strength: compression with heat and pressure, and the integration of topologically designed reinforcement within the mycelium matrix. While compression improves material strength and Young's modulus by increasing the density of the material [9], an embedded veneer lattice in mycelium is expected to increase the performance of the composite due to the combination of the compressive strength of mycelium and tensile strength of the internal fiber structure of wood. We investigate these two methods through physical prototyping and testing and compare them in terms of effectiveness, advantages and disadvantages. The possibility of combining the two methods and compressing a veneer-reinforced block is also explored. We develop a hybrid fabrication method suitable for this composite material system and test the samples structurally to assess the effect of compression and reinforcement on composite strength.

The focus of this research is to explore ways to improve the structural performance of this composite for architectural use cases, while maintaining satisfactory levels of suitable acoustic performance. The intended application is currently planned for interior use; therefore, the water and pest resistance of the resulting composite was not yet studied. The acoustic and fire performance will be subject to further studies.

2. State of the Art

2.1. Mycelium-Based Composites

The sustainability issues arising from the use of synthetic and non-renewable resins and binders in the engineered wood industry are well known. Thus, new solutions with bio-based resins with a lower environmental impact are being investigated globally [10]. Among the various materials used, mycelium has the potential to be a sustainable and more attractive alternative to most of the available binder matrices. Mycelium is the root part of fungi, composed of filamentous strands of fine white hyphae. When organic substrates, such as wood or natural fibers, are inoculated with specific fungi species, mycelium starts growing by using the substrates' nutrients [11]. By the time mycelium spreads through the whole substrate, a network structure is developed that binds the discrete particles of the substrate together. Therefore, a range of sustainable and green products can be manufactured in an environmentally friendly way without the need for any adhesives, potentially replacing various energy-intensive building materials. One advantage of mycelium-based composites over traditionally engineered wood-based materials is that they can be recycled or composted at their end of life without any negative impact on the environment [12]. No toxic substances or synthetic components are involved; therefore, mycelium-bound composite materials fit into the model of a bio-circular economy where there is no waste at the end of a product's lifecycle [6,13].

The properties of mycelium-bound materials can be customized to a certain extent by adjusting the parameters of the manufacturing process. A thorough framework of the main parameters influencing mycelium-based composites was presented in various studies recently and identified advantageous material properties, such as low thermal conductivity, high acoustic absorption, and fire protection properties [10–12]. However, challenges generally arise from research knowledge gaps [14] that limit the use of these materials only to non-structural or semi-structural applications.

2.2. Mycelium in Architecture

The use of mycelium-based composites as building materials has been explored in recent decades [15]. The most common approach has been to create mycelium building blocks that are assembled into larger structures. However, this application often relies on substructures, and is geometrically quite limiting due to the inherent properties of mycelium that only allows for structures in compression [7,16]. Studies have also been carried out on monolithic mycelium constructions, but these systems require either large scaffolds or extensive reinforcement systems that in most cases take over the structural functions and reduce the mycelium to a surface finishing, rather than a load-bearing material [17].

3D printing mycelium is an emerging research area that uses the mass-customization opportunity as the main research driver. While articulated surfaces created through 3D printing help mycelium growth [18], for large-scale structures, time-efficient additive manufacturing processes have not yet been developed, and the directional dependency of the fabrication method can cause the resulting components to display relatively low structural performance [19].

In order to compensate for the lack of tension and bending resistance of mycelium, research on reinforcing mycelium has been developed recently. Woven textiles, wood fibers, or 3D-printed spatial lattices are among the methods used [20–23]. However, these studies either heavily rely on manual production, or currently present very limited data about the effects of the reinforcement on the mycelium-based composite strength. For construction applications, to date only mycelium-based foam (MBF) and mycelium-based sandwich composites (MBSC) have been developed and investigated for their properties [24]. The latter uses natural fiber textiles on top and bottom of the components in order to increase bending resistance [25].

2.3. Additive Manufacturing with Timber

Current additive manufacturing technologies for producing high complexity objects are mainly based on inorganic materials. New processes that allow 3D printing with organic materials have been recently developed [26,27]. For the fused deposition modelling of timber, wood is ground to particles and mixed with various thermoplastics to create continuous printing filaments or pellets [28]. Both these materials cause timber to lose its natural material structure that provides strength, resulting in relatively weak printed structures [29].

In recent years, researchers proposed a fabrication method for architectural elements, using continuous natural timber fiber filaments through robotic fabrication [30,31]. The aim was to produce structural elements by combining the advantages of continuous fiber-based manufacturing with bio-based materials. This method can achieve highly controlled, sustainable, surface-like [32] and optimized geometries [33], as it can be seen in Figure 1.

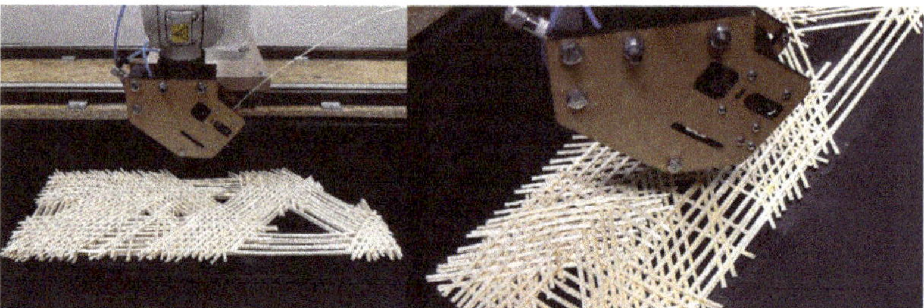

Figure 1. Robotic fiber laying process with processed willow strips from the research project TETHOK—Textile Tectonics for Wood Construction, University of Kassel.

2.4. Contribution

Combining the previously introduced mycelium composites and wood-based additive manufacturing processes, we propose a novel wood-veneer–mycelium bio-composite and its construction method for carbon neutral, circular building elements. As mycelium has excellent compression properties, but low tension and bending resistance, the integration of tailored continuous wood fibers in the composite is expected to increase the structural capabilities of mycelium-based components, while still being composed of exclusively natural materials. To demonstrate its potential in the context of architecture, we describe the material concept and its production process, and present results regarding its characterization with reinforcement strategies.

3. Materials and Methods
3.1. Selection of the Base Materials
3.1.1. Wood Veneer Species

We made a pre-selection of wood species indigenous to Germany, based on their availability at the time of the research, and data from the literature that proved their compatibility with mycelium growth: beech (*Fagus sylvatica*), maple (*Acer pseudoplatanus*), oak (*Quercus robur*), and spruce (*Picea abies*) veneers and willow branches of the genus *Salix americana*. Initial binding tests with these selected species were carried out at the University of Kassel. Maple demonstrated the best wood–wood bond with the selected binding method and was chosen as the reinforcement material.

H. Heitz Furnierkantenwerk from Melle, Germany supplied FSC (Forest Stewardship Council) certified maple veneer edge-bands that were 12 mm wide and 0.5 mm thick in spools. They are produced by lining up veneer sheets and joining them with fully glued finger joints. Non-woven cellulose-based fleece on one side of the roll is then added with PVAc dispersion glue to ensure that the material does not easily break during application. Due to the commercially available veneer rolls using a small amount of glue for their joining during production, the presented composites are not yet fully bio-based. However, custom veneer rolls made with bio-adhesives could be produced and utilized in future studies.

3.1.2. Substrates

We made a pre-selection of the substrates based on the availability of the raw materials mainly as waste stream in Europe: hemp fibers, hemp hurds, pine wood sawdust and shavings, and Silvergrass (Miscanthus) shavings. For the purpose of this study, only hemp hurds were used, which were collected from Bafa GmbH (Malsch, Germany), a local wood mill.

3.1.3. Mycelium Species

The mycelium mother culture of *Ganoderma lucidum* (*G. lucidum*) was purchased from Tyroler Glückspilze (Innsbruck, Austria) in the form of grain spawn and stored at 4 °C for up to four weeks. This selection was mainly made due to the already known faster growth rate on hemp hurds, and its availability in Europe. *Ganoderma lucidum* was grown on hemp hurds and subsequently reinforced with maple veneers to carry out a series of physical and mechanical tests on lightweight and dense veneer-reinforced mycelium-based composites.

3.2. Fabrication

3.2.1. Robotic Wood Fiber Laying

We developed a custom fabrication process to lay the continuous wood fibers robotically. The process consisted of the following sub steps: a single wood strip was extruded at a time, and the material was cut when a change in extrusion direction was needed. This allows for complex tool paths, the creation of multi-directional reinforcement patterns and controlled anisotropy. Two approaches can be used for the layering: placing the veneers with the same direction or similar directions at once, then moving on to the next layer (Figure 2); or printing one line from a different direction at a time, which results in a structure with interwoven fibers [31].

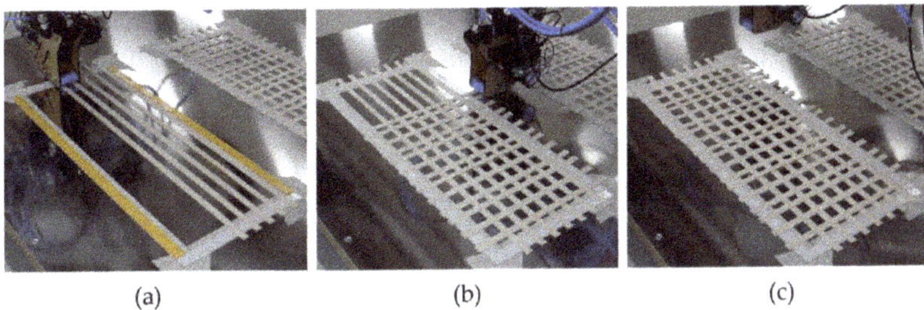

Figure 2. Robotic fiber laying process: (**a**) Fiber laying in direction one; (**b**) Fiber laying in direction two; (**c**) Completed 2D lattice.

We designed two types of 2D veneer lattices to reinforce the mycelium blocks/boards for this study: high- and low-density lattices (Figure 3). While the low-density lattice had two veneer strips in the longitudinal direction and four in the transversal direction, the high-density lattice had three and seven veneer strips, respectively. In each lattice, we placed the veneers to form a frame that was 19 cm × 8 cm in a total of two layers that were perpendicular to each other. The veneer strips were fixed at their ends using double-sided tape during printing. After all strips with the same direction were laid down, the second layer of strips perpendicular to the first layer was added, and the lattices were ready for the ultrasonic welding of the intersection points.

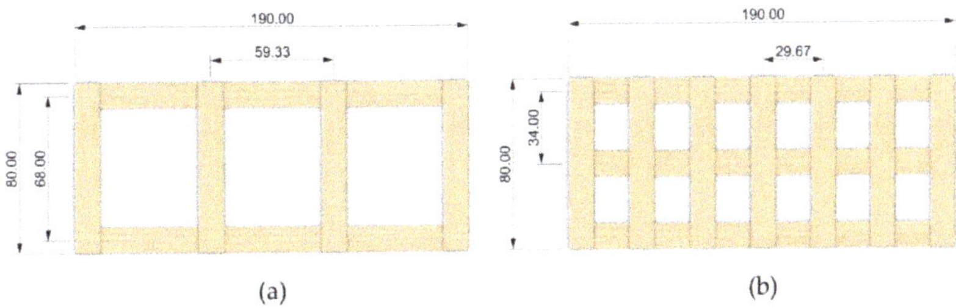

Figure 3. Veneer lattices produced: (**a**) Low-density lattice; (**b**) High-density lattice (dimensions in mm).

3.2.2. Ultrasonic Wood Welding for Wood–Wood Binding

Precedents of continuous wood fiber laying research have explored synthetic binders, such as UV-curing glue, contact glue and hot melt glue [31]. Since mycelium growth is incompatible with synthetic materials, and the goal of producing a 100% bio-based composite cannot be achieved with the binders investigated to date, it was necessary to research alternative binding methods.

Ultrasonic welding is a common adhesive-free joining method used in many industries, including automotive, electronic, and medical, due to its speed. It is performed by using ultrasonic energy at high frequencies that produce mechanical vibrations, which results in heat due to the friction between the two elements to be joined. Heat melts thermoplastic materials and binds the parts together after cooling [34]. In recent decades, this method has been used to weld thin woo, through heat softening and melting lignin in wood and binding the materials with entangled fibers [35]. Considering that no adhesives are needed for joining, this method was chosen as the wood–wood binding strategy for our custom manufacturing process.

The wood welding was performed with an ultrasonic welding horn, a generator that uses 20 kHz frequency, and a flat-ended sonotrode provided by Weber Ultrasonics (Karlsbad, Germany) mounted on a robotic arm (Figure 4).

(a)   (b)

Figure 4. (**a**) Robotic welding process; (**b**) Welded intersection point close-up.

As the veneer rolls have fleece on one side, initial welding tests were made comparing the welds of wood to wood, wood to fleece and fleece to fleece sides. The material was always placed with the wood side facing the welding horn to avoid the fleece from sticking onto the welding horn. Once the robot reached the intersection point to be welded, the pressure was applied by moving the robot arm down in the vertical direction. Then, the welding was performed by a signal of the digital control unit connected to the generator.

3.2.3. Mycelium-Based Composite Fabrication

Substrate Inoculation

Hemp hurds were collected from a local wood mill called Bafa GmbH (Malsch, Germany) and mixed with wheat bran to enhance the growth of mycelium, while calcium sulfate ($CaSO_4$) was added in a dry condition to adjust the pH of the mixture to the desirable threshold of 5 to 6, suitable for mycelium growth. The mixture was then blended with 60 wt% (weight percentage) of water, and eventually sterilized at 121 °C for 60 min. Subsequently, the mixture was cooled to room temperature before it could be inoculated with the selected *G. lucidum* grain spawn. Once cooled down, it was mixed with 1 wt% of colonized mycelium spawn. The colonized substrate was eventually left in the incubation room at 26–28 °C with 70–80% humidity for two weeks to develop the full mycelium network.

Molding and Sample Preparation

After two weeks of substrate colonization in the incubation room, the samples were taken out and transferred into molds prepared for compression, pull-out, and flexural tests. Following the filling of the molds with colonized substrates, they were transferred to the incubation room with similar conditions to the previous phase. The molds were kept there for an additional 3–6 days until the mycelium network was observed to have covered the substrate surface. Then, the mold was removed, and the samples were left in the incubation room for another 3–5 days to expedite the growth of the mycelium network inside and on the surface of the samples with better aeration. The samples´ preparation process and the final mycelium-based composites for each type of test are shown in Figure 5.

Figure 5. Mycelium composite samples' production process: (**a**) Compressive strength test cube; (**b**) One-side single veneer pull-out test cube; (**c**) Two-side, middle overlapped veneer with and without welding reinforced cube; (**d**) Lightweight block with and without low- and high-density lattices; (**e**) Pressed board with and without low- and high-density lattices.

For compression and pull-out tests, we prepared cubes of $5 \times 5 \times 5$ cm^3. A moistened and sterilized maple veneer strip with a length, width and thickness of about 16 cm, 1.2 cm, and 0.05 cm, respectively, was then placed in the center of the mold, parallel to one mold side for the pull-out tests. We prepared three types of pull-out samples: a series with a single veneer strip penetration of up to 75% of the height of the cubes, samples with unwelded overlapped veneer extending from both sides of the cubes, and lastly samples with overlapped and welded veneer. For the last two series of the above-mentioned samples, we placed the overlapped section of the veneer strips in the center of the cubes. The tests aimed to determine the interfacial shear strength between the mycelium matrix and veneer, and to evaluate the bonding mechanism that was developed at the interface of the veneer and mycelium matrix.

Flexural samples were prepared in molds of 19 cm × 8 cm × 7 cm, with and without veneer lattices (Figure 5d,e). First, we filled up half of the height of the mold with the colonized substrates before placing a veneer lattice, and afterwards finished filling up the rest of the mold with substrate. It was ensured that the density of all the samples would stay the same throughout the sample preparation. We used two types of veneer lattices for this study: high- and low-density lattices. Given the lack of prior research on the use of veneer reinforcement for mycelium-based composite materials, the size of the samples was chosen to suit the available testing facilities, while necessary references to ASTM and European standards were made. For comparison purposes, we prepared a series of flexural test samples with two layers of low-density veneer lattices embedded: one lattice at the top and one at the bottom of the molds with a 10 mm distance from the surface of the substrate. Further details are provided in Section 3.2.4.

Post-Processing

Once the growth cycle was completed, the samples were transferred to a drying oven and kept there at a temperature range of 60–70 °C for 2–3 days. The samples were weighed regularly during this period to ensure their weight was stabilized. When no change was observed, they were removed from the oven, and their final density was measured.

Compression and pull-out test samples were directly tested after drying, while the flexural test samples with and without veneer lattices were prepared for an additional pressing process to produce dense mycelium-based composites (DMC) as per the procedure explained in an earlier study [11]. The flexural test samples were placed in a hot press compression molding machine and pressed at a temperature of 120 °C, with the pressure set to 10 MPa for a duration of 15 minutes. The compressed samples were then moved to an oven with a temperature of 40 °C for 12 to 24 h to adjust to the room temperature and avoid any thermal stress shock within the samples.

3.2.4. Testing
Tensile Tests

Ten maple veneer strips were tested in order to determine the tensile strength of the reinforcement material used for this study. Each end of the veneer strip was fixed to the grip of a UTM with a 30 kN HBM load cell attached, and pulled by applying 1 N with a loading rate of 10 mm/min.

Similarly, the weld strength was also investigated through tensile tests. Two maple veneer strips that were 10 cm long were overlapped along their grains (in the same axis) and on the end points with an area of 1.2 cm × 3 cm, in which the 1.8 cm from the center of the overlap was welded. Twenty samples were prepared, and the ends were fixed to the grip of the testing machine with the same setup and pulled apart by applying 1 N with a loading rate of 10 mm/min.

A testing standard specific to wood veneers was not found. However, for the climate conditions of the testing, there are numerous standards, such as DIN 52377 (Testing of plywood—Determination of modulus of elasticity in tension and of tensile strength), DIN EN 302-x and DIN EN 205 (Adhesives—Wood adhesives for non-structural applications—Determination of tensile shear strength of lap joints) for wood, and DIN EN ISO 291 (Plastics—Standard atmospheres for conditioning and testing) for plastics with very similar conditions that can be considered as a baseline. Therefore, the climate recommendations from these standards were taken as the reference and the tests were carried out in circa 23 °C and at 50% humidity. While the welded samples were stored in 20 °C and at 60–65% humidity prior testing, single veneer strips were tested immediately in the recommended testing room conditions. The test setup was designed as per recommendations given by the DIN EN 205 testing standard, while the sample shape had to be adapted due to the material restrictions. Other specifications, such as clamping length and temperature, were followed.

Since almost all the welded samples demonstrated material failure rather than joint failure (see Section 4.2.1), tensile strength was evaluated instead of shear strength. The following formula was used for the calculation:

$$\sigma_t = \frac{F_{max}}{bt} \tag{1}$$

where F_{max} stands for the maximum load in N measured by the UTM at the failure, and b and t represent the specimen width and thickness in mm, respectively.

Compression Tests

The compression test samples had an average density of 145 kg/m^3, which places them in the range of flexible polyurethane foam products, due to their soft texture and low density. Given the lack of standard testing methods for lightweight mycelium-based materials, ASTM D3574:2017 (Standard Test Methods for Flexible Cellular Materials—Slab, Bonded, and Molded Urethane Foams), which is a commonly used standard for testing flexible cellular foam materials, was used as the reference for the testing and evaluation of the compressive properties. The samples were tested using a Universal Testing Machine (UTM) with a 5 kN HBM load cell (HBK, Germany). The cubes were compressed at a rate of 0.5 mm/min and tested until failure. The compressive strength and elastic modulus were calculated using the following formula:

$$\sigma_c = \frac{F_{max}}{bd} \tag{2}$$

where F_{max} is the failure load in N recorded during the test, and b and d are the width and depth of the cubes in mm, respectively. The elastic modulus was calculated using the slope of the stress–strain curves obtained from each individual test.

Pull-Out Tests

The bond between the veneer and mycelium matrix can be assessed with the pull-out tests, similar to the methods employed to measure the reinforcement-concrete matrix bond in steel-reinforced concrete elements. There are multiple testing standards with similar scenarios where the reinforcement (veneer strips in this study) is embedded in concrete. Using a UTM, the reinforcement is pulled out by applying a tension force with a defined loading rate, while the sample is restrained to avoid its movement. However, a testing standard for the exact type of material combination of timber and mycelium does not yet exist, since it is a novel composite material. Therefore, testing standards for other materials had to be followed. We selected the procedures explained in RILEM technical recommendation (RC6 Bond test for reinforcement steel. 2. Pull-out test) and ASTM D7913:2020 (Standard Test Method for Bond Strength of Fiber-Reinforced Polymer Matrix Composite Bars to Concrete by Pullout Testing) as the most relevant standards for this study. Thus, we designed the test setup as per the recommendations given by these two testing standards and made the necessary modifications to suit the available testing machines.

The interfacial shear strength (IFSS), or the bond strength, was measured by using a 5kN HBM load cell attached to a Universal Testing Machine (UTM). As explained earlier, three types of pull-out tests were performed in this study. For the pull-out tests where the veneer strip was extended only from one side of the cubes, the veneer strip was fixed to the grip of the tensile test setup and was pulled out on the fixed end. It was ensured that the cube could be held in place to prevent any movement or slipping during the tests. Subsequently, the IFSS was measured using the following formula:

$$\tau = \frac{F_p}{2l(t+b)} \tag{3}$$

where F_p is the pull-out force measured by the machine in N; t and b are the veneer thickness and width in mm, respectively; and l is the embedded veneer length (75% of the cube height) in the mycelium matrix in mm.

In the case of welded and unwelded overlapped veneers, which extended from the two sides of the pull-out samples, a similar overlapped area of 1.2 cm $\times$ 3 cm was used to compare the bonding properties. The veneers were overlapped and then embedded within the mycelium matrix and the two free ends were pulled out with the help of the UTM tensile grips from both sides.

Flexural Tests

To evaluate the flexural capacity of the samples, the recommendations of ASTM D1037:2020 (Standard Test Methods for Evaluating Properties of Wood-Base Fiber and Particle Panel Materials) were adopted. A three-point flexural test was used to find the modulus of rupture and to evaluate the flexural properties, including the elastic modulus in flexure. The support span was set to 140 mm, and a loading rate of 1 mm/min was used for the testing. The lightweight blocks and dense boards with and without veneer lattices were each tested for their flexural properties. The Modulus of Rupture (MOR) was calculated using the following formula, while the elastic modulus in flexure was calculated using the stress–strain curves obtained for each sample from the UTM:

$$MOR = \frac{3LF_{max}}{2bt^2} \tag{4}$$

where F_{max} stands for the maximum load in N measured by the UTM at failure, and b, t and L represent the specimen width, thickness, and distance between the support points in mm, respectively. L was set to 140 mm for all the samples, given the size of the specimens and the available testing machines. It should be noted that the lightweight blocks and the dense boards had a final size of 18.5 cm $\times$ 7.5 cm $\times$ 6.5 cm after drying and pressing.

4. Results

4.1. Fabrication

We successfully developed a custom robotic fabrication process consisting of two steps for this study: robotic wood fiber laying and ultrasonic wood welding. These individual processes were carried out, respectively, using different end effectors mounted on a robot arm. By carefully placing the wood fibers and binding them where necessary, we produced flat lattices made up of two orthogonal layers with this method and used them to reinforce mycelium blocks. Thanks to the mechanical properties and directionality of the material being preserved with this fabrication method, precise control over the reinforcement orientation was achieved.

Despite the geometric freedom of this additive manufacturing method, we encountered some challenges in the production: when placing the veneer, the start and end points of the strips should be fixed. Similarly, when more than one layer is deposited on one point, the strips must be bound together to stay in place until mycelium growth. Since only 2D lattices were produced for this study, this was solved by fixing the end points of the veneers with double sided tape on an aluminum plate. However, to produce more complex geometries with multiple layers, a more robust and local solution needs to be researched.

As a result of the initial welding tests, the most promising bond was obtained with the sample where the wood side of one veneer was welded to the fleece side of the other. Therefore, all samples were produced, while keeping the wood side on top; two layers of veneer were overlapped on their opposite sides (wood to fleece) and prevented the fleece from sticking to the sonotrode.

The welds on the intersection points were successful in keeping the lattice together during mycelium growth. However, not every point could be welded at once, or with the same quality due to slight pressure differences caused by using robot motion to apply pressure for welding. The welding setting used for most of the intersection points was

0.5 seconds with 100% amplitude. While this setting performed well in most welds, inconsistencies were observed, with burned welds and welding failures. In our robot setup, we used contact pressure through robot motion, which was not possible to be precisely controlled. This is therefore assumed to be one of the main reasons for welding inconsistencies. With the welding equipment being highly sensitive, if the sensors detect that the target welding time is not reached, or the system uses too much power to weld, the process is interrupted. In further studies, the integration of a force-controlled pneumatic cylinder and motion control system is planned to ensure constant pressure, prevent delays in production, and provide weld consistency.

4.2. Testing

4.2.1. Tensile Tests

Most samples showed wood tensile failure (Figure 6b), rather than welding area failure, proving that the weld strength is higher than the veneer tensile strength. The average tensile strength of the welded samples was measured as 63.6 ± 10.5 MPa, which is almost the same as the average tensile strength of a single maple veneer strip, 61.95 ± 10.83 MPa, and confirms the previous statement (Figure 6a).

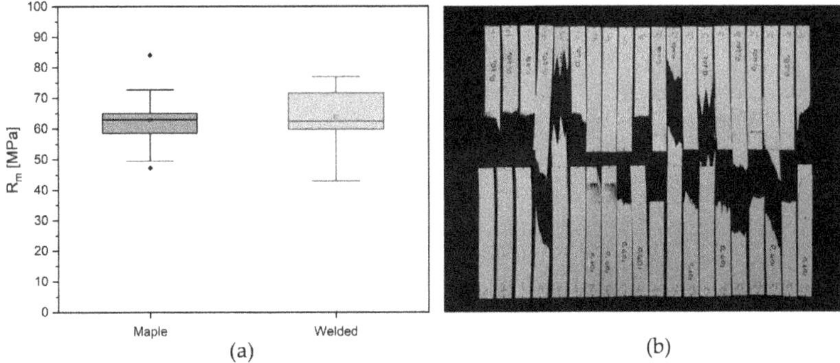

Figure 6. Tensile strength tests: (**a**) Comparison graph of maple veneer's tensile strength to welded joints' tensile strength; (**b**) Tested welded veneer samples.

4.2.2. Compression Tests

The results of compression, pull-out and flexural tests of all samples, together with their densities, are summarized in Table 1.

The average compressive strength of the samples was 1.2 MPa and the average elastic modulus in compression was measured to be around 4.1 MPa. As it can be seen in Figure 7a, the samples were compressed until reaching 50% of their original height. No cracking or breakage was observed during the tests. The samples showed deformation under compression load and were compressed until the end of the test. It was also observed, when the test continued beyond 50% of the sample's original height, that higher loads could be achieved. However, given the recommendations of the ASTM D3574, the test continued until the samples were compressed to 50% of their original height and showed similar behavior to conventional foams and flexible cellular materials.

Table 1. Summary table for physical and mechanical properties of mycelium-based composites with and without veneer lattices.

Test	Sample	Density (kg/m³)	Strength (MPa)	Elastic Modulus (MPa)
Compressive strength	Cube		1.2 ± 0.12	4.10 ± 0.67
Pull-out strength	Cube + one-side veneer Cube + two-side of unwelded veneer	145 ± 14	0.34 ± 0.04 0.36 ± 0.1	NA NA
Tensile strength	Cube + two-side of welded veneer		30.6 ± 3.6	NA
Flexural strength lightweight	Block Block + low-density lattice Block + high-density lattice * Block + 2 layers of low-density lattice *	140 ± 8	0.17 ± 0.04 0.19 ± 0.04 0.16 ± 0.05 0.13 ± 0.02	1.31 ± 0.33 1.32 ± 0.29 1.29 ± 0.16 0.89 ± 0.24
Flexural strength dense	Board Board + low-density lattice Board + high-density lattice	1180 ± 75	10.2 ± 1.73 21.99 ± 2.01 10.81 ± 3.18	2390.95 ± 444.91 6236.22 ± 322.2 3900.2 ± 1621.9

* Flexural failure was not observed for these samples; the failure mode was shear as explained in the text.

Figure 7. Samples during testing and after failure: (**a**) Compressive strength test; (**b**) One-side veneer pull-out test; (**c**) Two-side unwelded veneer pull-out test; (**d**) Two-side welded veneer tensile test.

4.2.3. Pull-Out Tests

The pull-out samples with one strip of maple veneer in the middle showed similar IFSS as those samples prepared with two strips of unwelded maple veneer. Mycelium growth was observed on the maple veneer surface that was embedded within the mycelium matrix (Figure 7b,c). The samples have shown a relatively good bonding: the bond strength between the mycelium matrix and the veneer strips was measured as 0.34 MPa and 0.36 MPa for a single veneer strip and overlapped veneer strips, respectively. The slight increase in the IFSS in samples with two unwelded overlapped veneer strips can be attributed to the better growth of the mycelium network around and between the layers of the veneer strips. A stronger bond was developed in these areas in comparison to the samples with a single strip of veneer where less surface area resulted in a lower mycelium growth density. For both series of samples, clear pull-out of veneers from the mycelium matrix was observed.

However, for samples prepared with welded overlapped maple veneers, the failure of the veneer strips due to tensile mode of failure was observed, before the veneer could be pulled out (Figure 7d). This can be explained by the higher failure load observed during the tests compared to the other pull-out samples. Furthermore, when the resulting stress is compared with the tensile strength of the maple veneer, it can further validate the hypothesis that the weld strength is relatively higher than the bond strength between the veneer strips and the mycelium matrix (Figure 6). However, further testing is required to find out the effect of mycelium growth combined with welded veneers on improving the bond strength in wood-veneer-reinforced mycelium-based composites.

4.2.4. Flexural Tests

Lightweight samples with one layer of high-density veneer lattice and the ones with top and bottom low-density veneer lattices showed shear mode of failure, while all the other samples, including dense boards with low- and high-density veneer lattices, and samples with no veneer lattices showed flexural mode of failure (Figure 8). The flexural strength of lightweight blocks increased slightly with the addition of one layer of low-density veneer lattice in the middle, compared to non-reinforced blocks. On the other hand, the use of high-density lattices resulted in shear failure and lower flexural strength. Similarly, samples with top and bottom low-density lattices also showed lower flexural strength. The results are summarized Figure 9.

The abovementioned behavior could be explained by the shear failure mode as a result of the potentially lower bonding strength between the veneer lattices and mycelium matrix. High-density lattices decrease the areas where the mycelium network would grow through the lattice holes and connect the two sides of the block divided by the lattice. This would create a weaker interlocking mechanism, which could result in a higher chance of de-bonding when exposed to flexural loads. Furthermore, the lower flexural strength of these samples compared to low-density lattices could also be attributed to the mycelium growth on the veneer lattices. In the case of high-density lattices, more surface area would result in a higher bonding strength between the veneer strips and substrates by forming a stronger mycelium network. However, as was observed from the pull-out tests of samples with unwelded overlapped veneer strips, the bond mechanism was not fully developed and all the samples showed clear pull-out failure, rather than the tensile failure of the veneer strips. Therefore, it is possible to state that weaker bonding areas between the mycelium matrix and the veneer strips increase the chance of de-bonding and interlaminar shear failure.

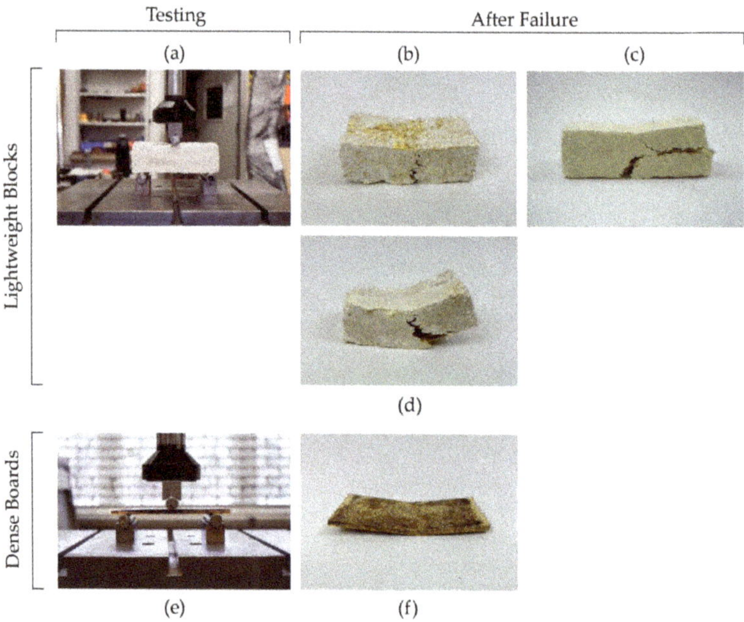

Figure 8. Samples during testing and after failure: (**a**) lightweight block under 3-point flexural test; (**b**) Block without or with low-density lattice in the middle; (**c**) Block with high-density lattice in the middle (shear failure); (**d**) Block with two layers of low-density lattices close to the top and bottom of the block; (**e**) Dense board under 3-point flexural test; (**f**) Dense board after failure without lattice reinforcement.

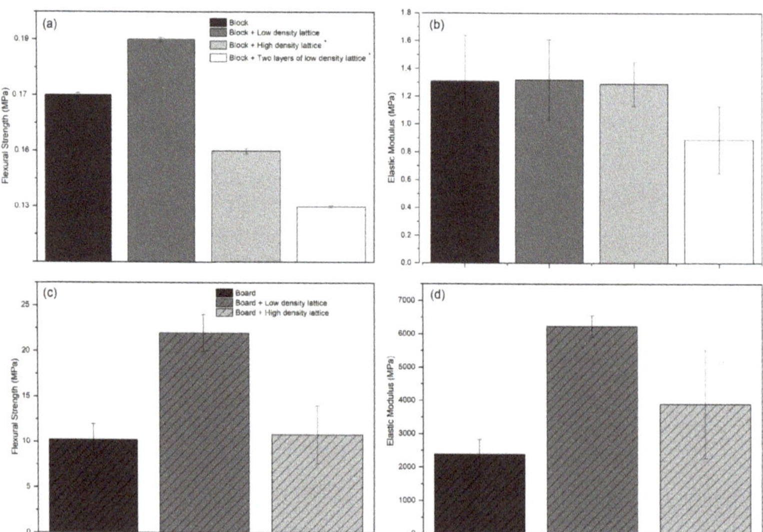

Figure 9. Flexural properties, including strength and elastic modulus: (**a**) Flexural strength of lightweight blocks; (**b**) Elastic modulus in flexure of lightweight blocks; (**c**) Flexural strength of dense boards; (**d**) Elastic modulus in flexure of dense boards. * Shear failure was observed, further explanation is given in the text.

Unlike the lightweight samples, dense boards showed clear flexural failure, which indicates a stronger bond mechanism between the veneer strips and mycelium matrix (Figure 8f). Even though no shear failure was observed, similar trends in flexural strength could be observed when the high-density veneer lattices were used in dense boards. No significant increase in flexural strength was detected within dense boards with high-density lattices compared to dense boards with no lattices. However, dense boards with low-density lattices showed a significant increase in both flexural strength and elastic modulus. In general, it was also observed that the increase in density helped to increase the flexural properties when the results of dense boards are compared with lightweight blocks.

Further testing is required to evaluate the impact of different veneer lattice densities and layouts in combination with different substrate densities on the flexural properties of wood-veneer-reinforced mycelium-based composites. Furthermore, the investigation of veneer placement within the samples along the height of the blocks should also be carried out to explore the bond mechanism developed between mycelium matrix and the veneer lattices with varying densities, and their impacts on flexural properties of the final samples.

Lightweight samples reinforced with either one layer of high-density lattice in the middle or two layers of low-density lattices on top and bottom of the block showed shear failure (Figure 8c,d). However, for lightweight samples reinforced with one layer of low-density veneer lattice and dense boards with low- and high-density veneer lattices, flexural failure was observed as the dominant mode of failure as expected (Figure 8b,f).

5. Discussion

The wood fiber laying process used in this study has the potential to become a resource-efficient, rapid production method, with material carefully placed in the structurally required areas. Fiber placement and binding were carried out with two different end effectors, as sequential steps of the process. In order to speed up production, further studies are planned to develop a single tool that can lay wood veneers and at the same time weld the intersections.

The novel wood-veneer-reinforced mycelium composites developed in this study were investigated for their mechanical properties, including compressive strength, pull-out strength and flexural properties. The suitability of 2D veneer lattices as a reinforcement system with welded and unwelded joints was also investigated separately through a series of tensile tests. The results of the investigation of welded joints show that they perform relatively well; in the majority of the tensile tests, no failure in the joints was observed, which indicates that the joints have a higher strength than the veneer itself. The bonding between the veneer strip and mycelium matrix was investigated through a series of pull-out tests and the results show that the bond might not have been developed fully, as most of the single veneer-strip-reinforced cubes showed purely pull-out failure modes rather than any failure in the veneer strip. However, the visual examination after completion of the tests confirmed the growth of mycelium network on the veneer strips. This again validates our hypothesis that the selected mycelium species (*Ganoderma lucidum*) can grow well when combined with the selected veneer species (maple) and hemp hurds as the main substrate.

Furthermore, the performance of the welded and unwelded joints was investigated through a series of pull-out tests with similar overlapping veneer areas embedded within the mycelium matrix. It was observed that the welded joints outperform the unwelded ones. The results show that, even though the mycelium growth was observed in both cases on the veneers, the interfacial shear strength developed within the unwelded veneer strips and mycelium matrix was lower than the strength of the welded joints. Further investigation on enhancing the mycelium growth on the veneer strip and improving the interfacial shear strength between the mycelium matrix and veneer strip is necessary to achieve a better bonding strength.

The results of the flexural tests on various samples once again strengthen the hypothesis that the bonding between the veneer reinforcement and mycelium matrix plays an important role in the structural integrity and mechanical properties of these composites.

Moreover, compressing the lightweight blocks into dense boards showed a significant improvement in flexural properties as a result of densification, and improved the bending mechanism between the veneer lattice and mycelium matrix. While samples with top and bottom veneer reinforcement did not show any significant increase in the overall flexural properties, samples with one layer of low-density veneer lattice before and after compression showed better flexural performance. The lower flexural strength and elastic modulus measured correspond to the shear mode of failure observed during the tests. Therefore, further investigation is necessary to identify the optimum design of the veneer lattices and to explore the effect of connecting the top and the bottom reinforcement lattices, namely 3D lattices.

Composed of one bottom and one top 2D lattice connected by an undulating layer of wood veneer, 3D lattice reinforcements could potentially improve the shear capacity of the mycelium composites and provide additional strength and stiffness via the spatial lattice system. Their design would be strictly connected to the design of the base 2D lattice. Following the production of the flat lattice, the layer that gives the structure its depth would be achieved through placing the material diagonally between the two opposite corners of a quadrilateral cell created by the 2D lattice, with a pre-calculated length. More material length would result in more structural depth.

This system could also achieve surfaces with varying depths through the management of the middle layer's height and the corresponding non-parallel and non-planar top and bottom surfaces. Similarly, through the gradual cell size modifications along the structure, heterogeneous reinforcement could be achieved. The parameters that can be adjusted on a cell level provide high flexibility, and opportunities for lightweight, optimized reinforcement based on local requirements.

However, certain limitations must be considered when designing 3D lattices with wood veneers: different veneer thicknesses and species allow for different bending radii. Due to the fibers being oriented in the direction of fiber laying, forcing the material into a very small bending radius would result in the filament breaking. Therefore, the minimum bending radius would be the main determining factor for the component height and as a result, the 3D lattice density. In Figure 10, some 3D lattice design studies based on the bending radius of maple veneer can be seen.

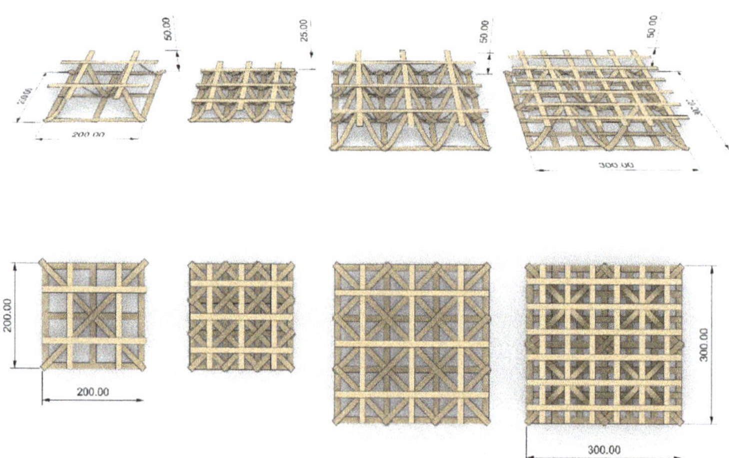

Figure 10. 3D lattice layout studies (dimensions in mm).

6. Conclusions

A novel wood-veneer-reinforced mycelium-based composite material was developed for this study as a sustainable and green alternative to traditional building materials with potential applications in the construction industry. Structural testing on physical prototypes was carried out to investigate the fundamental mechanical properties of this novel composite. The test samples were prepared with different variations of veneer lattices as reinforcement systems and tested for compressive strength, bond strength, and flexural properties. The tests provided an initial understanding of the mechanical behavior of the wood-veneer-reinforced mycelium composites in terms of densities, strength and stiffness at material scale. Both strategies, integrating a topologically designed veneer lattice and compression with heat and pressure, proved to be effective methods in increasing the bending resistance of the presented composites. It was shown that the effect of veneer lattices as reinforcement systems is strictly tied to the density and configuration of the lattice. For lightweight blocks, the most promising results were achieved with a single layer of low-density veneer lattice placed in the middle of the mycelium block. This configuration helped to increase the flexural strength of the block slightly (approximately from 0.17 MPa to 0.19 MPa), whereas the high-density lattice and two low-density lattices at the top and bottom of the block resulted in a lower flexural strength (approximately 0.16 MPa and 0.13 MPa, respectively) than that of the unreinforced block itself. The samples with two low-density lattices demonstrated a low flexural strength and elastic modulus and resulted in shear failure. Therefore, 3D lattice systems connecting top and bottom lattices are proposed to avoid shear failure in lightweight blocks for future studies.

The dense boards with one low-density lattice in the middle demonstrated a similar trend to the lightweight blocks and increased the flexural strength to more than double (from approximately 10 MPa to 25 MPa) of the unreinforced dense boards. On the other hand, the dense boards with one high-density lattice in the middle did not show a significant change in flexural strength compared to the unreinforced dense boards.

When the two methods are compared, dense boards have a better overall flexural strength. The dense boards reinforced with one low-density lattice are the most promising specimens, and would be appropriate for applications that require planar components and higher bending resistance. However, if more complex geometries that do not require high bending resistance are needed, lightweight blocks reinforced with one low-density lattice would be suitable.

The study provided the fundamental material inputs for the further development of the system at a larger scale. In the next steps, a digital model will be developed to integrate the material properties as design inputs and material constraints; geometrical variations as design variables; and structural Finite Element Analysis (FEA) and acoustic analyses as solvers to evaluate and optimize various design options within one digital computational framework.

Further studies, including the investigation of the growth compatibility between other wood veneers and mycelium species combined with a range of available organic waste by-products from wood and agricultural industries, will be carried out in the next steps of the research. Additional mechanical testing of the mycelium composites as larger panels reinforced with 3D lattice systems made of veneer are also planned to gain further insights into the materials' behavior, which will subsequently support the design and development of these composites. The initial results obtained show that wood-veneer-reinforced mycelium composites could be a promising environmentally friendly and sustainable substitute material to conventional building materials with potential applications in the context of architecture.

Author Contributions: Conceptualization, E.Ö., A.J. and N.S.; methodology, E.Ö., A.J. and A.R.; software, A.R. and E.Ö.; validation, N.S., A.R. and E.Ö.; formal analysis, A.J. and N.S.; investigation, A.J.; data curation, N.S.; writing—original draft preparation, E.Ö. and A.J.; writing—review and editing, N.S., A.R., N.N., S.R., A.D., I.A. and P.E.; visualization, E.Ö., N.S. and N.N.; supervision, P.E., D.E.H. and J.W.; project administration, E.Ö., A.R. and P.E.; funding acquisition, P.E., D.E.H. and J.W. All authors have read and agreed to the published version of the manuscript.

Funding: This research was funded by "Forschungsinitiative Zukunft Bau des Bundesinstitutes für Bau-, Stadt- und Raumforschung", grant number 10.08.18.7-21.48.

Institutional Review Board Statement: Not applicable.

Informed Consent Statement: Not applicable.

Data Availability Statement: Data are available upon request.

Acknowledgments: The authors would like to thank Jannis Heise from the Fachgebiet Trennende und Fügende Fertigungsverfahren (TFF) at the University of Kassel for conducting tensile strength tests on single maple veneer strips and welded maple veneer joints, and analyzing and visualizing the test results; Thomas Bierwirth and Marco Klocke from Heitz Furnierkantenwerk, Melle for supplying wood veneers for physical prototyping and testing; and lastly, Tobias Neff and Christian Ringwald from Weber Ultrasonics, Karlsbad for technical support and ultrasonic welding equipment supply.

Conflicts of Interest: The authors declare no conflict of interest. The funders had no role in the design of the study; in the collection, analyses, or interpretation of data; in the writing of the manuscript, or in the decision to publish the results.

References

1. Raut, S.P.; Ralegaonkar, R.V.; Mandavgane, S.A. Development of sustainable construction material using industrial and agricultural solid waste: A review of waste-create bricks. *Constr. Build. Mater.* **2011**, *25*, 4037–4042. [CrossRef]
2. Madurwar, M.V.; Ralegaonkar, R.V.; Mandavgane, S.A. Application of Agro-Waste for sustainable construction materials: A review. *Constr. Build. Mater.* **2013**, *38*, 872–878. [CrossRef]
3. Arıoğlu Akan, M.Ö.; Dhavale, D.G.; Sarkis, J. Greenhouse gas emissions in the construction industry: An analysis and evaluation of a concrete supply chain. *J. Clean. Prod.* **2017**, *167*, 1195–1207. [CrossRef]
4. Maraveas, C. Production of sustainable construction materials using Agro-Wastes. *Materials* **2020**, *13*, 262. [CrossRef]
5. Elsacker, E.; Vandelook, S.; Brancart, J.; Peeters, E.; De Laet, L. Mechanical, Physical and chemical characterisation of mycelium-based composites with different types of lignocellulosic substrates. *PLoS ONE* **2019**, *14*, e0213954. [CrossRef]
6. Jones, M.; Mautner, A.; Luenco, S.; Bismarck, A.; John, S. Engineered mycelium composite construction materials from fungal biorefineries: A critical review. *Mater. Des.* **2020**, *187*, 108397. [CrossRef]
7. Heisel, F.; Lee, J.; Schlesier, K.; Rippmann, M.; Saeidi, N.; Javadian, A.; Nugroho, A.R.; Mele, T.V.; Block, P.; Hebel, D.E. Design, Cultivation and application of load-bearing mycelium components: The MycoTree at the 2017 seoul biennale of architecture and urbanism. *Int. J. Sustain. Energy Dev.* **2017**, *6*, 296–303. [CrossRef]
8. Forest Products Laboratory Mechanical Properties of Wood. In *Wood Handbook-Wood as an Engineering Material*; United States of Department of Agriculture Forest Service: Madison, WI, USA, 2010; p. 508.
9. Appels, F.V.W.; Camere, S.; Montalti, M.; Karana, E.; Jansen, K.M.B.; Dijksterhuis, J.; Krijgsheld, P.; Wösten, H.A.B. Fabrication factors influencing mechanical, moisture- and water-related properties of mycelium-based composites. *Mater. Des.* **2019**, *161*, 64–71. [CrossRef]
10. Javadian, A.; Ferrand, H.L.; Hebel, D.; Saeidi, N. Application of Mycelium-Bound Composite Materials in Construction Industry: A Short Review. Available online: https://www.semanticscholar.org/paper/Application-of-Mycelium-Bound-Composite-Materials-A-Javadian-Ferrand/9070067ca5a3aed37daea39d88dd5dfa40ee009b (accessed on 10 March 2022).
11. Chan, X.Y.; Saeidi, N.; Javadian, A.; Hebel, D.E.; Gupta, M. Mechanical properties of dense mycelium-bound composites under accelerated tropical weathering conditions. *Sci. Rep.* **2021**, *11*, 22112. [CrossRef]
12. Elsacker, E.; Vandelook, S.; Van Wylick, A.; Ruytinx, J.; De Laet, L.; Peeters, E. A comprehensive framework for the production of mycelium-based lignocellulosic composites. *Sci. Total Environ.* **2020**, *725*, 138431. [CrossRef]
13. Soh, E.; Saeidi, N.; Javadian, A.; Hebel, D.E.; Le Ferrand, H. Effect of common foods as supplements for the mycelium growth of Ganoderma lucidum and *Pleurotus ostreatus* on solid substrates. *PLoS ONE* **2021**, *16*, e0260170. [CrossRef] [PubMed]
14. Attias, N.; Livne, A.; Abitbol, T. State of the Art, Recent advances, and challenges in the field of fungal mycelium materials: A snapshot of the 2021 mini meeting. *Fungal Biol. Biotechnol.* **2021**, *8*, 12. [CrossRef] [PubMed]
15. Sydor, M.; Bonenberg, A.; Doczekalska, B.; Cofta, G. Mycelium-Based composites in art, architecture, and interior design: A review. *Polymers* **2021**, *14*, 145. [CrossRef] [PubMed]
16. Heisel, F.; Hebel, D.E. Pioneering construction materials through Prototypological research. *Biomimetics* **2019**, *4*, 56. [CrossRef] [PubMed]

17. Dessi-Olive, J. Monolithic mycelium: Growing vault structures. In Proceedings of the 18th International Conference on Non-Conventional Materials and Technologies "Construction Materials and Technologies for Sustainability, Nairobi, Kenya, 24–26 July 2019.
18. Goidea, A.; Floudas, D.; Andréen, D.; Burry, J.; Sabin, J.; Sheil, B.; Skavara, M. Pulp Faction: 3d Printed Material Assemblies through Microbial Biotransformation. In *Fabricate 2020; Making Resilient Architecture;* UCL Press: London, UK, 2020; pp. 42–49, ISBN 978-1-78735-812-6.
19. Ulu, E.; Korkmaz, E.; Yay, K.; Burak Ozdoganlar, O.; Burak Kara, L. Enhancing the structural performance of additively manufactured objects through build orientation optimization. *J. Mech. Des.* **2015**, *137*, 111410. [CrossRef]
20. MycoKnit: Cultivating Mycelium-Based Composites on Knitted Textiles for Large-Scale Biodegradable Architectural Structures. Available online: https://somfoundation.com/fellow/davis-ghazvinian-gursoy-oghazian-pecchia-west/ (accessed on 9 March 2022).
21. Woven Mycelium. Available online: https://www.materialbalance.polimi.it/2021/05/21/woven-mycelium/ (accessed on 16 March 2022).
22. Mycelium Composite+Adaptive Lattice Structure. Available online: https://www.materialbalance.polimi.it/portfolio_page/mycelium-composite-adaptive-lattice-structure/ (accessed on 16 March 2022).
23. Adamatzky, A.; Ayres, P.; Belotti, G.; Wosten, H. Fungal architecture. *arXiv* **2019**, arXiv:1912.13262.
24. Girometta, C.; Picco, A.M.; Baiguera, R.M.; Dondi, D.; Babbini, S.; Cartabia, M.; Pellegrini, M.; Savino, E. Physico-Mechanical and thermodynamic properties of mycelium-based Biocomposites: A review. *Sustainability* **2019**, *11*, 281. [CrossRef]
25. Jiang, L.; Walczyk, D.; McIntyre, G.; Bucinell, R.; Tudryn, G. Manufacturing of biocomposite sandwich structures using mycelium-bound cores and preforms. *J. Manuf. Process.* **2017**, *28*, 50–59. [CrossRef]
26. Mogas-Soldevila, L.; Duro-Royo, J.; Oxman, N. Water-Based robotic fabrication: Large-Scale additive manufacturing of functionally graded hydrogel composites via Multichamber extrusion. *3D Print. Addit. Manuf.* **2014**, *1*, 141–151. [CrossRef]
27. Le Duigou, A.; Correa, D.; Ueda, M.; Matsuzaki, R.; Castro, M. A review of 3D and 4D printing of natural fibre biocomposites. *Mater. Des.* **2020**, *194*, 108911. [CrossRef]
28. Das, A.K.; Agar, D.A.; Rudolfsson, M.; Larsson, S.H. A review on wood powders in 3D printing: Processes, properties and potential applications. *J. Mater. Res. Technol.* **2021**, *15*, 241–255. [CrossRef]
29. Le Duigou, A.; Castro, M.; Bevan, R.; Martin, N. 3D printing of wood fibre biocomposites: From mechanical to actuation functionality. *Mater. Des.* **2016**, *96*, 106–114. [CrossRef]
30. Dawod, M.; Deetman, A.; Akbar, Z.; Heise, J.; Böhm, S.; Klussmann, H.; Eversmann, P. Continuous Timber Fibre Placement. In *Proceedings of the Impact: Design with All Senses;* Gengnagel, C., Baverel, O., Burry, J., Ramsgaard Thomsen, M., Weinzierl, S., Eds.; Springer International Publishing: Cham, Switzerland, 2020; pp. 460–473.
31. Eversmann, P.; Ochs, J.; Heise, J.; Akbar, Z.; Böhm, S. Additive timber manufacturing: A novel, wood-based filament and its additive robotic fabrication techniques for large-scale, material-efficient construction. *3D Print. Addit. Manuf.* **2021**. [CrossRef]
32. Ochs, J.; Akbar, Z.; Eversmann, P. Additive manufacturing with solid wood: Continuous robotic laying of multiple wicker filaments through micro lamination. In Proceedings of the DC I/O 2020, London, UK, 7–8 October 2020; Design Computation Ltd.: London, UK, 2020.
33. Lienhard, J.; Eversmann, P. New Hybrids–from Textile logics towards tailored material behaviour. *Archit. Eng. Des. Manag.* **2021**, *17*, 169–174. [CrossRef]
34. Chapter 2-Ultrasonic Welding. In *Handbook of Plastics Joining*, 2nd ed.; Troughton, M.J. (Ed.) William Andrew Inc.: Norwich, NY, USA, 2008; pp. 15–35, ISBN 978-0-8155-1581-4.
35. Tondi, G.; Andrews, S.; Pizzi, A.; Leban, J.-M. Comparative Potential of alternative wood welding systems, ultrasonic and Microfriction stir welding. *J. Adhes. Sci. Technol.* **2007**, *21*, 1633–1643. [CrossRef]

Article

Mycomerge: Fabrication of Mycelium-Based Natural Fiber Reinforced Composites on a Rattan Framework

Mai Thi Nguyen [1], Daniela Solueva [1], Evgenia Spyridonos [2,*] and Hanaa Dahy [2,3,4]

1   Faculty of Architecture and Urban Planning, University of Stuttgart, Keplerstrasse 11,
70174 Stuttgart, Germany; st163415@stud.uni-stuttgart.de (M.T.N.); st161479@stud.uni-stuttgart.de (D.S.)
2   BioMat Department of Bio-Based Materials and Materials Cycles in Architecture, Institute of Building
Structures and Structural Design (ITKE), University of Stuttgart, Keplerstrasse 11, 70174 Stuttgart, Germany;
hanaa.dahy@itke.uni-stuttgart.de
3   Department of Architecture (FEDA), Faculty of Engineering, Ain Shams University, Cairo 11517, Egypt
4   Department of Planning, Technical Faculty of IT & Design, Aalborg University, 2450 Copenhagen, Denmark
*   Correspondence: evgenia.spyridonos@itke.uni-stuttgart.de

Abstract: There is an essential need for a change in the way we build our physical environment. To prevent our ecosystems from collapsing, raising awareness of already available bio-based materials is vital. Mycelium, a living fungal organism, has the potential to replace conventional materials, having the ability to act as a binding agent of various natural fibers, such as hemp, flax, or other agricultural waste products. This study aims to showcase mycelium's load-bearing capacities when reinforced with bio-based materials and specifically natural fibers, in an alternative merging design approach. Counteracting the usual fabrication techniques, the proposed design method aims to guide mycelium's growth on a natural rattan framework that serves as a supportive structure for the mycelium substrate and its fiber reinforcement. The rattan skeleton is integrated into the finished composite product, where both components merge, forming a fully biodegradable unit. Using digital form-finding tools, the geometry of a compressive structure is computed. The occurring multi-layer biobased component can support a load beyond 20 times its own weight. An initial physical prototype in furniture scale is realized. Further applications in architectural scale are studied and proposed.

Keywords: bio-based materials; mycelium; mycelium-based composites; natural fiber reinforced polymers; NFRP; growing materials; rattan; lightweight structure

Citation: Nguyen, M.T.; Solueva, D.; Spyridonos, E.; Dahy, H. Mycomerge: Fabrication of Mycelium-Based Natural Fiber Reinforced Composites on a Rattan Framework. *Biomimetics* **2022**, 7, 42. https://doi.org/10.3390/biomimetics7020042

Academic Editors: Andrew Adamatzky, Han A.B. Wösten and Phil Ayres

Received: 15 March 2022
Accepted: 6 April 2022
Published: 8 April 2022

Publisher's Note: MDPI stays neutral with regard to jurisdictional claims in published maps and institutional affiliations.

1. Introduction

1.1. Relevance

Since the linkage between human consumption behavior and the rapidly increasing global warming becomes evident, life in the 21st century faces an ongoing environmental crisis. The use of conventional building materials in the global industry impacts largely on climate change by destructing more than 45 percent of global resources and emitting up to 40 percent of the energy-related carbon dioxide into the atmosphere [1,2]. Within an ever-growing society, there is and will be a constant need for materials, and consumption prevention is not the optimal choice. Awareness of already existing alternative systems is crucial for achieving sustainability. Waste can only be repurposed as a new resource if the majority of building components can be disassembled and returned to their original material cycles separately [3,4]. Fungal substrates are considered waste products and thus, can be used as compost and in a range of other applications [5]. By decreasing the cradle-to-gate manufacturing in the building sector, construction processes could be optimized to meet the social, ecologic, and economic values of the future generation. While the composting process was absent during the non-regenerative production line of the 18th and 19th centuries, at the present time, reorientation processes take place towards the cultivation of natural resources by breeding, raising, or growing materials. With sufficient improvement of the current fabrication methods, which are biologically driven and technologically supported, designing and manufacturing sustainable objects are goals within reach [3,6,7].

1.2. Recyclability of Composites

With an increasing demand for lightweight and durable building materials, fiber-reinforced polymer composites are considered a reliable alternative to conventional building materials such as concrete and steel. Determined by their components and fabrication methods, the structural and functional performance of fiber-reinforced polymers (FRC) can be adjusted to match the preferred application. Natural fiber-reinforced polymer (NFRP) composites consist of a high-strength reinforcement and a high-ductility matrix. Cellulosic fibers are highly sustainable and commonly used as reinforcement in composites. They are commonly agricultural residues; hence they enhance the ecological role of renewable resources, can be found in nature, are non-toxic, renewable, cost-effective, and allow bonding with different matrices. Bio-based composites with mineral and petrochemical matrices are widely used. However, their full biodegradability is costly to achieve due to the complex separation of the composites into their initial components, causing limited end-of-life options. Recyclability of composites with bio-based matrices is also limited, as degradation can only take place in specific industrial composting conditions [4,6,8]. Alternatively, mycelium-based matrices are organic matter and fully biodegradable, fulfilling the requirements of the circular material life cycle [8,9]. Since the main constituents of mycelium composites are fibrous substrates, lignocellulosic agricultural or forestry by-products and wastes such as straw and hemp, or porous substrate, e.g., sawdust, the costs of mycelium composites are low and enable waste upcycling [10,11]. The results of the first methods for the disintegration of mycelium-based composites (Ganoderma resinaceum and hemp fibers) in soil have strengthened their biodegradability, with a maximum weight loss of 43% after 16 weeks [12].

1.3. Mycelium Based Composites

Mycelium is the root of fungi, building large thread-like networks, which are made of individual hyphae. Hyphae grow from mycelium fungal strain spores and consume feedstock containing carbon and nitrogen [13]. To create mycelium-based composites, fast and robust colonization of the substrate is required. Among the numerous subordinates of fungi, Dikarya build large and complex structures. These fungi have two special characteristics: Septa—transverse cell wall opening which can close—decreases damage caused to the colony by a rupture; and Anastomosis—the ability of two hyphae fusing together to build large networks and distribute nutrients from high to low concentrated areas. In the presence of hosting materials from agricultural waste products such as hemp or flax, mycelium merges with its environment and absorbs its host. Colonization and growth are highly dependent on the amount of cellulose in the given hosting material, as the nutrition of fungi consists of glucose. Mycelium can break down cellulose into glucose, which means that a high cellulose environment can improve its growth. Apart from compatibility, some natural fibers offer additional protection by a waxy outer layer, preventing contamination by other microorganisms. Generally, the hosting material must be sterilized through the processes of pasteurization, hydrogen-peroxide treatment, and natural composting [8]. The substrate can then be inoculated with the preferred fungal species. There are crucial growing conditions for successful cultivation, including low light, high humidity, medium temperature, and access to oxygen. After sufficient growth, the growing phase can be interrupted or stopped by exposing the cultivated composite to high temperatures over 80 degrees [8,9,13].

1.4. Previous Studies

Mycelium-based biocomposites are perceived as a sustainable and competitively performing material alternative in several application fields, including thermal and acoustic insulation, or as a replacement to standard expanded polystyrene packaging [10]. Because of mycelium's high compressive mechanical properties, previous studies have focused on compression-only structural applications. Further applications of mycelium in design and construction have also been a topic of further research studies [14].

A study carried out by Jiang (2017) [15] examines the performance of mycelium-bound sandwich composites by measuring the flexural stiffness of the composites' core and skin layers. During this experiment, a discontinuous composite core is placed in between two skin layers of continuous and randomly oriented cellulosic fibers. Interfacial bonding between the core composite and the outer binding layers was possible because of mycelium's ability to grow through the fiber matrix and effectively bind with the fibers. The compressive properties of the composite are mainly dependent on the core stiffness and can be improved through highly cellulosic skin materials [10].

Two mycelium composite prototypes were developed during the Material Matter Lab IV at the BioMat Department in the ITKE Institute of the University of Stuttgart. This is a seminar practicing validation through small-scale structural demonstrators in the form of chairs and stools [16]. A timber veneer mold (Figure 1a) and a soft cotton fabric mold (Figure 1b) were used to develop the two alternatives. The timber mold bends because of the moisture being held throughout the entire growth process, interfering with the overall shape and causing separation. In addition, due to the outer skin's density, there is insufficient ventilation, which leads to mold contamination. Separation and inconsistent formability occurred in the cotton fabric option prototype, just as in the previous experiment, because of the fabric's stretchability. In industrial treatment methods such as bleaching, nutrition loss occurs in natural cellulose material, making it less suitable for the mycelium to grow on.

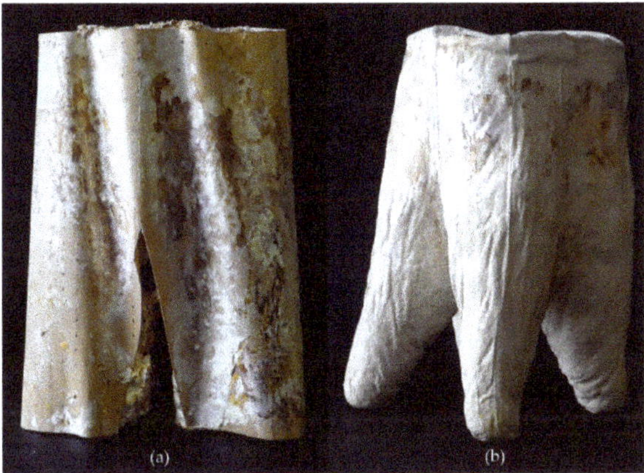

Figure 1. Mycelium-based prototypes: (**a**) using a timber veneer mold, and (**b**) using cotton fabric mold, by F. Milano, K. Antorveza, L. Kiesewetter, G. Lochnicki, Materials Matter Lab IV, 2020.

2. Materials and Methods

2.1. Workflow

After obtaining sufficient knowledge about the environmental preferences and growing behavior of mycelium, the initial step of the practical process is the proper cultivation of the organism. The aim is to prevent creating an external mold that will have to be discarded. The bonding qualities of mycelium and outer skin materials are investigated through small-scale samples. Digital form-finding and optimization tools, specifically Rhino Vault 2, a plugin for Rhino McNeel that concentrates on funicular form-finding, are used to calculate design possibilities. A rattan framework is used as a supporting skeleton to enable fabrication without the usage of an external mold while also being able to form double-curved surfaces. A workflow diagram in Figure 2 presents the basic steps of this bottom-up process.

Figure 2. Bottom-Up workflow diagram.

2.2. Cultivation of Homegrown Substrate

To inoculate the substrate, wood plugs already infused with *Pleurotus ostreatus* cultures are utilized. Short, chopped fibers such as wood chips and long continuous fibers such as hemp are compared as a hosting environment (Figure 3). The hosting materials are sterilized via pasteurization. Then, the sterilized materials must cool down to a temperature of 28 °C before adding nutrients and the mycelium-infused wood plugs. The growing period takes place at an ambient temperature of approximately 20–25 °C over the course of three weeks (Figure 4).

(a)   (b)   (c)

Figure 3. Mycelium-infused wood plugs (**a**), hosting materials: hemp fibers (**b**), wood chips (**c**).

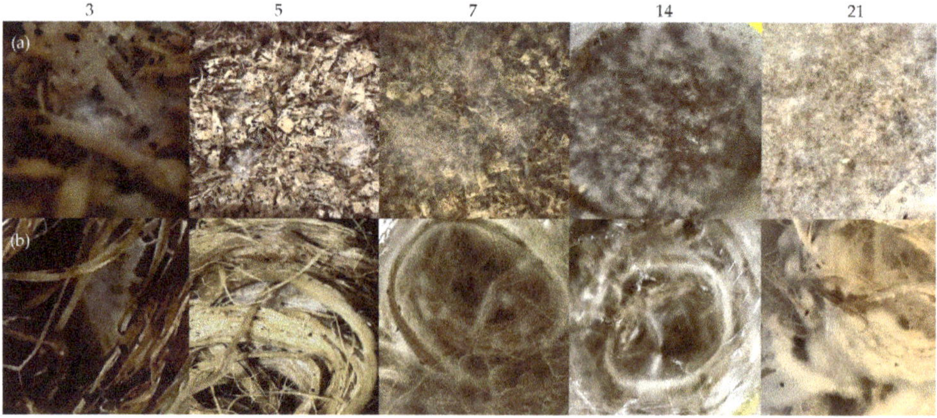

Figure 4. Growth process in days: (**a**) wood chips, (**b**) hemp fibers.

2.3. Compatibility with Skin Materials

The pre-grown substrate was purchased from the market due to the lack of a sterile environment throughout this study, as well as the time constraints imposed by the long growing period required. The substrate's merging capabilities are examined using three natural fiber materials: continuous bidirectional woven jute fabric, discontinuous randomly oriented compressed hemp sheets, and a continuous unidirectional hemp rope knitted outer layer. An easily detachable framework is necessary to keep the sterilized soft textiles in position (Figure 5). Wooden frames are CNC cut and then assembled. After the material growth is completed, each component of the framework can be detached and reused.

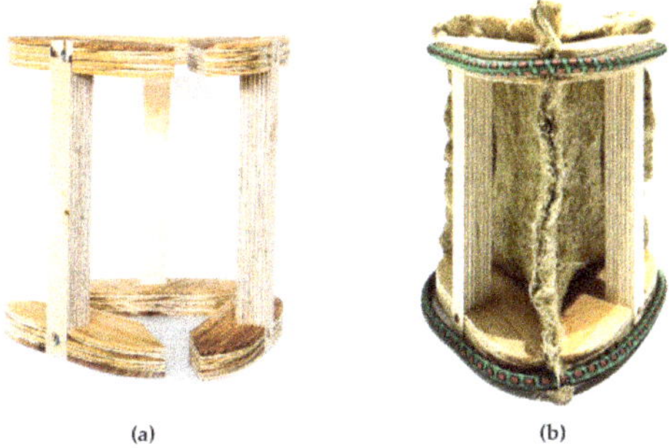

Figure 5. (**a**) Wooden framework, (**b**) hemp mat.

2.4. Results of Growth on Skin Materials

2.4.1. Hemp Sheets

The first sample contains mycelium substrate, compressed into a soft mold of randomly oriented hemp fibers (Figure 6a). Due to the high density of the hemp sheets, water absorption levels are high and ensure constant moisture levels throughout the growth process. Because of that, the mycelium grew beyond the geometrical restrictions of the soft mold and along the outer edges of the hemp sheets. The high growth density prevents separation during the shrinking process and results in the stiffest sample and most successful binding outcome. The concept of a "soft mold" suggests the use of a natural frame that is integrated during the fabrication process and stays embedded in the end-product. The outcome of this initial test was successful, resulting in new ideas for alternate molding methods.

Figure 6. Results of grown on different skin materials: (**a**) hemp sheets, (**b**) jute sheets, (**c**) knitted hemp rope.

2.4.2. Jute Sheets

In the second experiment, pre-woven bidirectional jute sheets are used as a skin alternative (Figure 6b). The low thickness and density of the fibers do not contribute to containing sufficient moisture levels, which results in the sample drying before enough growth is achieved. Consequently, uneven shrinkage and separation between substrate and skin layer occurred.

2.4.3. Knitted Hemp Rope

Throughout the third experiment, knitted hemp rope is used as an alternate outer skin (Figure 6c). While compressing the mycelium substrate into the mold, the too loosely knitted skin resulted in a deformed overall shape. Similar moisture deficiency as in the previous observation occurs. The sample's final state is less rigid and hardly successful, due to the uneven distribution of the substrate.

2.5. Results on the Growth of Multi-Layer Samples

Based on the successful growing outcome of the first sample (Section 2.4.1), two further experiments with hemp were carried out.

2.5.1. Hemp Sheet Sandwich

As an alternative to filling up a voluminous mold, in this experiment, the mycelium substrate is pressed between two hemp mat layers to form a thin and rigid sandwich element. As in the above-described example, the density of the hemp fibers ensures a constantly moist growth environment. No separation between substrate and hemp sheets is visible. This sandwich results in the stiffest sample (Figure 7b).

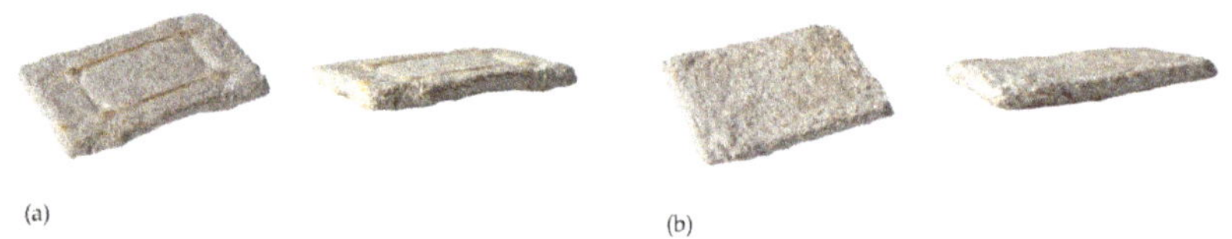

(a) (b)

Figure 7. (**a**) Results of multilayer composite, (**b**) hemp mat sandwich.

2.5.2. Multilayer Composite: Rattan, Loose Hemp Fibers, Mycelium Substrate with Chopped Hemp Fibers

In the final material test, loose hemp fibers act as a substitute for the mechanically compressed hemp sheets, and rattan reinforcement is introduced in between the mycelium substrate. Due to the greater airflow between the randomly oriented loose hemp fibers, there is proportionally more space for the mycelium to spread, still resulting in a stiff sample but also exhibiting higher elasticity. Rattan acts as an integral structural reinforcement, as it successfully merges with the mycelium. This compatibility leads to a significant increase in the overall stiffness (Figure 7a).

2.6. Form-Finding

Based on the findings of Sections 2.4 and 2.5, considering the volume of small-scale samples is sufficient to hold a person's weight of approximately 80 kilos, a prototype in the form of a stool, named Mycomerge, is designed and built. The geometry is developed through form-finding procedures using Rhino Vault 2. The digital form-generating methods are used to create the entire geometry as well as for basic optimization of the structure [17]. The structure's skeleton, in this case, the rattan framework, is first generated, starting with single lines and the core of the geometry. Then three-dimensional surfaces are integrated to form the rest of the shape (Figure 8). The main parameters of the computational model are associated with the grid density for the skeleton and the overall dimensions. A "funnel-shaped" structure is generated, which is only supported centrally, forming a canopy that cantilevers without the need for additional supports at its edge [18,19]. In this case, the resulting canopy acts as the seating area, with the center of gravity meeting at the central support. The purpose of this design is to maximize material efficiency while achieving the appropriate load-bearing capacities using the least amount of material. The stool is designed with a seating height of 45 cm to provide comfortable seating. The funnel shell of the resulting Thrust Diagram (Figure 8) is used for developing the arrangement of the rattan skeleton, which serves as an integrated structural element on which the fibers and mycelium substrate are placed.

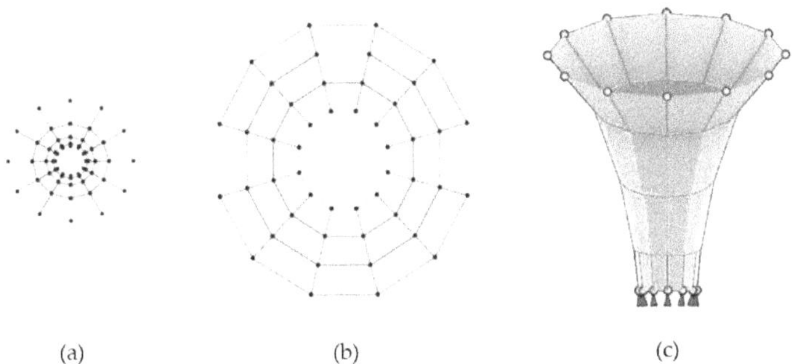

(a)   (b)   (c)

Figure 8. (**a**,**b**) Form and Force Diagram, (**c**) Thrust Diagram.

2.7. Prototyping

A full-scale paper model is first produced to verify that the structure is self-supporting and also to serve as a guide for positioning the rattan rods along with the desired shape. Rattan serves as the framework in this fabrication approach since it is incorporated with the structural system and, as with all other components, is fully compostable. The number of reinforcement rods in the computed design is doubled to ensure the proper positioning of the hemp fiber and mycelium composite layers (Figure 9).

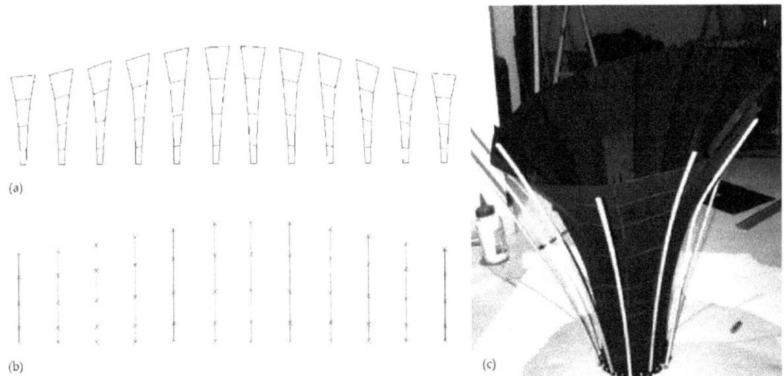

Figure 9. (**a**) Layout for paper strips, (**b**) goal lengths, (**c**) assembled paper model.

Based on the results of Section 2.5, the rattan is the outer supporting skeleton; loose hemp fibers are flexible sub-layers for the mycelium to grow through and bind with the skeleton. Mycelium pre-grown substrate forms the core, which is subsequently covered with loose hemp fibers (Figure 10). Before assembly, fibers and rattan rods must be sterilized either by steaming or boiling. In addition, flour is added to the fibers to improve mycelium growth. Table 1 presents an overview of the materials used in the two experiments.

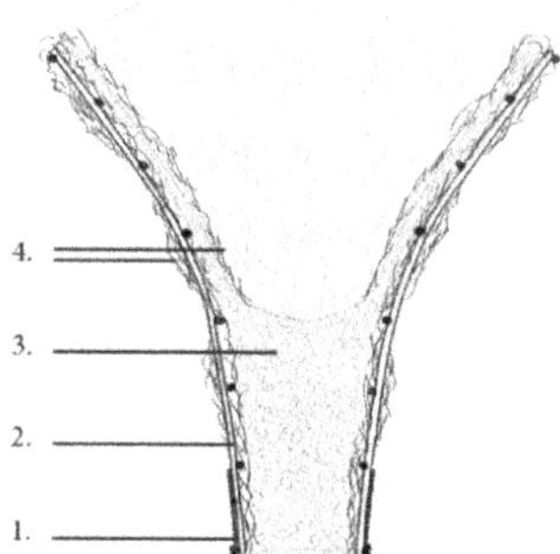

Figure 10. Section: 1. Rattan ⌀ 2 mm, 2. Rattan ⌀ 5 mm, 3. Mycelium substrate, 4. Hemp fibers.

Table 1. Material overview.

Material	First Prototype	Final Prototype
Hemp fibers (g)	180	420
Mycelium substrate (l)	3	7
Psyllium husk (g)	-	360
Jute rope (m)	100	100
Rattan (g) ⌀ 5 mm, 2 mm	250, 125	250, 125
Weight (kg)	2.1	3.7

2.8. Assembly

All materials must be sterilized before the assembly process begins. Figure 11a presents a material overview. After cooling down to room temperature, flour must be added to the wet hemp fibers. The rattan should be still wet so that one can bend the rods into their initial shape. The connection of vertical and horizontal members is secured with sterilized jute rope using traditional square knot techniques (Figure 11b). When the skeleton is assembled, the fibers can be placed on top so that no gaps emerge during the substrate placement (Figure 12a). This layer is approximately 1 cm thick. The pre-grown substrate is then mixed with psyllium husk until reaching a clay-like texture. Afterwards, the mixture is evenly distributed on top of the wet hemp fibers with a thickness of 3 cm (Figure 12b). An additional centimeter of wet hemp fibers follows. The multi-layer composite is then wrapped in perforated plastic foil to sustain moisture but also provide air circulation. To ensure constant moisture and nutritional levels, occasional spraying with a water-flour solution takes place. The assembled piece needs to be kept in a sterile environment for a minimum of 5 days while sufficient growth density can be reached (Figure 12c). Baking of the prototype at 80 degrees is then necessary to improve its compressive strength and to stop the growth process until the sample does not lose any further weight. While baking, a color change from white to a darker beige or brown is expected due to the hemp fibers.

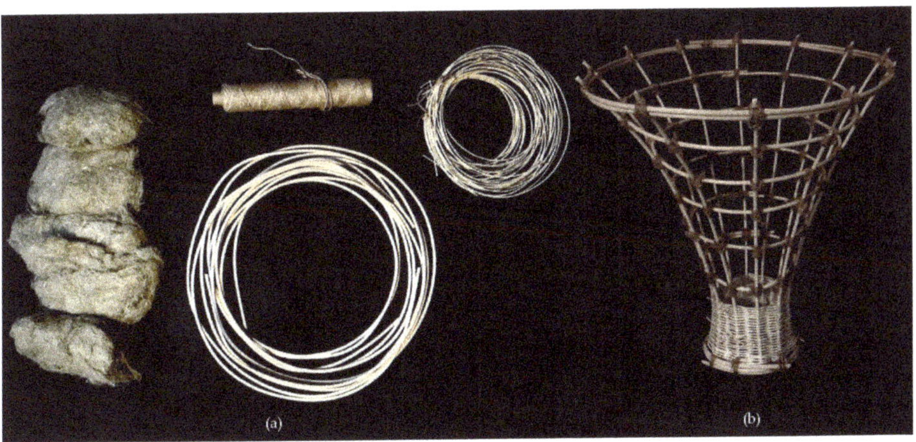

Figure 11. Material overview: (**a**) hemp fibers, rattan rods, jute fabric, (**b**) assembled rattan frame.

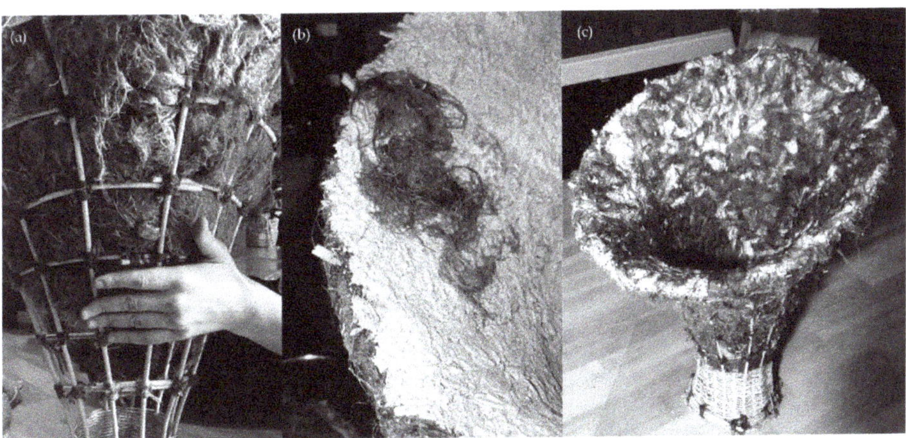

Figure 12. Assembly: (**a**), placing of fibers, (**b**) substrate and cover with additional layer of fibers, (**c**) assembled stool left to grow.

3. Results

3.1. Comparison

Three scenarios are explored in order to determine the importance of including a rattan framework in the specified system:

3.1.1. Composite without Rattan Reinforcement (Assumption)

Since a full prototype without a rattan skeleton was not created during this study, this scenario is an assumption. Although the soft mold samples indicate excellent merging capabilities with the mycelium-based core, the whole composite might not withstand the applied forces without the core breaking or splitting. A framework to fix the soft fabric to fill in the substrate, or an external mold, would be required to construct a composite without a rattan skeleton. During the growing and drying phases, the composite might deform unevenly due to the randomly oriented fibers.

3.1.2. Composite with Rattan Reinforcement

The tensile capabilities of the mycelium-based rattan-reinforced composite have improved due to higher water content within the rods and wet hemp fibers, while the whole composite performs best under compression. Rattan's load-bearing capacity is advantageous not only when merging with mycelium but also during the growing and drying phase to minimize uneven shrinking. The bottom support's finely woven rattan maintains the core in place and prevents breakage.

3.1.3. Rattan without Mycelium Matrix

Since rattan rods perform best under bending, this research demonstrated that rattan might be used as reinforcing rods in a mycelium composite. In the first attempt, the behavior of the rattan framework was tested through a seating test before placing fibers and substrate, which resulted in severe, irreversible deformation of the framework (Figure 13).

Figure 13. First Prototype.

3.2. Physical Prototypes

In the first attempt, with an insufficient amount of 3 L of the substrate, the stool is able to hold the needed load but is still relatively unstable. The shell thickness varies from 0.5 cm to 1.5 cm. To improve its structural performance, the bottom radius is upscaled by 3 cm, which also prevents the stool from slipping. Additionally, in the second attempt, the amount of mycelium substrate is doubled, and 2.3 times more fibers are used. Psyllium husk is added to the substrate for improved material distribution, giving it a clay-like texture. In both attempts, mycelium binds effectively with all the elements. Significant growth has been observed in the vertical rattan members, particularly through the capillaries of the rattan. This is caused by the capillary effect, transporting the water, nutrients, and mycelium throughout the whole length. Due to increasing water content within the rods and wet hemp fibers, the tensile properties of the mycelium-based rattan-reinforced composite have improved, whereas the whole composite performs best under compression. In addition to rattan's load-bearing capabilities when merging with mycelium, it is also beneficial in preventing uneven shrinking throughout the growing and drying process. In comparison to earlier research (Figure 1), in Mycomerge, the substrate entirely binds to the skin materials, leaving no evidence of separation. The densely woven rattan at the bottom support keeps the substrate in place and prevents breakage. The final prototype (Figure 14) weighs 3.7 kg and can support more than 20 times its own weight, demonstrating high structural capabilities and upscaling possibilities (Figure 15).

Figure 14. Final Prototype.

Figure 15. Seating tests (44 to 90 kilos).

4. Architectural Application

Possible interior applications of the developed system can be in partition walls or sound insulation panels. Large elements can be fragmented, or the design can be developed in modular pieces able to fit in an industrial oven. In comparison with other fabrication techniques, working with rattan as a structural and form-giving framework eliminates the necessity of an external mold. Forming double-curved geometries can also be achieved. To showcase the potential of a full-scale structural application of the developed system, an initial design is developed (Figure 16). Since this structure is intended to be placed outside, it will not be baked, and the mycelium cultures will continue to grow in their natural environment until it decomposes. To prevent the growth of undesirable mold or other species, the growth phase must be interrupted, for instance, by exposing the structure to a high temperature of above 80 degrees or cooling it to below 0 degrees. Because an oven the size of an architectural building is unrealistic, assembly and growth are suggested to take place during the wintertime for the growth to stop naturally by simply being kept outside. However, this way of stopping the growing process may significantly influence the structural behavior and performance. Furthermore, this method is completely dependent on weather conditions and lacks consistency and applicability in various locations and seasons. The core material itself is not waterproof, and it will lose rigidity by being exposed to water and weathering. However, in between rainfalls, the material can dry and stabilize back again. By letting the structure air dry, mycelium can grow further on top of the surface, which will be covered with a pure mycelium layer. The foam-like mycelium layer does not absorb water, as one can see in several mycelium leather products found on the market. Keeping the structure fully waterproof is yet not possible, with an additional coating being necessary. The proposal of an exterior application is expected to last for two up to three months, similar to already developed mycelium temporary structures such as the Hy-Fi towers at MoMa in 2014. Further research on this topic is needed.

Figure 16. Architectural proposal.

5. Discussion

Although the developed physical prototype has proven mycelium's load-bearing and binding capacities, there is still a lot of space for further research on the structural capabilities of mycelium-based composites in large-scale applications. The necessity for interruption of the growth process presents limitations in the manufacturing on an architectural scale, due to the need for an industrial oven with a restricted size. That obstacle can be overcome through segmentation of the structural object into separate pre-grown and assembled on-site modules. For interior applications, Mycomerge presents a successful concept of material efficiency and load-bearing capacity of mycelium-based structures. Water content and moisture during growth are critical for the successful bonding of the rattan rods; otherwise, the mycelium will not grow onto the rattan's surface. As an outcome, the tensile characteristics of the rattan rods and the mycelium composite will degrade, resulting in possible separation. In this study, rattan is utilized as an exterior skeleton; however, given the common reinforcement methods, such as steel rebar in concrete, more testing of layering, rattan binding, and reinforcement capabilities are required.

The materials used in this prototype are only agricultural waste products, which can be sourced regionally. Because all the components can be grown, this approach has no limitations in terms of resources. Mycomerge is fully biodegradable and hence promotes an eco-friendly alternative to the commonly used conventional materials, aiming toward sustainability in the building industry.

Author Contributions: M.T.N. and D.S. contributed equally to this paper. Conceptualization, methodology, design development, prototyping and testing were handled by M.T.N. and D.S. This paper was supervised and reviewed by H.D. (BioMat Department director) and E.S. (BioMat research associate) and both gave prerequisite knowledge, support, and guidance in the field of biocomposites and sustainable architectural building elements. All authors have read and agreed to the published version of the manuscript.

Funding: This research received no external funding.

Institutional Review Board Statement: Not applicable.

Informed Consent Statement: Not applicable.

Data Availability Statement: Not applicable.

Acknowledgments: The project was developed in the seminar Material Matter Lab (Material and Structure, winter semester 2020/2021), offered by BioMat (The Department of Biobased Materials and Materials Cycles in Architecture) at ITKE at the University of Stuttgart under the supervision of Hanaa Dahy and tutoring of Evgenia Spyridonos. M.N and D.S are especially grateful for the support and knowledge provided throughout the whole process.

Conflicts of Interest: The authors declare no conflict of interest.

References

1. Dahy, H. Towards Sustainable Buildings with Free-Form Geometries: Development and Application of Flexible NFRP in Load-Bearing Structures. In *Biocomposite Materials: Design and Mechanical Properties Characterization*; Composites Science and Technology; Hameed Sultan, M.T., Majid, M.S.A., Jamir, M.R.M., Azmi, A.I., Saba, N., Eds.; Springer: Singapore, 2021; pp. 31–43. ISBN 978-981-334-091-6.
2. Dahy, H. Biocomposite Materials Based on Annual Natural Fibres and Biopolymers—Design, Fabrication and Customized Applications in Architecture. *Constr. Build. Mater.* **2017**, *147*, 212–220. [CrossRef]
3. Hebel, D.E.; Heisel, F. *Cultivated Building Materials: Industrialized Natural Resources for Architecture and Construction*; Birkhäuser: Berlin, Germany; Boston, MA, USA, 2017; ISBN 978-3-0356-0892-2.
4. Rajak, D.K.; Pagar, D.D.; Menezes, P.L.; Linul, E. Fiber-Reinforced Polymer Composites: Manufacturing, Properties, and Applications. *Polymers* **2019**, *11*, 1667. [CrossRef] [PubMed]
5. Grimm, D.; Wösten, H.A.B. Mushroom Cultivation in the Circular Economy. *Appl. Microbiol. Biotechnol.* **2018**, *102*, 7795–7803. [CrossRef] [PubMed]
6. Dahy, H. Natural Fibre-Reinforced Polymer Composites (NFRP) Fabricated from Lignocellulosic Fibres for Future Sustainable Architectural Applications, Case Studies: Segmented-Shell Construction, Acoustic Panels, and Furniture. *Sensors* **2019**, *19*, 738. [CrossRef] [PubMed]
7. Dahy, H. 'Materials as a Design Tool' Design Philosophy Applied in Three Innovative Research Pavilions Out of Sustainable Building Materials with Controlled End-Of-Life Scenarios. *Buildings* **2019**, *9*, 64. [CrossRef]
8. Lelivelt, R.J.J. The Mechanical Possibilities of Mycelium Materials. Master's Thesis, Eindhoven University of Technology, Eindhoven, The Netherlands, 2015.
9. Jiang, L. A New Manufacturing Process for Biocomposite Sandwich Parts Using A Myceliated Core, Natural Reinforcement and Infused Bioresin. Ph.D. Thesis, Rensselaer Polytechnic Institute, Troy, NY, USA, 2015.
10. Girometta, C.; Picco, A.; Baiguera, R.M.; Dondi, D.; Babbini, S.; Cartabia, M.; Pellegrini, M.; Savino, E. Physico-Mechanical and Thermodynamic Properties of Mycelium-Based Biocomposites: A Review. *Sustainability* **2019**, *11*, 281. [CrossRef]
11. Jones, M.; Mautner, A.; Luenco, S.; Bismarck, A.; John, S. Engineered Mycelium Composite Construction Materials from Fungal Biorefineries: A Critical Review. *Mater. Des.* **2020**, *187*, 108397. [CrossRef]
12. Van Wylick, A.; Elsacker, E.; Yap, L.L.; Peeters, E.; de Laet, L. Mycelium Composites and Their Biodegradability: An Exploration on the Disintegration of Mycelium-Based Materials in Soil. *Constr. Technol. Archit.* **2022**, *1*, 652–659. [CrossRef]
13. Heisel, F.; Lee, J.; Schlesier, K.; Rippmann, M.; Saeidi, N.; Javadian, A.; Nugroho, A.R.; Mele, T.V.; Block, P.; Hebel, D.E. Design, Cultivation and Application of Load-Bearing Mycelium Components: The MycoTree at the 2017 Seoul Biennale of Architecture and Urbanism. *IJSED* **2017**, *6*, 296–303. [CrossRef]
14. Kırdök, O.; Akyol Altun, D.; Dahy, H.; Strobel, L.; Hameş Tuna, E.E.; Köktürk, G.; Andiç Çakır, Ö.; Tokuç, A.; Özkaban, F.; Şendemir, A. Chapter 17—Design Studies and Applications of Mycelium Biocomposites in Architecture. In *Biomimicry for Materials, Design and Habitats*; Eggermont, M., Shyam, V., Hepp, A.F., Eds.; Elsevier: Amsterdam, The Netherlands, 2022; pp. 489–527, ISBN 978-0-12-821053-6.
15. Jiang, L.; Walczyk, D.; McIntyre, G.; Bucinell, R.; Tudryn, G. Manufacturing of Biocomposite Sandwich Structures Using Mycelium-Bound Cores and Preforms. *J. Manuf. Processes* **2017**, *28*, 50–59. [CrossRef]
16. Rihaczek, G.; Klammer, M.; Başnak, O.; Petrš, J.; Grisin, B.; Dahy, H.; Carosella, S.; Middendorf, P. Curved Foldable Tailored Fiber Reinforcements for Moldless Customized Bio-Composite Structures. Proof of Concept: Biomimetic NFRP Stools. *Polymers* **2020**, *12*, 2000. [CrossRef] [PubMed]
17. Sippach, T.; Dahy, H.; Uhlig, K.; Grisin, B.; Carosella, S.; Middendorf, P. Structural Optimization through Biomimetic-Inspired Material-Specific Application of Plant-Based Natural Fiber-Reinforced Polymer Composites (NFRP) for Future Sustainable Lightweight Architecture. *Polymers* **2020**, *12*, 3048. [CrossRef] [PubMed]
18. Block, P.; Mele, T.V.; Rippmann, M. Geometry of Forces: Exploring the Solution Space of Structural Design/Geometrie der Kräfte. Untersuchungen zum Lösungsraum des Tragwerksentwurfs. In *GAM 12: Structural Affairs*; Birkhäuser: Berlin, Germany; Boston, MA, USA, 2016; pp. 46–55, ISBN 978-3-0356-0984-4.
19. Rippmann, M.; Block, P. Funicular Funnel Shells. In Proceedings of the Design Modeling Symposium, Berlin, Germany, 30 September 2013.

biomimetics

Article

A Study on the Sound Absorption Properties of Mycelium-Based Composites Cultivated on Waste Paper-Based Substrates

Natalie Walter * and Benay Gürsoy

Department of Architecture, Penn State University, University Park, PA 16802, USA; bug61@psu.edu
* Correspondence: nvw5160@psu.edu

Abstract: Mycelium-based composites have the potential to replace petrochemical-based materials within architectural systems and can propose biodegradable alternatives to synthetic sound absorbing materials. Sound absorbing materials help improve acoustic comfort, which in turn benefit our health and productivity. Mycelium-based composites are novel materials that result when mycelium, the vegetative root of fungi, is grown on agricultural plant-based residues. This research presents a material study that explores how substrate variants and fabrication methods affect the sound absorption properties of mycelium-based composites grown on paper-based waste substrate materials. Samples were grown using *Pleurotus ostreatus* fungi species on waste cardboard, paper, and newsprint substrates of varying processing techniques. Measurements of the normal-incidence sound absorption coefficient were presented and analyzed. This paper outlines two consecutive acoustic tests: the first round of experimentation gathered broad comparative data, useful for selecting materials for sound absorption purposes. The second acoustic test built on the results of the first, collecting more specific performance data and assessing material variability. The results of this study display that cardboard-based mycelium materials perform well acoustically and structurally and could successfully be used in acoustic panels.

Keywords: mycelium; acoustic materials; bio-fabrication; sound absorption

Citation: Walter, N.; Gürsoy, B. A Study on the Sound Absorption Properties of Mycelium-Based Composites Cultivated on Waste Paper-Based Substrates. *Biomimetics* 2022, 7, 100. https://doi.org/10.3390/biomimetics7030100

Academic Editors: Andrew Adamatzky, Han A. B. Wösten and Phil Ayres

Received: 30 June 2022
Accepted: 20 July 2022
Published: 22 July 2022

Publisher's Note: MDPI stays neutral with regard to jurisdictional claims in published maps and institutional affiliations.

1. Introduction

Increasing urban populations, scarce resources, and climate change will force a paradigm shift in our material use and approaches to construction. Our current framework of construction is unsustainable; we rely on fleeting systems of resource extraction, waste management, and energy consumption. By relying on man-made polymers and petroleum-based components in our built environment, our building materials either cannot naturally decay or take centuries in a landfill to degrade. Biodegradable materials and biologically derived materials present an alternative to this traditional construction framework. Mycelium-based composites, a bio-material derived from fungi, have the potential to successfully replace plastic-based materials in our building systems without the extraction of non-renewable resources. Instead, mycelium, the vegetative root of fungi, is grown on agricultural plant-based residues, resulting in a new compound material. This research aims to further understand the characteristics of the material and the potential for implementation as acoustic architectural components. Specifically, this research began with systematic material tests, assessing the acoustic properties of mycelium-based components grown on local and accessible paper-based waste products. These material tests then inform the development of mycelium-based sound absorption panels. Using *Pleurotus ostreatus* fungi species, commonly known as the oyster mushroom, this research tested how substrate variants and fabrication methods affect acoustic absorption.

1.1. Noise Control through Sound Absorption

Exposure to prolonged environmental noise is associated with several negative effects that can be mitigated with proper sound treatments. Chepesiuk addresses the health problems associated with hazardous noise, including tinnitus, elevated blood pressure, cardiovascular constriction, and hearing loss [1]. These effects, in turn, lead to social handicaps, reduced workplace productivity, and decreased student–teacher communication. The Centers for Disease Control and Prevention (CDC) even declares that "occupational hearing loss is one of the most common work-related illnesses in the United States" [2]. Addressing this problem requires the implementation of noise control treatments in architectural systems to reduce the negative effects of noise.

Noise control and architectural acoustics are a growing sector of the design field, given the importance of maintaining acoustic comfort. Aletta and Kang argue that while noise can be hazardous, the pursuit of "silence" from a health standpoint is not what defines a successful acoustic environment [3]. They point out challenges in architectural acoustics but suggest that we move away from total noise control and instead embrace a certain threshold of environmental sound. Therefore, thoughtful consideration must be made to regulate acoustic quality rather than just reducing all sound.

Regulating interior acoustic quality is performed through environmental assessment and sound treatment, pending the spatial and programmatic requirements. All building materials either reflect, transmit, or absorb incident sound, and thus to manage acoustic comfort, materiality must be designed with acoustic intent [4].

1.1.1. Sound Absorption

Sound absorption is one method of acoustic treatment in which the energy of a sound wave is converted into low-grade heat, reducing the strength of reflected sound [4]. This reduces the amount of sound perceived as well as the effects of acoustic discomfort. Sound absorptive materials have many different applications within architectural, studio, automotive, and industrial acoustics. They can be used as interior lining in vehicles, aircraft, ducts, industrial equipment, and buildings/interiors. These materials are notably used within performance spaces to control unwanted echo, work environments to quiet the reverberant field, and restaurants to improve users' communication [5]. A measurement of a material's sound absorption is called the sound absorption coefficient, which is the ratio of energy absorbed to the incident energy. The higher the sound absorption coefficient, the more absorptive the material [4].

There is a need to develop sustainable alternatives for conventional synthetic sound-absorbing materials (i.e., glass wool, stone wool, and polystyrene). Both Arenas and Sakagami [6] and Desarnaulds et al. [7] address the environmental impacts of conventional sound-absorbing materials. Arenas and Sakagami mentioned that sound absorbing materials began with asbestos-based materials but were replaced with mineral-based fibrous materials once asbestos was linked to human health hazards. These fibers are most commonly made from glass and rock wool fibers, but their use is associated with negative environmental effects. The researchers suggest the use of sustainable alternatives, such as "eco-materials elaborated from residues" [6]. Desarnaulds et al. added to this by assessing the environmental performance of sustainable acoustic materials [7]. In this article, they specified that glass and rock wools are unsustainable because they are disposed of in a non-inert waste landfill. They also release airborne fibers that are harmful to contractors, laborers, and future occupants.

1.1.2. Factors Influencing Sound Absorption

Sound absorptive materials are generally fibrous or porous in nature. Their absorption behavior is dependent on physical material characteristics, such as the following: Fiber size, porosity, material thickness, and material density [5].

Fiber Size: Fiber diameter affects sound absorption because of the fiber's movement when sound waves travel through the material. Fibers act as frictional elements, which convert sound energy into heat as they move. Thinner fibers have a higher sound absorption coefficient for two reasons. First, thin fibers move more easily than thicker fibers. Second, more fibers are needed to reach the same volume density as a material with thicker fibers, which creates more tortuous paths for sound waves, thus increasing airflow resistance [8]. Thus, having many thin fibers in a material rather than a few thick fibers creates greater frictional resistance.

Porosity: Porosity deals with the number, size, and type of pores/voids existing in a material through which sound waves travel through and become dampened. When sound waves enter pores, the air molecules within the channels vibrate, converting part of the sound energy into heat [9]. Continuous channels are more successful at absorbing sound than shorter, closed pores.

Material's Thickness: The thickness of a sound-absorbing material has a direct relationship with low-frequency sounds (100–2000 Hz), while it has no effect on high-frequency sounds. As the material becomes thicker, the sound absorption increases. Studies show that effective sound absorption for low-frequency sounds is achieved when the thickness is approximately one-tenth of the wavelength of the incident sound [5].

Material's Density: The sound absorption coefficient increases for middle and high-frequency sounds as the density of the material increases. Less dense materials absorb low frequencies (500 Hz), while denser structures absorb higher frequencies (2000 Hz) [5].

The relationship between material characteristics and acoustic performance is also relevant with regard to musical instruments. Wegst [10] addressed why the physiological properties of bamboo and wood make them ideal materials for instrument manufacturing. An important point made is that the loss coefficient (acoustic energy dissipated due to friction) is dependent on the temperature and moisture content within a sample.

Understanding the physical material characteristics that determine acoustic performance is relevant to this research because the growth factors of mycelium-based composites can be curated to achieve optimal acoustic performance. Since mycelium-based composites characteristics are highly variable, understanding what outcome is preferred enables narrowing down the growth parameters.

1.1.3. Testing Sound Absorption

Testing sound absorption can be performed using different methods depending on the desired result. In order to test the sound absorption of a specific material, an impedance tube is often used. The two-microphone transfer-function method is a common method when using an impedance tube. This is when a sound source sends broadband sound waves at a sample, which reflect off the sample. The sound waves generate a pattern of forward and backward traveling waves inside the tube. Digital frequency analyzers then measure the sound pressure at specific locations to determine the sound absorption and acoustic impedance of the material.

1.2. Mycelium-Based Composites as Biodegradable Alternatives for Sound Absorption

The construction industry generates a significant amount of waste with undeniable negative environmental impacts. The use of biodegradable materials as building components can reduce the amount of building waste generated and the ensuing environmental consequences. Transporting waste is associated with resource consumption and pollution emissions, landfills are associated with land use and ground contamination, and waste incineration produces contaminated ash, air pollution, and greenhouse gas emissions [11]. According to the Environmental Protection Agency (EPA), approximately 600 million tons of construction and demolition debris were generated in 2018, which amounts to more than twice the amount of municipal solid waste generated in the same year [12]. There is a clear need to reduce the amount of waste generated from building construction and demolition, and biodegradable materials offer a low waste alternative.

1.2.1. Cultivating Mycelium-Based Composites

Mycelium-based composites result when fungal growth is stopped during colonization of the substrate, and a resulting compound material is created [13]. Mycelium grows in search of food and spreads through the substrate in a network colony. During this growth, mycelium produces enzymes that convert the substrates' biomass into nutrients while simultaneously binding the substrate. The organic matter decomposes over time as the plant polymers are replaced with fungal biomass. Fabrication of mycelium-based composites involves the growth of mycelium on organic substrates. The composites' properties and performance are highly variable; factors include fungal species, substrate type, environmental conditions during growth (temperature, humidity), and forming/processing techniques [14]. The resulting materials differ immensely in their density, tensile and compressive strength, morphology, and insulative/acoustic performances [15]. Figure 1 illustrates the typical stages of mycelium-based composite cultivation.

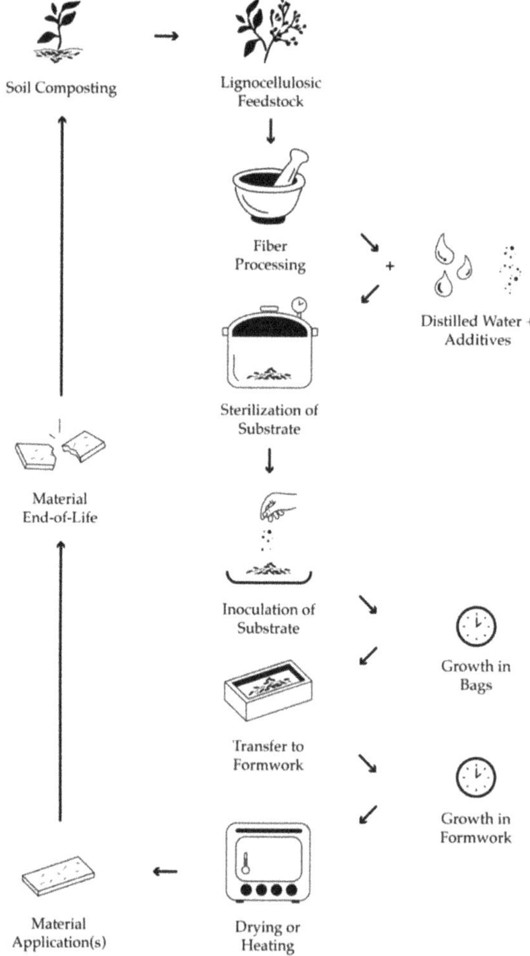

Figure 1. Typical Stages of Mycelium-Based Composite Cultivation. Diagram reworked from [15].

There is a growing field of knowledge on mycelium-based composites as more researchers are testing the characteristics of different growth methodologies. It is important to note that because the material constitution and mechanical properties of mycelium-based composites vary immensely, it is difficult to establish set protocols for growth and fabrication methods. The compressive strength of one composite material, for example, may be drastically different than another composite because of the different growth protocols. That said, below are some prominent experiments that assess the physical, chemical, and mechanical properties of mycelium-based composites.

Appels et al. [16] experimented with the growth and fabrication techniques of mycelium-based composites by growing *Trametes multicolor* and *Pleurotus ostreatus* on beech sawdust and rapeseed straw. Appels et al. found that the different fungal strains and substrate compositions cause differing mechanical and physical characteristics of the resulting composite. One finding, for example, is that *Trametes multicolor* grown on rapeseed straw resulted in flexible and soft skin, while *Pleurotus ostreatus* also grown on rapeseed straw resulted in firm and rough skin.

Elsacker et al. [15] grew *Trametes versicolor* on five different fiber types (hemp, flax, flax waste, softwood, and straw). They also varied the fiber processing techniques into four categories: loose, chopped, dust, pre-compressed, and tow. The resulting materials were then tested for dry density, Young's modulus, compressive stiffness, stress–strain curves, thermal conductivity, and water absorption rate. One finding that Elsacker discovered was that the mechanical properties of the composites are dependent on fiber types. The fiber condition (loose vs. chopped) had a large impact on the compressive stiffness, and the samples grown were dense.

1.2.2. Mycelium-Based Composites as Sound Absorbers

There is limited research and literature existing on the acoustic performance of mycelium-based composites. Moreover, since the resulting material characteristics are variable, the results of one study may not correlate with another. It is difficult to conclude that all mycelium-based composites are successful acoustic absorbers based on the few studies that exist.

Mogu [17] is a company selling mycelium-based interior acoustic wall panels. The company, however, does not disclose its growth methodologies. One prominent study that reported on the experiments on the acoustic properties of mycelium-based composites is [18]. This study tested how substrate variants affect sound absorption. Their substrates were rice straw, hemp pith, kenaf fiber, switch grass, sorghum fiber, cotton bur fiber, and flax shive, and they assessed sound pressure levels. The results found that mycelium-based composites are successful absorbers, but the acoustic performance varies between samples depending on the substrate material. It was also noted that even the low performer, the 100% cotton bur fiber, still yielded higher than 70% acoustic absorption at 1000 Hz. In a subsequent study that built upon this research, the team, instead of testing rigid composites, tested the acoustic properties of mycelium foam [19]. They used *Ganoderma* as the fungal species and a combination of ground corn stover, grain spawn, maltodextrin, and other nutrients as the substrate. They also used a specifically designed growth chamber to grow the foam. These two studies were the main experiments published regarding the acoustic properties of mycelium-based composites, and to gather a further understanding of the acoustic potentials of the mycelium-based composites, more experiments are needed.

Another approach to using mycelium as an acoustic material was seen in the development of the biotech violin [20]. Schwarze and Morris developed a mycelium-based material, coined mycowood, using *Physisporinus vitreus* and *Schizophyllum commune* fungi. This material was developed and manufactured into violins that match the tone of a Stradivarius, an extremely high-quality violin.

Additionally, while not fungal-based, there is a growing field of research regarding alternative natural acoustic materials. Putra et al. [21] analyzed the utilization of natural waste fibers from paddy as an acoustic material. Similarly, Rachman et al. [22] assessed the acoustical performance of a particleboard made of coconut fiber and citric acid solution.

2. Materials and Methods

The following experiment consisted of three stages: (1) the cultivation of mycelium-based composites, (2) the assessment of the cultivated samples' acoustic performance, and (3) the cultivation of mycelium-based acoustic panel prototypes.

Material cultivation began with substrate selection and preparation. The prepared substrates were then sterilized in an autoclave chamber to mitigate contamination. Once sterilized, the materials were inoculated with *Pleurotus ostreatus* spawn. These samples were left to grow in a controlled growth environment, first in autoclavable bags for 12 days and then in sterile formworks for 16 more days. Once grown, the samples were dried and heated in an oven to kill the mycelium and stop the cultivation process. The samples were then shaped to fit into an impedance tube to test for sound absorption. Table 1 shows the samples that were tested in the impedance tube.

Table 1. Cultivated Samples.

Sample	Abbr.	Substrate	Substrate Treatment	Sample Size
Shredded Cardboard High freq.	SCH	cardboard	shredded	29 mm
Shredded Cardboard Low freq.	SCL	cardboard	shredded	100 mm
Fine Cardboard High freq.	FCH	cardboard	pulverized	29 mm
Fine Cardboard Low freq.	FCL	cardboard	pulverized	100 mm
Shredded Paper High freq.	SPH	paper	shredded	29 mm
Shredded Paper Low freq.	SPL	paper	shredded	100 mm
Fine Paper High freq.	FPH	paper	pulverized	29 mm
Fine Paper Low freq.	FPL	paper	pulverized	100 mm
Shredded Newsprint High freq.	SNH	newsprint	shredded	29 mm
Shredded Newsprint Low freq.	SNL	newsprint	shredded	100 mm
Fine Newsprint High freq.	FNH	newsprint	pulverized	29 mm
Fine Newsprint Low freq.	FNL	newsprint	pulverized	100 mm
Ecovative Mixture High freq.	EMH	undisclosed	undisclosed	29 mm
Ecovative Mixture Low freq.	EML	undisclosed	undisclosed	100 mm

The cultivated samples were tested in an impedance tube, following standard ASTM E1050-12, to compare sound absorption in the 500 Hz to 6.4 kHz frequency range.

2.1. Cultivation of Mycelium-Based Composites

The following methodology for the growth of these mycelium-based samples was conducted following an initial growth experiment. In the initial experiment, failure to consider material shrinkage resulted in the inability to test for acoustic absorption. The mycelium mixtures were grown in Petri dishes that were the exact size necessary to test for sound absorption. Once dried, they shrunk and warped considerably and would not permit accurate results. The following experiment was executed with shrinkage in mind.

2.1.1. Lignocellulosic Substrate Materials

The selected substrate materials are paper-based waste products, specifically sorted office paper, cardboard, and newsprint. These paper-based materials are all lignocellulosic materials, meaning they provide the lignin and cellulose for fungi to feed. The office paper and cardboard were obtained from recycling bins in the Stuckeman School of Architecture at Penn State University, University Park Campus. The newsprint was similarly obtained from recycling bins across campus and local recycling centers. All materials were sorted to ascertain unsoiled samples.

In order to maintain the cyclical nature of biodegradable materials, the importance of waste and recycled materials was stressed in this study. Thus, strictly local paper-based waste products were used for substrate materials/feedstock. According to the EPA, paper-based materials are largely recycled, yet still, 4.2 million tons of paper were combusted in 2019, making up 12.2 percent of all combusted municipal solid waste (MSW) that year. Additionally, 17.2 million tons of paper-based MSW landed in landfills, making up 11.8 percent of MSW landfilled in 2018 (Environmental Protection Agency, n.d.). This study addressed the accessibility of paper-based waste products and the need to reduce the amount combusted/landfilled.

2.1.2. Substrate Preparation

Six substrate mixtures were prepared using: (a) shredded cardboard (SCL and SCH), (b) fine cardboard (FCL and FCH), (c) shredded paper (SPL and SPH), (d) fine paper (FPL and FPH), (e) shredded newsprint (SNL and SNH), and (f) fine newsprint (FNL and FNH) seen in Figure 2a. For all samples, the materials (cardboard, newsprint, paper) were first shredded using an office shredder [23]. The three materials were then split in half to make 6 separate sample mixtures, and half of each was ground to make a fine cottony material. All 6 mixtures were supplemented with 10% (w/w) wheat bran and mixed thoroughly. Wheat bran was used as a supplementary substance to induce mycelial growth and increase cultivation speed by adding nitrogen to the substrate mixtures. The prepared substrates were then adjusted to 65% moisture content by adding water. Each prepared substrate mixture contained 100 g of dry weight material, 185 g of water, and 18 g of wheat bran.

Figure 2. Growth Process of Low-Frequency Samples and Resulting Materials: (**a**) Prepared Substrates (80 mm × 80 mm square); (**b**) Mycelium Mixtures in Formworks (250 mm × 125 mm × 38 mm); (**c**) Composite Materials After Drying (250 mm × 125 mm × 38 mm); (**d**) Composite Materials After Drying—Side View.

To compare against commercially available mycelium-based composite materials, Ecovative Design's Grow-It-Yourself Mushroom® Material was also cultivated (EML and EMH) (see Section 2.1.8). The substrate material of these samples was hemp hurd, as seen in Figure 2a.

2.1.3. Sterilization

The 6 substrate mixtures were placed in polypropylene autoclavable bags [24], 200 mm × 125 mm × 480 mm, and stored overnight in a cold room. The bags were then autoclaved for 45 min at 121 °C. This sterilization process assured the substrate was not contaminated with other organisms, making the material unlikely to grow mold. The bags were then cooled down in a clean, room-temperature room overnight.

2.1.4. Inoculation

Each substrate mixture was inoculated with *Pleurotus ostreatus* spawn. The mycelium spawn is purchased from Lambert Spawn [25] (*Strain 123 Pleurotus ostreatus*) in a pre-spawn bag. These prepared bags were made of supplemented cotton seed hulls and straw. A total of 10% of the dry weight of the substrate was added to spawn. The spawn was added directly into the autoclavable bags and thoroughly mixed and compressed. The mycelium was left to grow in the bags for 12 days. The bags are kept in an environmentally controlled growth room, with 99% relative humidity and a temperature of 24 ± 1 degree Celsius.

2.1.5. Cultivation in Formworks

After 12 days of growth in bags, the cultivated mycelium mixtures were transferred to rectangular acrylic formworks, as seen in Figure 2b. Before transferring the cultivated mycelium, the formworks were sterilized with ethanol solution (70%). The transfer from bags to formworks was cautiously performed in a sterile environment. The formworks are then covered with plastic wrap and left to grow for an additional 16 days in the same environmentally controlled growth room.

2.1.6. Heating and Drying

After 16 days, the samples were taken out of the formworks and left to dry with a fan. After two days, the samples were placed in an oven at 90 °C for 24 h, resulting in the rectangular composite materials shown in Figure 2c,d. Drying the samples caused the material to lose 2/3 of its water content and fully kill the mycelium.

2.1.7. Sample Shaping

In order to ascertain whether the samples would fit into the impedance tube to test the sound absorption, the rectangular samples (thickness 38 mm) had to be shaped into 100 mm and 29 mm circles. Therefore, the materials were cut on a band saw, seen in Figure 3, and sanded using a belt sander.

(**a**)　　　　(**b**)　　　　(**c**)　　　　(**d**)

Figure 3. Sample Shaping of Fine Cardboard Material: (**a**) Large Formwork Dried Fine Cardboard Sample; (**b**) FCL Samples (100 mm diameter, 38 mm thickness); (**c**) Small Formwork Dried Fine Cardboard Sample; (**d**) FCH Samples (29 mm diameter, 38 mm thickness).

2.1.8. Commercial Mycelium Comparison

In order to compare against commercially available mycelium-based composite mixtures, Ecovative Design's Grow-It-Yourself Mushroom® Material [26] was grown in the same two formworks and cut to the same sample circles. Ecovative is one of the pioneers in utilizing mycelium-based composites in industrial applications. This start-up began by producing packaging and insulation materials as an alternative to polystyrene-based (Styrofoam) materials and has developed into a large biotechnology company, making myco-leather, mycelium meat alternatives, and beauty industry alternatives [27]. The company sells Grow-It-Yourself bags with their own mycelium mixture. Samples grown using their mixture were also tested in this study (EML and EMH).

2.2. Testing and Assessing Sound Absorption of Mycelium-Based Composites

The following experiment outlined two sets of acoustic tests. The first round of tests was useful in selecting appropriate sound-absorbing materials for acoustic panels. The second set of tests builds on the results of the first by testing the best performing materials again using a larger sample size.

2.2.1. Preliminary Testing for Sound Absorption

As a preliminary study, first, two replicates for each of the samples (material thickness: 38 mm) are tested three times using an impedance tube, specifically the two-microphone transfer-function method, illustrated in Figure 4, following the standard ASTM E1050-12. Brüel and Kjær's Impedance Tube Kit (50 Hz–6.4 kHz) Type 4206 was used in this experiment. Type 4206 consists of:

1. 100 mm diameter tube (large tube)

    a. Frequency range: 50 Hz to 1.6 kHz;
    b. Material sample size requirements: 100 mm diameter, 200 mm max sample length.

2. 29 mm diameter tube (small tube)

    a. Frequency range: 505 Hz to 6.4 kHz;
    b. Material sample size requirements: 29 mm diameter, 200 mm max sample length.

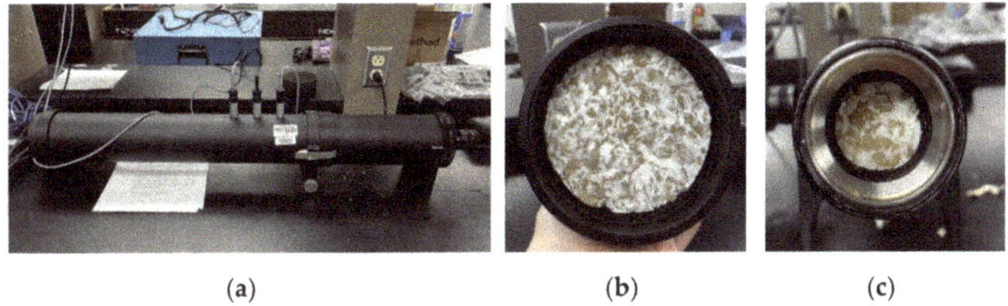

(a)   (b)   (c)

Figure 4. Impedance tube Testing: (**a**) Impedance Tube; (**b**) FCL Sample in the impedance tube; (**c**) FCH sample in the impedance tube.

2.2.2. Testing with Larger Sample Size

The results of the preliminary study informed the second stage of acoustical testing. Substrates that resulted in the structural failure of the samples were omitted. Two of the most promising substrates from the preliminary study were determined for both low-frequency and high-frequency sound absorption. These were SCL and FCL and SCH and FCH, respectively. Six replicates were created for each of the low-frequency samples (SCL and FCL), and 9 replicates were created for each of the high-frequency samples (SCH and FCH). These are listed on Tables 2 and 3. All replicates' thicknesses were 38 mm. These samples were each tested again, three times, using an impedance tube following the same standard (ASTM E1050-12).

2.2.3. Statistical Analysis

There are usually six frequencies used to determine whether a material is sound absorbing. These are: 125 Hz, 250 Hz, 500 Hz, 1000 Hz, 2000 Hz and 4000 Hz. If the average sound absorption coefficient to the above-stated six frequencies α is bigger than 0.2, the material is called a sound absorbing material [28]. For comparison of the two selected sample groups, the sound absorption coefficients at the following low-frequency levels are used for the 100 mm samples: 125 Hz, 250 Hz, 500 Hz, and 1000 Hz; additionally, the sound absorption coefficients at the following high-frequency levels are used for the 29 mm samples: 2000 Hz and 4000 Hz. The mean sound absorption coefficients of the sample groups at the given frequency levels were compared using the Mann–Whitney U test with the SPSS software (IBM Corp. Released 2015. IBM SPSS Statistics for Windows, Version 23.0. IBM Corp: Armonk, NY, USA). The Mann–Whitney U test was used to determine whether there is a difference in the dependent variable for two independent groups and to compare whether the distribution of the dependent variable is the same for two groups [29]. A p-value of <0.05 was considered statistically significant.

2.3. Paneling Experiments

Initial experiments were conducted regarding the design and fabrication of acoustic panels using the best-performing materials presented in this study. As seen in Figure 5, the fabrication of the panels was performed by first CNC-milling a positive wooden form of 380 mm × 380 mm × 50 mm, and then thermoforming the wooden form with PVC sheets to create a reusable plastic negative formwork. This formwork was then filled with a fine cardboard substrate mixture, and panels were grown using the same procedure presented in Section 2.1.

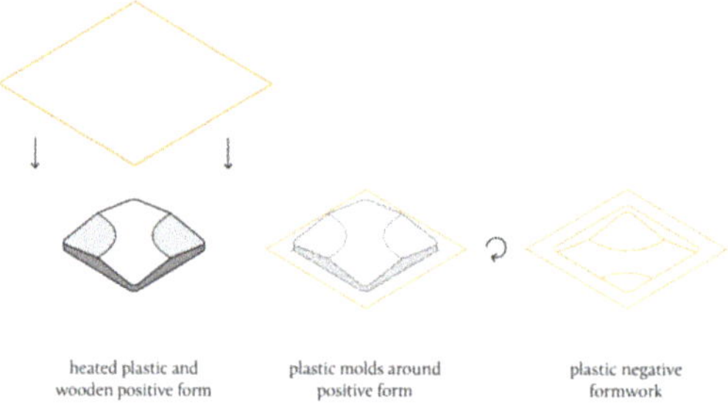

Fill Formwork with Mycelium Mixture

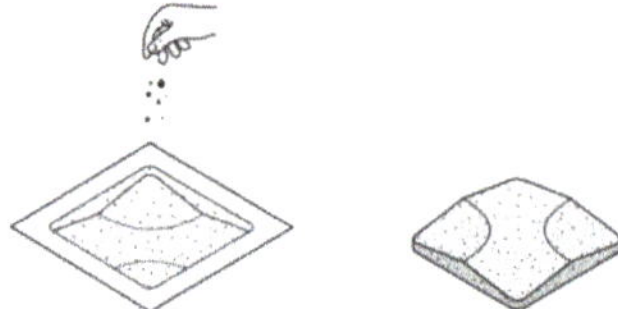

Figure 5. Acoustic Panel Fabrication Diagram.

3. Results

3.1. Physical Characteristics of the Cultivated Mycelium-Based Composites

A visual inspection was conducted to determine initial growth conclusions. This first analysis was useful in determining which mycelium mixtures were unsuitable for the impedance tube tests and therefore unsuitable as acoustic paneling materials.

3.1.1. Warpage

The materials, once dried, shrank and warped from their original form, as seen in Figure 2d. The material with the most warpage was the Sample FPL (fine paper), and the material with the least warpage was the Sample EML (Ecovative mixture), followed by the Sample FCL (fine cardboard), and the Sample FNL (fine newsprint). In order to ascertain accurate test results from the impedance tube, the sample surface must be flat. As a result, the formworks were made larger than the sample size, so the inaccuracy due to warpage was minimized. For the use of mycelium-based composites as acoustic panels, warpage becomes a challenge for form-to-performance accuracy and mounting purposes. Further research is necessary to predict warpage for specific mycelium mixtures. One possible solution could be to add weights to the corners of the materials as they dry.

3.1.2. Structural Integrity

The structural integrity of the composite material largely relies on the structure of the substrate material and how well mycelium can grow throughout the substrate. The Sample FNL (fine newsprint) did not hold together once dried, cracked, and crumbled, as seen in Figure 6a. A possible explanation for the deterioration is the structure of the fine newsprint. When pulverized, the newsprint became very fine dust, while the pulverized paper and cardboard maintained more of their structure. After shaping the materials using industrial tools, the materials' structural integrity was affected, as the fungal skin is a large component holding the material together. The Sample SPH (shredded paper), once cut with a saw, could not yield accurate results in the impedance tube because of its structural integrity, as seen in Figure 6b. Therefore, no further tests were conducted with the FNL, FNH, SPL, and SPH Samples.

(a) **(b)**

Figure 6. Structural Integrity of Mycelium Samples: (**a**) FNL (**b**) SPH.

Once all the samples were cut open, the inner growth of the samples was analyzed. The samples were observed to have higher mycelial growth on the outer surface and less mycelial growth internally. This could be related to the absence of light and air and the heat produced by mycelium during growth [15]. While the shaping of the samples was necessary for the impedance tube tests, it is worth noting that for the purpose of acoustic panels, shaping/cutting the samples negatively affects their durability and structural integrity.

It is possible that a combination of different substrate materials would lead to more mycelium growth and stronger material. Further research can be conducted to assess the mechanical and physical characteristics of mycelium-based composites grown on different combinations of paper-based waste substrates.

3.2. Sound Absorption of the Cultivated Mycelium-Based Composites

3.2.1. Results of the Preliminary Acoustic Absorption Testing

The results of the impedance tube for each sample were recorded and graphed. For each material mixture, two replicates grown in a single formwork were tested, and their results were averaged. It is important to note that the surface of the samples varied depending on the growth of the mycelium. These tests were useful in determining which mixtures performed better than others and informed the second stage of acoustical testing with additional replicates.

Of the low to mid-frequency samples, the fine cardboard samples (FCL) showed the best absorption, as can be seen in Figure 7, though none of the samples showed very high absorption in the low-frequency range (50 Hz to 500 Hz). It was noted that the sound absorption results shown do not include the effect of an air gap behind the material. The introduction of an air cavity between the material and the rigid backing surface can increase the sound absorption performance at low frequencies [30]. Of the mid-range frequencies (500 Hz to 2 kHz), the fine cardboard samples FCL had the best acoustic absorption performance.

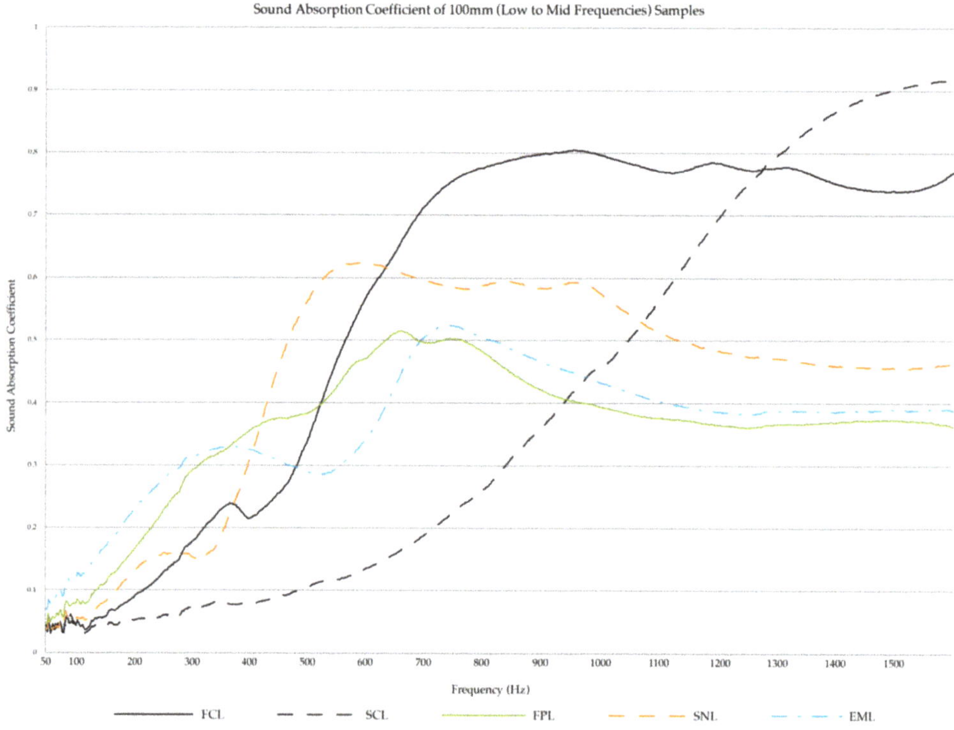

Figure 7. Sound Absorption Coefficient of low-frequency samples (100 mm).

Of the high-frequency samples, the shredded cardboard samples (SCH) had the highest sound absorption from the 2 kHz to 6.4 kHz frequency range, as can be seen in Figure 8. Samples SCH is followed by the fine cardboard samples (FCH), then the fine paper samples (FPH). The lowest absorption is from the shredded newsprint samples (SNH). EMH does not show to be a successful absorber.

The graphs in Figures 7 and 8 display the results for the 100 mm and 29 mm samples, respectively. The first graph (Figure 7) represents the materials' absorption from the 50 Hz to 1.6 kHz frequency range. The second (Figure 8) is from 500 Hz to 6.4 kHz.

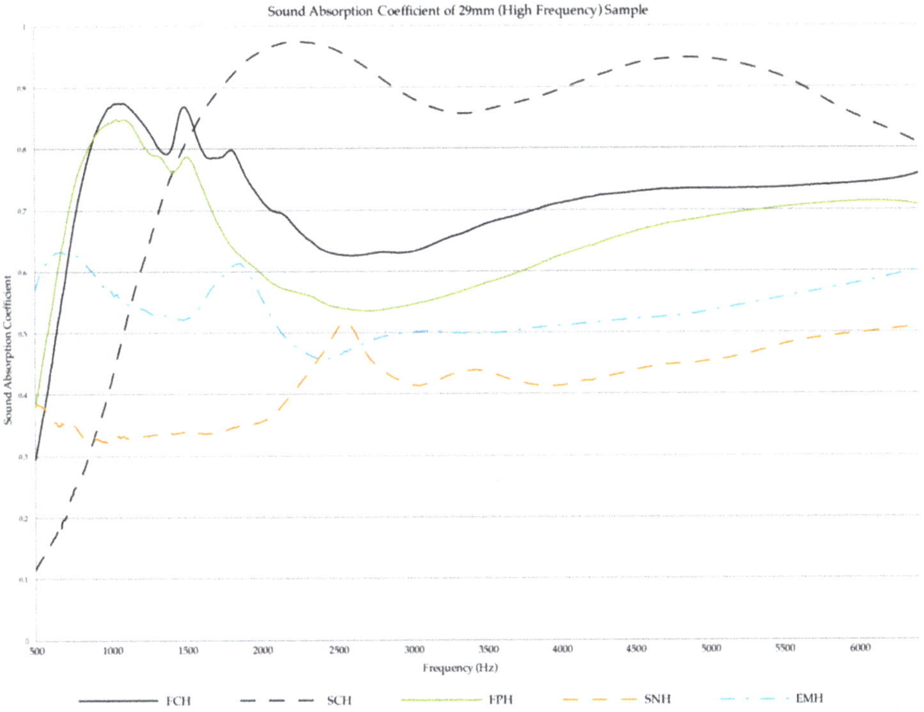

Figure 8. Sound Absorption Coefficients of high frequency samples (29 mm).

3.2.2. Results of the Acoustic Absorption Testing with Larger Sample Size

Low-frequency sound absorption coefficients: To obtain a better understanding of how the two best-performing samples in low-frequency sound absorption, SCL and FCL, compare with each other, we created six replicates for each sample group and tested their acoustic absorption. Three formworks were filled for each substrate mixture, resulting in two replicates per formwork, thus six replicates per substrate mixture (Table 2). Figure 9 presents the test results of the six SCL replicates grown in three separate formworks. Figure 10 presents the test results of the six FCL replicates grown in three separate formworks.

Table 2. Samples Tested for Low-Frequency Sound Absorption.

Abbr.	Sample	Formwork Number	Number of Replicates	Sample Size
SCL	Shredded Cardboard Low freq.	1–2–3	6	100 mm
FCL	Fine Cardboard Low freq.	4–5–6	6	100 mm

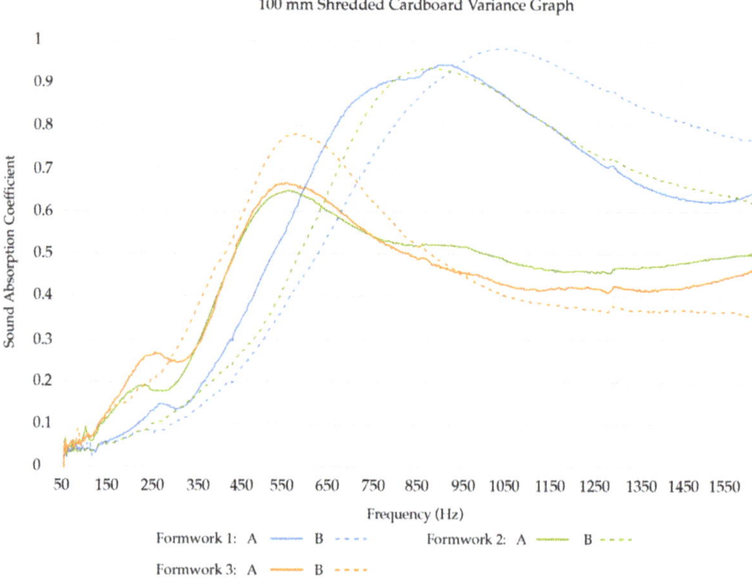

Figure 9. Sound Absorption Coefficients of SCL replicates (100 mm).

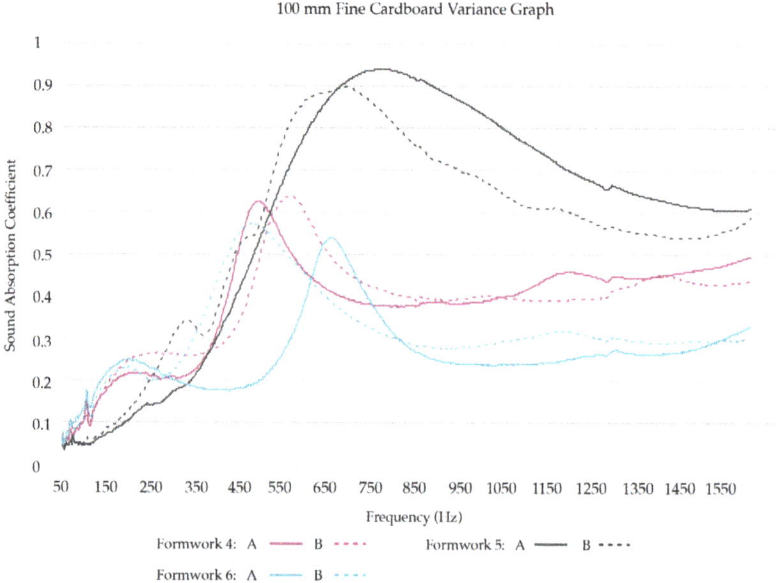

Figure 10. Sound Absorption Coefficients of FCL replicates (100 mm).

Of the low to mid-frequency samples, the shredded cardboard samples (SCL) follow two general trends. Half of the replicates' sound absorption coefficients peak between 450 Hz and 650 Hz and then begin to drop, while the other half has a much higher absorption rate, and the absorption peak shifts to between 750 Hz and 1050 Hz. Of the mid-range frequencies (500 Hz to 2 kHz), SCL performs well, with half of the replicates reaching over a 0.9 sound absorption coefficient at some frequency.

The fine cardboard samples (FCL) similarly follow two general trends in the low to mid-frequency ranges. Four of the replicates' sound absorption coefficients peak between 400 Hz and 700 Hz, then drop and remain constant, while the other two have a much higher absorption rate, and the absorption peak shifts to between 550 Hz and 850 Hz. Of the mid-range frequencies (500 Hz to 2 kHz), some of the FCL also perform well, with two of the replicates reaching a 0.9 sound absorption coefficient at some frequency.

As can be seen in Table 4, the test results show that in the selected low to mid frequencies (125 Hz, 250 Hz, 500 Hz, 1000 Hz), the sound absorption trends of both low-frequency sample groups (SCL and FCL) are statistically similar ($p > 0.05$).

High-frequency sound absorption coefficients: To obtain a better understanding of how the two best performing samples in high-frequency sound absorption, SCH and FCH, compare with each other, we created nine replicates for each sample group and tested their acoustic absorption. Three formworks were filled for each substrate mixture, resulting in three replicates per formwork, thus nine replicates per substrate mixture (Table 3). Figure 11 presents the test results of the nine SCH replicates. Figure 12 presents the test results of the nine FCH replicates.

Table 3. Samples Tested for High-Frequency Sound Absorption.

Abbr.	Sample	Formwork Number	Number of Replicates	Sample Size
SCH	Shredded Cardboard High freq.	7–8–9	9	29 mm
FCH	Fine Cardboard High freq.	10–11–12	9	29 mm

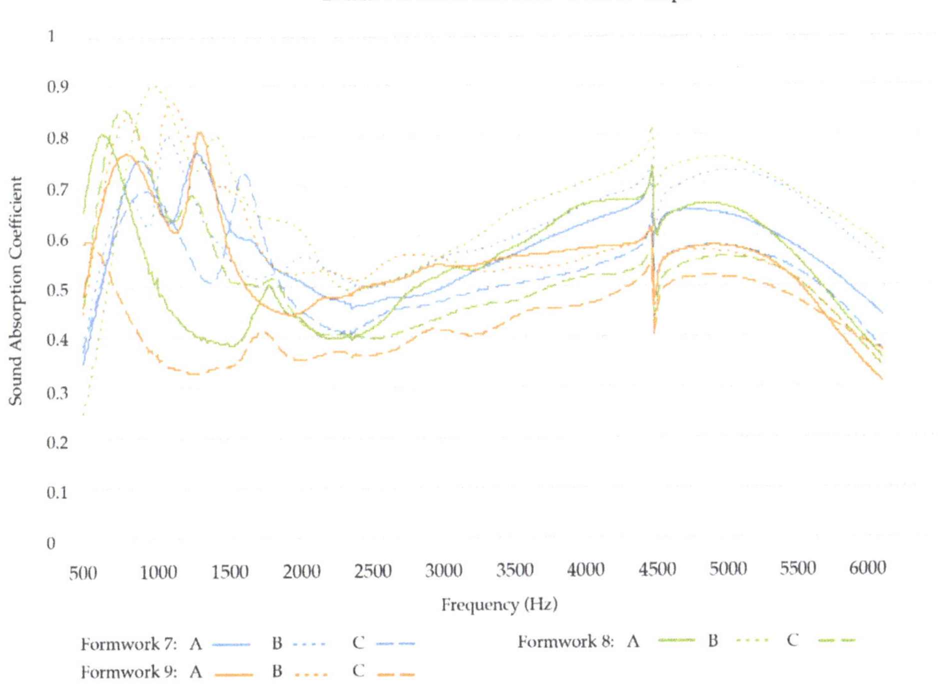

Figure 11. Sound Absorption Coefficients of SCH replicates (29 mm).

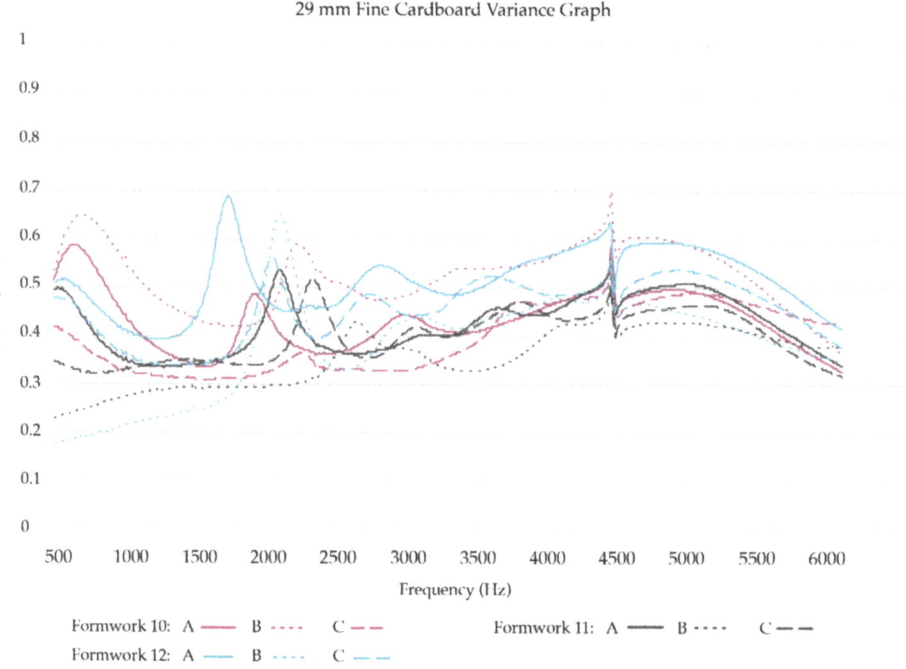

Figure 12. Sound Absorption Coefficients of FCH replicates (29 mm).

Of the high-frequency samples, the shredded cardboard samples (SCH) follow a single trend. The samples generally peak between 500 Hz and 1.5 kHz, dip and then gradually rise again. The samples have a medium to high absorption rate from 500 Hz to 1.5 kHz and 4 kHz to 5.5 kHz. The fine cardboard samples (FCH) also follow a single trend. However, as can be seen in Table 4, in the selected high frequencies (2000 Hz, 4000 Hz), high-frequency sample groups (SCH and FCH) have different sound absorption trends (p = 0.019 and p = 0.011, respectively). Shredded cardboard samples (SCH) had better sound absorption performance than fine cardboard samples (FCH).

Table 4. Sound absorption coefficients of shredded and fine cardboard samples (mean $\pm$ Standard Deviation).

	Frequencies (Hz)	Shredded Cardboard (SCL + SCH)	Fine Cardboard (FCL + FCH)	p-Value
Low Freq.	125	0.0698 $\pm$ 0.02	0.1218 $\pm$ 0.05	0.070
	250	0.1609 $\pm$ 0.07	0.2074 $\pm$ 0.04	0.240
	500	0.5116 $\pm$ 0.17	0.5096 $\pm$ 0.15	0.810
	1000	0.6891 $\pm$ 0.26	0.4697 $\pm$ 0.23	0.070
High Freq.	2000	0.4934 $\pm$ 0.07	0.3949 $\pm$ 0.08	0.019
	4000	0.5731 $\pm$ 0.07	0.4703 $\pm$ 0.06	0.011

3.3. Mycelium-Based Acoustic Panel Prototypes Cultivated with Fine Cardboard Substrates

The results of the acoustic panel prototypes revealed that mycelium growth is still consistent even in larger formworks (Figure 13). However, the durability of the material proves to be a problem on a larger scale. After dying and handling, the edges of the panels began to show signs of deterioration. In order to ensure durability with the fine cardboard material, additional support may be necessary. This could potentially be remedied with additional substrate materials, internal support, or external backing.

Figure 13. Acoustic panel prototype cultivated with Fine Cardboard substrate.

Concurrently with panel fabrication, a customizable panel system was generated using parametric modeling software (Rhinoceros 3D, Version 7.0. Robert McNeel & Associates, Seattle, WA, USA). Figure 14 shows a custom acoustic wall configuration generated using this system and illustrates how the panel configuration can be altered. The use of the parametric system aids in random wall configurations. Using a three-dimensional truchet tile in the parametric system allows for a number of wall configurations with only one panel, thus reducing the need for different formworks.

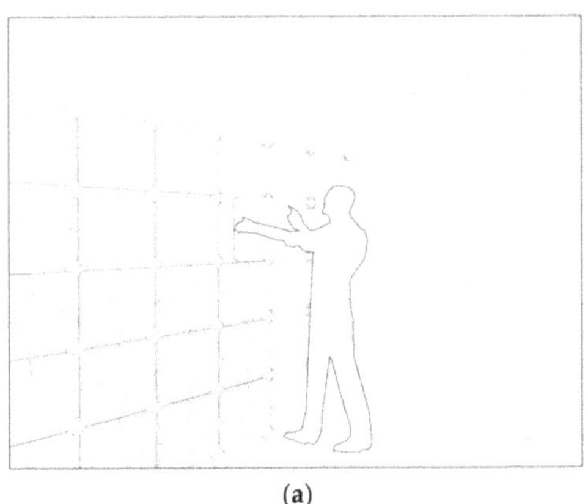

(a)

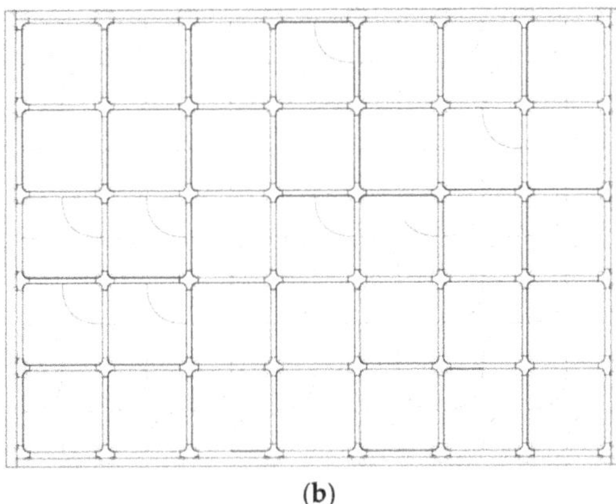

(b)

Figure 14. Parametric acoustic panel wall prototype design: (**a**) Acoustic panel installation illustrated (**b**) Example wall configuration.

4. Discussion

The results of this study indicate that mycelium-based composites grown on waste shredded and fine cardboard show potential as sound-absorbing materials, specifically in the mid to high-frequency ranges. Shredded cardboard samples (SCH) slightly outperform fine cardboard samples (FCH) in high-frequency ranges.

4.1. Comparison to Commercial Sound Absorbing Materials

The impedance tube test results show that mycelium-based composites cultivated on shredded cardboard and fine cardboard can both be considered sound-absorbing materials ($\alpha > 0.2$) and have the potential to compete with the performance of synthetic sound absorbers (Table 5). The sound absorbing coefficients of three types of commercially available synthetic sound absorbing products made with fiberglass, polypropylene, and plaster were compared with the fine cardboard samples (FCL + FCH) and shredded cardboard samples (SCL + SCH). The comparison revealed that the fiberglass insulation board shows better sound absorption than both sample groups in all the frequencies except 125 Hz. When compared with the polypropylene product, both sample groups have better sound absorption at low frequencies (125 Hz, 250 Hz, 500 Hz). Compared with plasterboard, one of the most common interior wall finishes, the absorption coefficients of both samples are significantly higher in the mid to high-frequency ranges. These comparisons are helpful in discussing the potential of the two sample groups as sound absorbers. For more accurate comparisons, commercially available sound-absorbing materials need to be tested by the authors for each frequency range using the same testing model.

Table 5. Sound Absorption Coefficients comparing commercial sound absorbing materials with the fine cardboard samples (FCL + FCH) and shredded cardboard samples (SCL + SCH).

Product/Sample	Octave Band Center Frequencies, Hz						Average
	125	250	500	1000	2000	4000	α
Fine Cardboard Samples (FCL + FCH)	0.12	0.21	0.51	0.47	0.40	0.47	0.36
Shredded Cardboard Samples (SCL + SCH)	0.07	0.16	0.51	0.69	0.49	0.57	0.42
Type 706 Series Fiberglas™ Insulation Board (Fiberglass) [31]	0.01	0.22	0.67	0.97	1.05	1.06	0.66
Quiet Board™ Acoustic Panel (Polypropylene) [32]	0.05	0.06	0.21	0.8	0.65	0.75	0.42
Plasterboard (1/2″ paneling on studs) [33]	0.29	0.1	0.06	0.05	0.04	0.04	0.10

4.2. Effects of Surface Texture and Porosity on Sound Absorption

The results of the impedance tube tests revealed significant variances between replicates of the same material (see Table 4). This is hypothesized to be the result of inconsistent mycelia growth, the size of the substrate material, and the random substrate filling technique. The following graphs in Figure 15 present the sound absorption coefficients of SCH replicates cultivated within three different formworks, alongside close-up images of the replicates cultivated within the same formwork. A visual inspection revealed that the replicates with more bumps and pores at the surface have higher sound absorption coefficients; however, further tests are needed to validate this hypothesis.

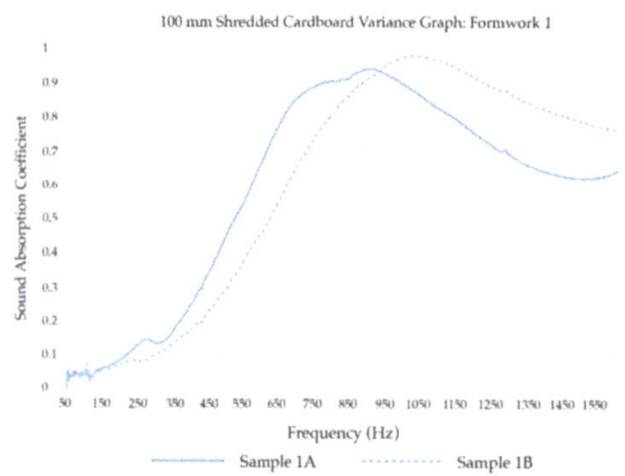

Figure 15. *Cont.*

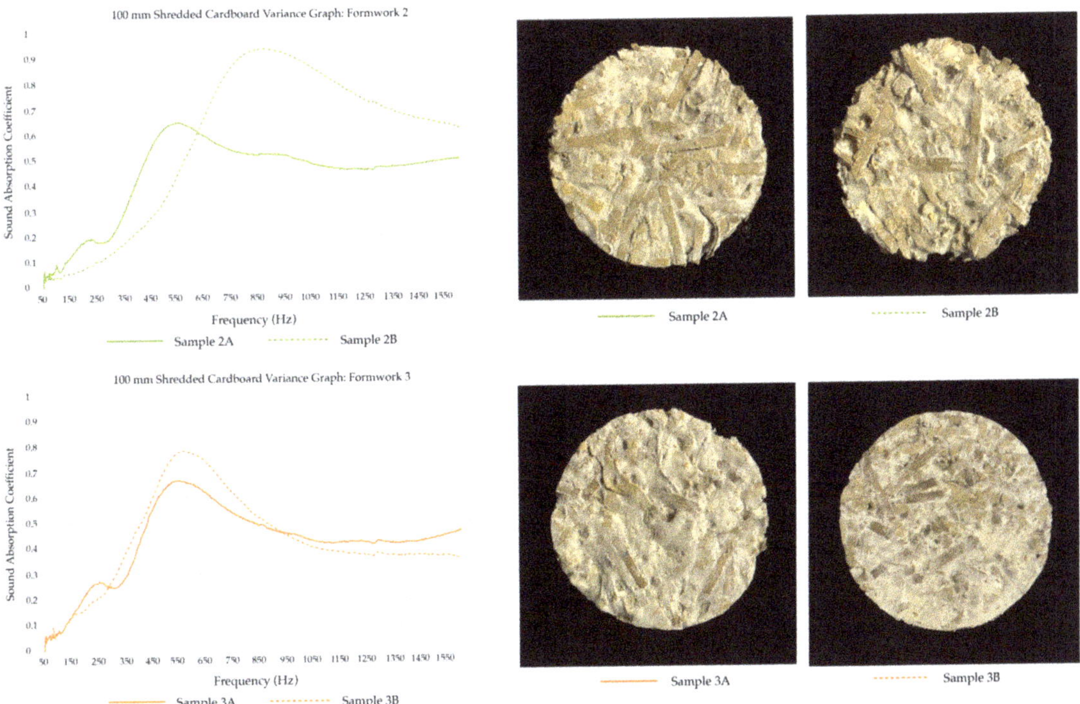

Figure 15. Sound Absorption Coefficient Graphs and the Shredded Cardboard Samples Tested. A visual inspection to compare surface texture and porosity with the respective material performance.

4.3. Limitations and Strengths of the Study

There are two main limitations to this study. The first limitation addresses Section 4.1. The data for the commercially available synthetic sound absorbers were collected from the existing literature [31–33]. While the sound absorption coefficients for these materials are validated numbers provided in their data sheets, to be able to ensure accurate results and have a meaningful comparison, commercially available sound absorbing materials need to be tested for each frequency range using the same testing model, with samples that have the same material thickness, density, and porosity.

The second limitation addresses Section 4.2. As can be seen in Table 4, the sound absorption coefficients for both sample groups in 500 Hz and 1000 Hz frequencies show significant variances. This limitation can be overcome by creating larger subgroups within each sample group through visual inspection of the replicates and testing these subgroups' sound absorption coefficients independently.

The strength of the study was initially performing preliminary tests with multiple waste paper-based samples. This enabled accurately deciding which substrates fit in the testing model and eliminating the ones that did not work.

5. Conclusions

Of the tested samples from the preliminary acoustic tests, the shredded and fine cardboard-based samples show the best acoustic performance. In addition to this, the fine newsprint and shredded paper substrates are not considered to be applicable for paneling purposes due to their (lack of) structural integrity. Due to these findings, the shredded and fine cardboard samples were regrown with larger sample size and tested again. The results show that both shredded and fine cardboard-based mycelium composites do show potential as sound absorbing materials, with shredded cardboard samples slightly performing better in high-frequency sound absorption. However, the inherent nature of bio-fabricated materials causes a variance in performance, even between samples of the same material.

The next steps of this research are to investigate how material thickness, density, and porosity affect sound absorption of shredded and fine cardboard-based mycelium composites. This will be performed by cultivating additional replicates and creating larger subgroups within each sample group by controllably varying their material thickness, density, and porosity. Along with their sound absorption properties, their mechanical properties (compression, bending, torsion, and tension and impact damping) and morphological characteristics (i.e., pore size, porosity, density), as well as the growth mechanisms of mycelium, will be studied. The main objective is to understand how the growth of mycelium at microscopic levels, the morphological characteristics at both mesoscopic and macroscopic levels, and the acoustic absorption performance of the composites interact with one another. Another follow-up study could be to test various commercially available synthetic acoustic absorbers using the same testing model, with samples that have the same material thickness, density, and porosity as the mycelium-based sample groups. This would enable a more thorough comparison of mycelium-based composites' acoustic absorption performance with synthetic absorbers.

Once a holistic understanding and more comprehensive data about the composites' acoustic, mechanical, and morphological characteristics are gathered, the next steps involve the applications of the shredded and fine cardboard-based composites as acoustic paneling. The material itself, though sound absorbing, has physical limitations such as structural integrity and warping when cultivated on larger scales. More experiments must be conducted to ensure the durability of the material. Concurrently with durability assessments, analyses regarding form-to-performance will be conducted. These experiments will be used to determine how the form of the acoustic panels affects the sound absorption performance. Therefore, full-scale prototypes will be built and tested alongside computer simulation models in reverberant chambers. These results will inform parametric iterations of panel systems.

Incorporating mycelium-based composites into architectural systems is significant because of their ability to reduce waste generated and energy consumed during material manufacturing compared to conventional building materials. Mycelium-based composites recycle waste materials for growth, require little energy to manufacture, and completely decompose at the end of their product life. This research is relevant in order to establish protocols for material use and implementation within acoustic systems.

Author Contributions: Conceptualization, N.W. and B.G.; methodology, N.W. and B.G.; software, N.W.; formal analysis, N.W. and B.G.; investigation, N.W. and B.G.; writing—original draft preparation, N.W.; writing—review and editing, N.W. and B.G.; visualization, N.W.; supervision, B.G.; project administration, B.G.; funding acquisition, N.W. and B.G. All authors have read and agreed to the published version of the manuscript.

Funding: This research was partially funded by the Pennsylvania State University's Erickson Discovery Grant and Institute of Energy and Environment Flower Grant.

Data Availability Statement: Data are available upon request.

Acknowledgments: We thank John Pecchia at the Mushroom Research Center at Penn State University for their expertise and for providing access to their facilities and equipment. We also thank Yun Jing, Hyeonu Heo, and Jun Ji for assisting with the impedance tube tests, as well as Ali Ghazvinian and Alale Mohseni for their help in cultivating the mycelium-based composite samples.

Conflicts of Interest: The authors declare no conflict of interest.

References

1. Chepesiuk, R. Decibel Hell: The Effects of Living in a Noisy World. *Environ. Health Perspect.* **2005**, *113*, A34–A41. [CrossRef] [PubMed]
2. National Institute for Occupational Safety and Health. Noise & Hearing Loss Prevention. CDC. 6 February 2018. Available online: https://www.cdc.gov/niosh/topics/noise/default.html (accessed on 2 May 2022).
3. Aletta, F.; Kang, J. Promoting Healthy and Supportive Acoustic Environments: Going beyond the Quietness. *Int. J. Environ. Res. Public Health* **2019**, *16*, 4988. [CrossRef] [PubMed]
4. Amares, S.; Sujatmika, E.; Hong, T.W.; Durairaj, R.; Hamid, H.S.H.B. A Review: Characteristics of Noise Absorption Material. *J. Phys. Conf. Ser.* **2017**, *908*, 012005. [CrossRef]
5. Seddeq, H.S. Factors Influencing Acoustic Performance of Sound Absorptive Materials. *Aust. J. Basic Appl. Sci.* **2009**, *3*, 4610–4617. Available online: https://www.semanticscholar.org/paper/Factors-Influencing-Acoustic-Performance-of-Sound-seddeq/f6e0dd4a9b34074b0beb25217d5415aa31ee90ca (accessed on 2 May 2022).
6. Arenas, J.; Sakagami, K. Sustainable Acoustic Materials. *Sustainability* **2020**, *12*, 6540. [CrossRef]
7. Desarnaulds, V.; Costanzo, E.; Carvalho, A.; Arlaud, B. Sustainability of acoustic materials and acoustic characterization of sustainable materials. In Proceedings of the 12th International Congress on Sound and Vibration 2005, ICSV 2005, Lisbon, Portugal, 11–14 July 2005; Volume 1. Available online: https://www.researchgate.net/publication/37649558_Sustainability_of_acoustic_materials_and_acoustic_characterization_of_sustainable_materials (accessed on 2 May 2022).
8. Lee, Y.; Joo, C. Sound Absorption Properties of Recycled Polyester Fibrous Assembly Absorbers. *Autex Res. J.* **2003**, *3*, 78–84. Available online: https://www.researchgate.net/publication/267680225_Sound_absorption_properties_of_recycled_polyester_fibrous_assembly_absorbers (accessed on 2 May 2022).
9. Arenas, J.P.; Crocker, M.J. Recent Trends in Porous Sound-Absorbing Materials. *Sound Vib.* **2010**, *44*, 12–17. Available online: https://www.researchgate.net/publication/272151761_Recent_Trends_in_Porous_Sound-Absorbing_Materials (accessed on 2 May 2022).
10. Wegst, U.G. Bamboo and Wood in Musical Instruments. *Annu. Rev. Mater. Sci.* **2008**, *38*, 323–349. [CrossRef]
11. Chen, K.; Wang, J.; Yu, B.; Wu, H.; Zhang, J. Critical evaluation of construction and demolition waste and associated environmental impacts: A scientometric analysis. *J. Clean. Prod.* **2020**, *287*, 125071. [CrossRef]
12. Environmental Protection Agency. Construction and Demolition Debris: Material-Specific Data. EPA. (n.d.). Available online: https://www.epa.gov/facts-and-figures-about-materials-waste-and-recycling/construction-and-demolition-debris-material (accessed on 8 October 2021).
13. Manan, S.; Ullah, M.W.; Ul-Islam, M.; Atta, O.M.; Yang, G. Synthesis and applications of fungal mycelium-based advanced functional materials. *J. Bioresour. Bioprod.* **2021**, *6*, 1–10. [CrossRef]
14. Ghazvinian, A.; Farrokhsiar, P.; Vieira, F.; Pecchia, J.; Gursoy, B. Mycelium-Based Bio-Composites for Architecture: Assessing the Effects of Cultivation Factors on Compressive Strength. *Blucher Des. Proc.* **2019**, *2*, 505–514. [CrossRef]
15. Elsacker, E.; Vandelook, S.; Brancart, J.; Peeters, E.; De Laet, L. Mechanical, physical and chemical characterisation of mycelium-based composites with different types of lignocellulosic substrates. *PLoS ONE* **2019**, *14*, e0213954. [CrossRef] [PubMed]
16. Appels, F.V.W.; Camere, S.; Montalti, M.; Karana, E.; Jansen, K.M.B.; Dijksterhuis, J.; Krijgsheld, P.; Wösten, H.A. Fabrication factors influencing mechanical, moisture- and water-related properties of mycelium-based composites. *Mater. Des.* **2018**, *161*, 64–71. [CrossRef]
17. Mogu Acoustics. Available online: https://mogu.bio/acoustic/ (accessed on 12 March 2022).
18. Pelletier, M.G.; Holt, G.A.; Wanjura, J.D.; Bayer, E.; McIntyre, G. An evaluation study of mycelium based acoustic absorbers grown on agricultural by-product substrates. *Ind. Crops Prod.* **2013**, *51*, 480–485. [CrossRef]
19. Pelletier, M.G.; Holt, G.A.; Wanjura, J.D.; Greetham, L.; McIntyre, G.; Bayer, E.; Kaplan-Bie, J. Acoustic evaluation of mycological biopolymer, an all-natural closed cell foam alternative. *Ind. Crops Prod.* **2019**, *139*, 111533. [CrossRef]
20. Schwarze, F.W.M.R.; Morris, H. Banishing the myths and dogmas surrounding the biotech Stradivarius. *Plants People Planet.* **2020**, *2*, 237–243. [CrossRef]
21. Abdullah, Y.; Putra, A.; Effendy, H.; Farid, W.M.; Ayob, R. Investigation on natural waste fibers from dried paddy straw as a sustainable acoustic absorber. In Proceedings of the 2011 IEEE Conference on Clean Energy and Technology (CET), Kuala Lumpur, Malaysia, 27–29 June 2011; pp. 311–314. [CrossRef]
22. Rachman, Z.A.; Utami, S.S.; Sarwono, J.; Widyorini, R.; Hapsari, H.R. The usage of natural materials for the green acoustic panels based on the coconut fibers and the citric acid solutions. *J. Phys. Conf. Ser.* **2018**, *1075*, 012048. [CrossRef]
23. Fellowes Powershred 12-Sheet Cross-Cut Shredder, 12C. Available online: https://www.walmart.com/ip/Fellowes-Powershred-12Sheet-Cross-Cut-Paper-Shredder/862121144?wl13=1640&selectedSellerId=0 (accessed on 2 May 2022).
24. Polypropylene Autoclavable Bags. Available online: https://www.amazon.com/gp/product/B07P7CD7HQ/ref=ppx_yo_dt_b_asin_title_o03_s01?ie=UTF8&th=1 (accessed on 2 May 2022).
25. Lambert Spawn Pre-Spawned Oyster Bags. Available online: https://www.lambertspawn.com/documents/products/LAMB_Specialty.pdf (accessed on 2 May 2022).
26. Ecovative Design Grow-It-Yourself Mushroom®Material. Available online: https://grow.bio/products/grow-it-yourself-material (accessed on 12 March 2022).
27. Ecovative. Available online: https://ecovative.com/ (accessed on 12 March 2022).

28. Li, Y.; Ren, S. (Eds.) Acoustic and Thermal Insulating Materials. In *Building Decorative Materials*; Woodhead Publishing Series in Civil and Structural Engineering; Woodhead Publishing: Sawston, UK, 2011; pp. 359–374. [CrossRef]
29. Karadimitriou, S.M.; Marshall, E. Mann-Whitney U Test. (n.d.) Available online: https://www.sheffield.ac.uk/polopoly_fs/1.71 4552!/file/stcp-marshall-MannWhitS.pdf (accessed on 2 May 2022).
30. Arenas, J.; Del Rey, R.; Alba, J.; Oltra, R. Sound-Absorption Properties of Materials Made of Esparto Grass Fibers. *Sustainability* **2020**, *12*, 5533. [CrossRef]
31. Type 706 and Type 707 Series Fiberglas™ Insulation Boards. Available online: http://commercial.owenscorning.com/assets/0/321/333/9d503f63-6274-49d3-b00a-d74cf4b4d5f1.pdf (accessed on 2 May 2022).
32. Quiet Board™ Acoustic Panel. Available online: https://www.soundproofcow.com/skin/common_files/pdf/quiet_board_pds.pdf (accessed on 2 May 2022).
33. Absorption Coefficients of Common Building Materials and Finishes. JCW Acoustic Supplies. (N.D.). Available online: https://www.acoustic-supplies.com/absorption-coefficient-chart/ (accessed on 2 May 2022).

Article

Strategies for Growing Large-Scale Mycelium Structures

Jonathan Dessi-Olive 

MycoMatters Laboratory, University of North Carolina at Charlotte (UNCC), Charlotte, NC 28223, USA; jdessiolive@uncc.edu; Tel.: +1-952-334-1775

Abstract: Fungi-based materials (myco-materials) have been celebrated and experimented with for their architectural and structural potential for over a decade. This paper describes research applied to assembly strategies for growing large building units and assembling them into efficiently formed wall prototypes. A major concern in the development of these two fabrication strategies is to design re-usable formwork systems. La Parete Fungina demonstrates two undulating wall units standing side-by-side, each composed of seventeen myco-welded slabs. L'Orso Fungino revisits the in situ monolithic fabric forming of units that are repeated, stacked, and post-tensioned. Although the design and research presented in this paper focuses on overcoming the challenges of growing large-scale building components, this work also touches on issues of accessibility and technology, economic and logistical systems needed for building-scale applications, and material ethics of energy and waste associated with emerging biomaterial production.

Keywords: mycelium; myco-materials; myco-fabrication; sustainable buildings; sustainable structures; architectural design; structural design; material ethics

Citation: Dessi-Olive, J. Strategies for Growing Large-Scale Mycelium Structures. *Biomimetics* **2022**, *7*, 129. https://doi.org/10.3390/biomimetics7030129

Academic Editors: Phil Ayres, Andrew Adamatzky and Han A.B. Wösten

Received: 13 July 2022
Accepted: 7 September 2022
Published: 11 September 2022

Publisher's Note: MDPI stays neutral with regard to jurisdictional claims in published maps and institutional affiliations.

1. Introduction

Within Euro-centric traditions of architecture, the significance of a building is often tied to its permanence. The Pantheon in Rome, for example, is a nearly 2000-year-old cementitious dome structure, whose resilience to time elevates it to a monumental status. Notwithstanding the significance of cultural and economic factors associated with the need for permanent buildings and structures, must all buildings be assembled with the goal of being permanent? Globally, the lifespans of buildings are rapidly decreasing. The average lifespan of buildings in China was recently reported to be 34 years [1], and 25 years for residential buildings in Japan [2]. To great detriment, buildings are more than ever being demolished prematurely and yet, use materials that are manufactured with energy-intensive processes and are expensive or impractical to recycle. In the United States alone, the Environmental Protection Agency (EPA) reported there was 600 million tons of construction and demolition waste generated in 2018 [3]. Structural materials, including wood, and architectural metals, such as steel, copper, and brass, are valuable commodities that can be reused and recycled. However, in present-day architectural assemblies, these materials nearly ubiquitously inter-face with expanded foams, plastics, and resins, sometimes in irreversible composites. For example, wood is widely treated with synthetic resins and glues to increase its resistance to decay or structural performance.

Fossil-fuel-based materials are versatile and economical. They are used to create building products such as floor and wall finishes, furniture, conduits, structural reinforcements, insulation, and sealants, to name a few. From their manufacture to their end-of-life, synthetic materials require significant amounts of energy and produce emissions that are harmful to environmental and human health. Plastics, such as polyvinyl chloride (PVC), use a known carcinogenic monomer (vinyl chloride) in their production [4], and are often manufactured to be more ductile using phthalate plasticizers, a known class of toxins posing risks to the immune response, reproductive health, and embryonic development [5].

Particularly in Europe, sorting programs are improving, and assessments of recycling products, such as PVC from window frames [6], have demonstrated successful programs for those contexts. Still, only 3 percent of PVC is diverted from the waste stream in Europe [4]. Expanded polystyrene (EPS), commonly used as a packaging material, is fully recyclable, but due to its low density, the cost of transporting it to be recycled quickly outweighs the benefit if performed over long distances [7]. The EPA reports that only 0.6 percent of EPS waste produced in the United States is recovered [8]. While the championing of recycling has kindled examples of robust systems that produce high recycling rates in Germany and Singapore [9], the fate of most foams, plastics, and fossil-based composites is disposal in landfills, elimination through thermal incineration, or pyrolysis [10].

At a time when buildings can be expected to have short, non-permanent lifespans that commonly result in landfill disposal, new building materials are needed that can help challenge our traditional perceptions of significance and building permanence, rethink what materials we use to build, and gain awareness of where those materials go when we are finished with using them. Wood has recently been championed for its potential as a low-cost and affordable building material, but a labor shortage during the COVID-19 pandemic caused the cost of wood to increase by nearly four times [11], exposing the fragility of existing supply chains. In the face of material insecurity, there is a critical need to explore and test alternate low-energy and rapidly renewable building materials that contribute to circular material economies and lessen the impact of the architecture, engineering, and construction industries on climate change. Adopting new materials into the standards of contemporary and future construction is challenging, but necessary. Importantly, the way such new materials are used to design and build at the architectural scale cannot be assumed. Innovation is possible, and presenting physical demonstrations at the building scale is an important aspect of research needed to prove that an emerging material is viable for future building construction.

1.1. Mycelium Composite Materials

Fungi-based materials are among a class of biotechnologies showing promise in vastly offsetting the impact of the short lifespans of buildings in the modern era. In their most common form, lignocellulosic fibers sourced from agriculture or forestry material streams are bound together with an entangled web of *mycelia*, the root-like structures of fungi [12]. Commonly known as "myco-materials", they are produced similarly to commercial mushroom farming, and can be composted at end-of-life. Myco-materials have become an international enterprise and are produced at an industrial scale. Companies such as Ecovative [13], Mycoworks [14], and Mogu [15] have explored their unique and variable properties to create products through different forms of production. Products finding commercial success include packaging materials [16–18], interior products such as lampshades and planters [19], and acoustical panels [15]. Mushroom leather products that serve as a sustainable alternative to animal leather are demonstrating increasing commercial success [14,20,21], and are created through the use of different solid- and liquid-state techniques [22].

Growing myco-materials involves propagating fungal hyphae (often from the phylum Basidiomycota) into a fibrous substrate for several days under correct environmental conditions until it forms a composite mass. Mycelium biomass is formally agnostic, having the capacity to be grown into nearly any shape by packing fibers inoculated with a living fungus into a formwork composed of a breathable non-cellulose-based material (usually plastic) to avoid the mycelium from permanently adhering to the mold. The limitations for growth are biological and environmental. Important precautions are proper sterility to avoid the contamination of unwanted organisms, access to food and nutrients, maximal darkness, and access to warm, humid air. Depending on the region, the fungal species being grown, and the scale of production, growth chambers may need to be actively controlled to maintain an optimal temperature and humidity, representing a likely demand for energy resources. A common issue myco-material growers face is the emergence of contaminants, sometimes dangerous molds, and other organisms that thrive in similar environmental conditions. Typically, the fibrous substrates into which mycelia are grown need to be steam-sterilized or pasteurized, which can also be prohibitively expensive due to the equipment and energy needed for such processing. Another important precaution that relates to design with myco-materials is that at certain thicknesses, mycelia do not grow sufficiently due to a lack of oxygen, presenting a chance for contamination.

Once fully grown, parts are typically actively dried to stop growth [23], resulting in a material that resembles expanded polyurethane or polystyrene foam with a flame spread resistance comparable to gypsum and low thermal conductivity. The numerous complexities associated with growing myco-materials make it difficult to control the associated material properties (whether mechanical, thermal, acoustical, or other) and are understood to be a reported average. Different combinations of mycelium strains and fibrous substrates yield varying properties of structural integrity, density, thermal conductivity, moisture resistance, and visual quality [24]. Studies have reported on mechanical qualities [25,26], the impact of moisture [27], acoustical properties based on mycelial growth [28], fire resistance [29], and their biodegradability [30], and their aesthetic capacities [31], among several others.

One of the most significant challenges of using mycelium in large-scale structural applications is that it is an inherently weak material (0.1–0.2 MPa of compressive stress on average without mechanical compaction) and assumed to work best in compression. Despite this limitation, myco-materials are also very lightweight, giving them advantageous strength-to-weight ratios compared to concrete. This suggests that through advantageous material placement large-scale and even long-span structures are possible. In the last decade, several large-scale pavilion structures have demonstrated the potential of myco-materials to be used for building structures. An important distinction must be determined between those which use mycelium in a load-bearing capacity, and those which use the material as a surface or cladding application. Pavilions such as "Shell Mycelium" in India [32], the "Living Pavilion" in the Netherlands [33], and the pavilion at the Rensselaer Polytechnique Institute, Troy, NY, USA [34], used mycelium cladding panels or units over wooden frame structures. Ecovative used mycelium panels as the insulation of a tiny house [35]. While these serve as examples of the building-scale use of myco-materials, they are definitively non-structural applications. Curiously, there has been little diversity in approaches to building with myco-materials, with fabrication techniques used to assemble myco-structures remaining canonically familiar to architecture and engineering. These include logical adaptations of assembly systems with bricks or blocks, monolithic castings, 3D printing-based, and hybrid techniques, which are described below.

1.2. Brick and Block Myco-Structures

The most common approach is based on the production of bricks or blocks grown in custom-made molds, actively dried in ovens, transported to the site and assembled, typically with the assistance of a temporary formwork and scaffolding structures. An early structural application of myco-materials was the "Myco-tectural Alpha" [36], a small catenary barrel vault built from bricks grown from reishi. The largest, and perhaps most widely publicized mycelium structure was the "Hi-Fi" [37], a 40-foot tower installation by David Benjamin and The Living in 2014, engineered by ARUP. The mycelium bricks sourced from Ecovative were stacked atop of a wood and steel supporting structure. The "MycoTree" exhibited at the 2017 Seoul Biennale [38] demonstrated how the structural capacity of mycelium can be exploited maximally by placing it in compression-only configurations. In each previous example, the structures were formed with the assumption that the material would only work in compression, with dome/vault, tower, and column structural forms dominating the literature. The masonry units themselves were grown in plastic formworks. Three-dimensional printing techniques for myco-materials have also been explored, with much attention being paid to the formulae of viscous living pastes to be extrude with techniques adopted from digital ceramics [39]. Unit-based column structures have been demonstrated by teams in Europe at Lund University, Lund, Sweden [40], and by Blast

Studio, London, the UK [41]. Among the numerous exciting prospects of 3D printing myco-materials, a significant benefit is that custom-designed building units can be produced without needing a plastic formwork.

1.3. Monolithic and Bio-Welded Myco-Structures

Though much weaker and lighter than concrete, grow-in-place monolithic mycelium techniques can inherit many of the advantages (and challenges) of cast-in-place concrete techniques, including the use of traditional board, plank, sheeting, and flexible fabric formwork techniques. Without some means of aeration, beyond a certain thickness (150 mm or so), there is a risk that the fungi die prematurely from a lack of oxygen. Beyond assemblies of discrete element techniques, other research has focused on stereotomic approaches and monolithically growing large colonies of myco-materials in situ.

1.3.1. Monolithic Myco-Structures

Monolithic mycelium requires the design and fabrication of complex formworks that permit the fungi to fully grow. Due to such challenges associated with the cultivation of large volumes of live myco-materials and the constructing of formworks to facilitate such growth, very little work on monolithic mycelium has been accomplished in the context of architecture and structural design. In 2016, a master's of science thesis on civil engineering at Miami University, in Coral Gables, FL, USA [42] suggested analytical methods for mycelium-based monolithic domes, but did not validate them through physical means. At a small scale, Dutch artist Eric Klarenbeek demonstrated structural monolithic growth [43] in combination with 3D printing to create furniture. Ecovative experimented with monolithic mycelium and exhibited a chair in 2018 [44] that used a proprietary process that aerated the growing colonies of myco-materials, allowing them to be grown at greater thicknesses. A dissertation from the University of Newcastle in Newcastle upon Tyne, the UK, explored the potential of monolithic mycelium chair structures [45] grown in a conventional plastic formwork. Another interesting application of monolithic mycelium was a functional canoe [46] that was over 2 m long, grown by a student at Wayne State College in Wayne, NE, USA, in 2020.

Beyond these examples in product and furniture design, very few examples of architectural structures have been attempted. A series of three prototype structures was previously presented by the author of this paper [47], proving that grow-in-place monolithic mycelium structures were feasible through novel constructive approaches. Two arch structures (Figure 1) brought to light crucial considerations for successfully growing monolithic mycelium structures. First, the external formwork must be strong enough to support the weight of a wet substrate while maintaining its precise form, it must be composed of removable non-cellulose materials, and must be sufficiently porous to allow promoting the mycelia to breathe. Second, internal reinforcing strategies are advisable to handle eccentric loadings and formal accuracy, and must be composed of a cellulose-based material to permit the mycelia to bind and grow through the reinforcing structure.

A third prototype structure, called the Monolito Micelio (Figure 2), was an architectural-scale monolithic mycelium structure, grown in early 2018 from a one-ton colony of mycelium-stabilized hemp procured from Ecovative. The structure was designed and executed in the context of a graduate research seminar at the Georgia Tech School of Architecture. The vaulted pavilion was a critical response to the observed monotony of brick/block-based myco-fabrication methods and built upon the constructive principles of structures before it. The pavilion demonstrated that myco-materials could inherit fabrication logics from cast-in-place concrete techniques, including traditional board formwork and flexible fabric formwork techniques. Importantly, the structure showed that much more work was needed to uncover new and previously unimaginable construction logics that go beyond the architectural cannon of traditional materials.

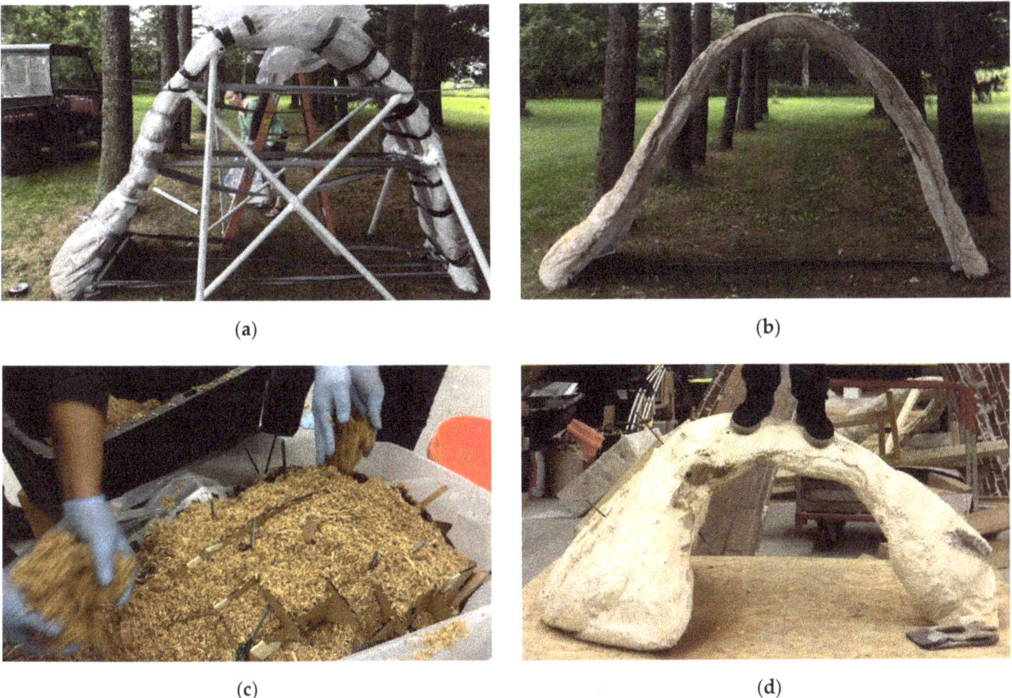

(a)

(b)

(c)

(d)

Figure 1. (**a**) Formwork for the "Mycoarch" composed of active bent PVC and plastic sheeting; (**b**) completed arch (late 2017, since renamed the "Diamond A Arch"), which collapsed due to inaccurate form and a myco-material matrix that had not sufficiently dried; (**c**) packing the internal reinforcing for the "Thick and Thin Arch" composed of recycled cardboard; (**d**) complete "Thick and Thin Arch" (early 2018) held seventy-five kilograms. Photos by the author.

The success of the project was also met with numerous failures, which provided the grounds for such a future inquiry. Notably, as part of a super-structure, myco-materials are highly susceptible to expansion and contraction in the face of external elements, making them unsuitable for external use, unless for temporary structures where the lifespan of the structure is understood to be short. Temperature swings and precipitation caused the material matrix of the Monolito Micelio to crack, decay, and become infested by other unfavorable organisms, including potentially dangerous mold (Figure 3). Furthermore, the materials used for the internal reinforcing system were much stronger and rigid than the myco-materials, which further exacerbated the cracking and decay of the structure.

While, in many regional contexts, there are minor active energy inputs needed to grow myco-materials, their reliance on plastics and molds that have limited reusability presents an ethical dilemma. For example, the plastic-lined plywood and woven nylon fabric formwork system used for the Monolito Micelio was a waste byproduct that resulted in land-fill disposal. The issue of formwork resulting in waste is an issue that has since been taken up by researchers interested in monolithic mycelium. A prototype structure by the multi-disciplinary collaboration in Europe called the FUNGAR project [48] provided early evidence that woven Kagome structures are an advantageous replacement for the polymeric in-situ formworks and molds typically needed to grow myco-materials. Such weaving crafts are globally ubiquitous, formally flexible, and often use natural lignocellulosic materials that are readily available. Such strong porous surfaces allow the fungi to breathe, provide a humid environment, and serve as a source of nutrition for the fungi. In contrast to plastic formworks, myco-weaves encourage mycelia to grow into the formwork and integrate into the biomass. More recently, the author of this paper grew a two-meter-tall monolithic mycelium column [49] along with students at Kansas State University that used basket weaving techniques. The woven formwork both participated in the visual expression of the column and potentially strengthened the assembly due to the deep bonds between the myco-materials and exoskeleton (Figure 4a).

(a) (b) (c)

Figure 2. The Monolito Micelio, grown in early 2018 with students at Georgia Institute of Technology. (**a**) Construction of the wooden internal reinforcing; (**b**) in a manner resembling cast-inplace concrete, mycelilum composite materials were processed on-site with water and nutritional additives and immediately packed into the plywood and geo-textile formwork; (**c**) finished structure, used as a stage and pavilion for a choir performance and exhibited at the School of Architecture. Photos by the author.

(a) (b)

Figure 3. Decay of the Monolito Micelio. (**a**) Cracking and decay of the structure after three months caused by expansion and contraction of the material matrix against the internal reinforcing structure; (**b**) cracking, decay, and infestation of the structure after four months. Photos by the author.

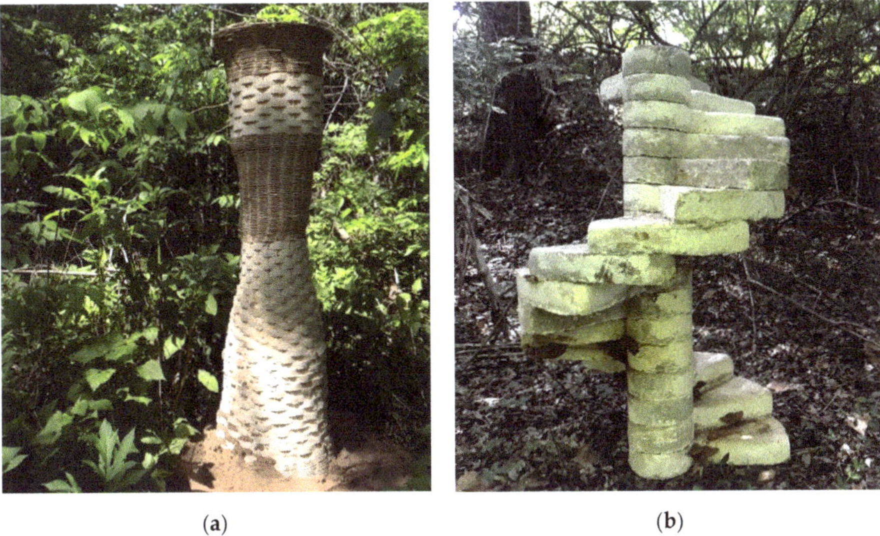

(a)   (b)

Figure 4. Monolithic and myco-welded structures grown by the author and students at Kansas State University in spring 2021, shown in their final installation sites. The structures were both larger than the available resources for actively drying the structures to stop growth, resulting in the emergence of fruiting bodies on the structures. (**a**) Two-meter-tall woven monolithic mycelium column; (**b**) half-scale myco-welded staircase with visible fruiting bodies that resulted from the two-stage growing process inherent to the myco-fabrication technique. Photos by the author.

1.3.2. Bio-Welded Myco-Structures

An increasingly popular technique called "bio-welding", or "myco-welding", involves assembling structures with discreet living parts and growing them together into monolithic wholes. Myco-welding is challenging because it requires two stages of growth. First, individual units are grown from loose inoculated substrates in molds. Second, assemblies of living units are kept in an intended formal configuration for several days, while maintaining necessary sanitary and environmental conditions. Drying and stopping the growth of large assemblies is also a challenge inherent to myco-welding large assemblies. If not completed quickly enough, fruiting bodies often grow on the structure (Figure 4b), which, depending on the application or context, may or may not be desirable. The technique has been demonstrated for small arch structures [50], furniture [51], for making monolithic blocks for use with robotic-controlled abrasive wire cutting [52], and a load-bearing half-scale spiral staircase recently grown by the author and their students [49]. At the large scale, the technique was demonstrated in the form of a triumphal arch at a short-term art installation in Europe [53].

1.4. Aims and Scope of This Research

The applied research described in this paper seeks to expand upon fabrication techniques using myco-materials, with the primary motivation being the excessive waste produced by contemporary construction practices. Among the numerous challenges and limitations associated with the application of myco-materials in architecture, this work focuses on overcoming (1) the challenge of cultivating large colonies of living myco-materials into precise forms and (2) the need for intuitive and re-usable formwork systems that reduce waste byproducts from growing and fabrication processes. The myco-fabrication strategies presented here were developed through the production of prototype structures that demonstrate growing large blocks of myco-materials and assembling them into efficiently formed wall structures. The prototypes share an underlying serpentine geometry deployed into assemblies that are categorically hybrids between monolithic and brick/block-based. One wall prototype demonstrates units created from myco-welded slabs, while the other revisits the in situ monolithic fabric forming of units that are repeated, stacked, and post-tensioned. Both structures were produced in academic contexts in collaboration with students from the University of Virginia (UVA) and Kansas State University (K-State), under the direction of the author. The prototypes were exhibited publicly in early 2022 at the Biomaterial Building Exposition (BBE) [54].

2. Context, Design, and Methods

The BBE gathered five teams of architect scholars from across the United States to develop and exhibit novel approaches for architectural-scale biomaterial research alongside students at their respective universities. There were three components to the BBE: a collaborative fabrication workshop with students in January 2022, full-scale installations outdoors on the UVA grounds, and an accompanying indoor gallery exhibition at the UVA School of Architecture. An opening symposium fostered discussion between the organizers and exhibitors on how renewable, carbon-sequestering biomaterials could be utilized in contemporary construction, while establishing a multi-institutional scholarly discourse that raised public awareness of novel biomaterial construction.

The UVA's academical village (now a UNESCO World Heritage Site) provided potent inspiration behind the geometry of the two prototype structures presented in this paper. Established and designed by Thomas Jefferson, a prominent feature of the grounds are the brick "serpentine" walls (Figure 5) enclosing the gardens behind each residence pavilion. The structures served as barriers to between the enslaved people and the white university community and to mask the use of slave labor visually and acoustically [55]. The history of serpentine walls at the UVA and their connection to slavery is inescapable. However, serpentine walls are not Jefferson's invention. Straight masonry walls, unless very thick or reinforced, cannot resist lateral loads [56]. Undulating walls can have a much wider footprint, which helps resist lateral loads and can be much thinner than straight walls. Therefore, the motivation to deploy serpentine wall technology for the BBE was to recontextualize the serpentine geometry from its connections to slavery. The prototypes intended to foreground that the inherent properties of myco-materials can through their flexibility, stability, and material efficiency, also act as a means of promoting environmental and human justice.

Figure 5. Serpentine walls designed by Thomas Jefferson and built by slave labor that enclose the gardens at the rear of the residences of the historical academical village at the University of Virginia located in Charlottesville, VA, USA. Photos by the author.

2.1. Parametric Design for Serpentine Walls

To facilitate the generation of an expansive and diverse family of undulating wall geometries, a custom computational design script was developed in Rhino/Grasshopper [57].

The script generated a range of three-dimensional forms for serpentine walls. The geometries were generated from a periodic base curve that informed later design decisions, including the generation of digital fabrication protocols for making the formworks. First, a curve was generated using design variables that included the number of control points that composed a V or U-shaped "unit", the length and width of the unit, whether the curve was generated with poly-lines or poly-curves, and how many units composed the length of the curve. Figure 6a shows three examples of such basic walls in top view. Next, the underlying three-dimensional geometries of the wall units were represented as a planer ruled surface. The geometries used for the two prototypes were generated from the base curve and its mirror, according to a specified height. The final design stages consisted of a set of thickening and geometrical extraction protocols (Figure 6b) that helped generate the formwork schema specific to the myco-fabrication technique being tested.

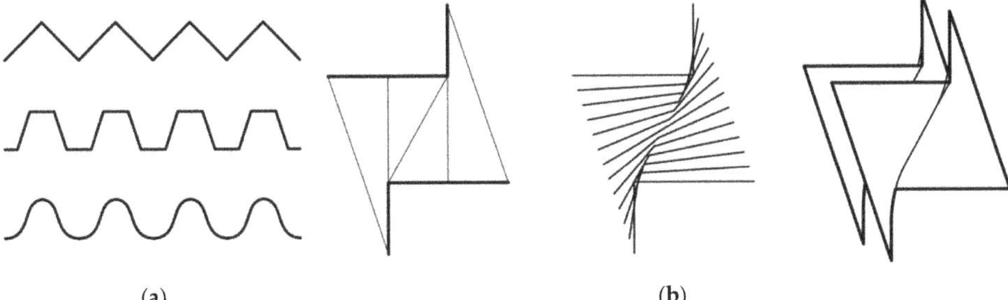

(a)   (b)

Figure 6. (a) Examples of poly-line- and poly-curve-based periodic curves generated with the parametric design script for designing serpentine walls; (b) three-dimensional extractions and transformations afforded by the script. Vertical slicing and thickening were both used to design formwork schemes and to estimate material volume requirements.

2.2. Summary of Myco-Fabrication Methods

The experimental structures grown for the biomaterial building exposition tested myco-welding and fabric-forming techniques for growing large monolithic blocks and assembling them into efficiently formed wall structures. Due to their inherent lightness, assemblies of large elements were not only possible, but also offered potential advantages over other previously demonstrated methods of building with myco-materials. Taking inspiration from pre-cast concrete traditions, the goal for both prototypes was to demonstrate re-usable formwork systems that produced large myco-material building components offsite in semi-controlled working conditions. The strategies intended to reduce the demand of long labor hours, reduce the risk of contaminating large colonies of myco-materials, and reduce uncertainty during on-site assembly. Numerous practical and contextual considerations had to be determined, which ultimately influenced the specific designs and techniques used to complete them. These considerations included if the structure was going to be exhibited indoors or outdoors, the location the structure's parts were going to be grown, the materials and fabrication resources on-hand for fabricating the formwork, how many students were available to contribute to the project, and if the author would be present for the various stages of growing and assembly. While the two structures shared a common underlying formal logic for undulating "serpentine" walls, the formal character and complexity of each serpentine wall prototype was intimately related to its respective method of myco-fabrication. The two structures were grown in two different geographic locations in the United States. At their core, these were academic projects, whereby the complexities in the form and technique had to remain accessible to UVA and K-State students both at undergraduate and graduate levels.

2.2.1. Myco-Welding Slabs into Monolithic Building Units

The structure exhibited outdoors on the grounds was intended to be cultivated and assembled locally by UVA students. As such, the scale of the structure, the complexity of its form, and the accessibility of the fabrication and growing techniques were precisely selected. As a base technique, myco-welding offered numerous advantages that better aligned with the number of students involved and how much time they could contribute. In devising the proposed methodology, a driving consideration was that most of the physical effort was during a week-long workshop with participating UVA students with design and engineering backgrounds. Myco-welding was advantageous in this context because the two phases of growth de-concentrated continuous labor hours needed for large in situ monolithic mycelium casting techniques.

The prototype wall structure, later named La Parete Fungina, was created from two wall units, each built from seventeen V-shaped slabs myco-welded into three "chunks". Nine different V-shaped formworks were needed (Figure 7a). Noting the labeling scheme in the figure, the palindromic sequence <a, b, c, d, e, f, g, h, i, h, g, f, e, d, c, b, a> described a complete wall unit, with eight of the forms being repeated in each unit. The formworks were intended to function as re-usable slip-molds to address the ethical dilemma of plastic or other non-cellulosic material being ubiquitously used in the production of myco-material objects. These formworks were intended to be simple to make, and because they had very limited contact with growing materials, they could be created from wood and fabricated with basic tools. For assembly on-site, a friction-based connection system (Figure 7b) was developed, so the structure could easily be disassembled when the Exposition was taken down. A unique byproduct of prolonged growth inherent to myco-welding was that it produced overgrowth: a thick layer of pure mycelium grew on the surfaces of the units. The overgrowth was like a layer of hydrophobic defense for the fungal colony, both while it was alive and after being dried and immobilized. Having such a performative benefit meant that myco-welding was the logical choice for a structure being exhibited outdoors.

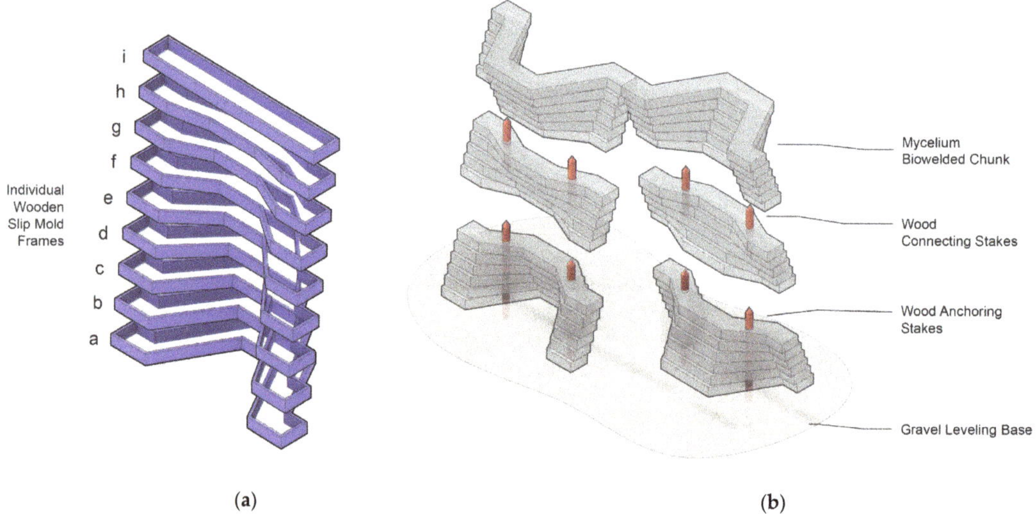

Figure 7. Design diagrams for La Parete Fungina. (**a**) Diagram of the nine different wooden slip-form frames (a–i) designed for growing slabs; (**b**) axonometric assembly diagram of the unit chunks highlighting the friction-based connection system of the wooden stakes. Drawings by Emmett Lockridge.

2.2.2. Fabric-Forming Monolithic Units

The structure exhibited indoors in the UVA's School of Architecture gallery was grown in spring 2022 at K-State in the context of a research seminar instructed by the author on myco-materials and myco-fabrication. In the introduction of the course, the graduate architecture students were immersed into the question of myco-fabrication techniques for large monolithic blocks assembled into efficiently formed wall structures. They were challenged to work collaboratively and contribute efforts toward an alternative expression of myco-fabrication. In contrast to the structure being grown at the UVA, the prototype later named L'Orso Fungino leveraged the lightweight properties of myco-materials using large monolithic elements that were cast in re-usable wood and fabric formwork. The complexity and fabrication methods chosen were tuned to the available resources, the skill levels of the students involved, and the short 6-week timeline for all the design and production. In situ monolithic mycelium casting techniques were deemed advantageous, because most of the physical effort available from participating students and research assistants in the lab was during a weekly four-hour session. The custom formwork apparatus for growing the wall units (Figure 8a) was designed to be quickly assembled, collapsible, and re-usable. Vertical perforated cardboard tubes were grown into the matrix of the units to provide air into the thickest parts of the colony during growth and to later serve as a conduit for a post-tension connection system (Figure 8b). The top and bottom surfaces of each unit were detailed such that there were interlocking adjacencies between the two units, the top compression plates, and the base. Due to the anticipated lightness of each unit, a post-tensioning system was a key feature of this prototype. It was hypothesized that loading the units in compression with cables running through the cardboard tubes would bring additional strength and stability to the assembly.

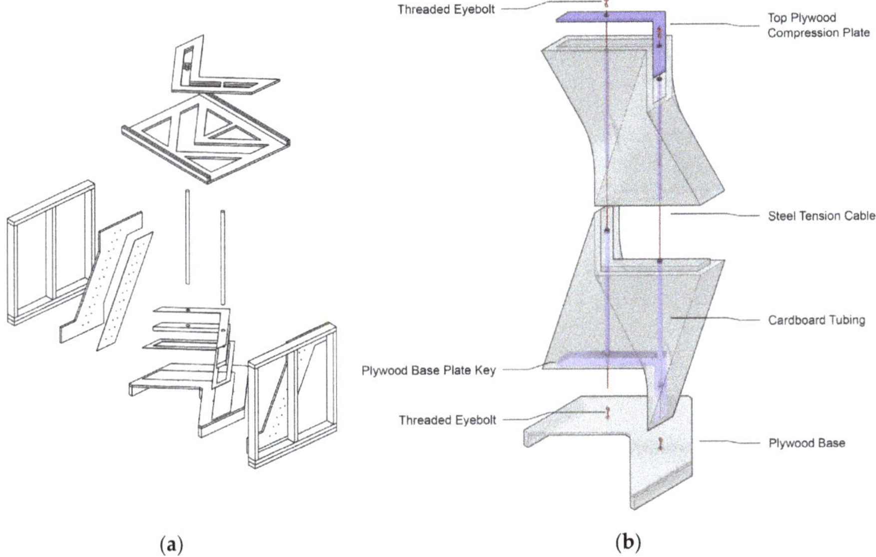

Figure 8. Design diagrams for L'Orso Funigno. (**a**) Exploded axonometric diagram of the formwork apparatus created to grow the monolithic wall units; (**b**) axonometric diagram of the monolithic wall units highlighting the post-tension system that applied a compressive force on the units with a mechanically tightened cable between plywood base and top plate. Drawings by Emmett Lockridge.

2.3. Materials

The myco-materials used for both prototypes presented below were procured from Ecovative [13] and paid for with funds provided by the exposition. Within the budget, each prototype structure could be grown from at most one pallet of myco-materials, weighing roughly 325 kg. For these prototypes, one pallet held sixty-five 5 kg bags or 0.6 cubic meters in total volume of wet living material. Ecovative's patented material was a hemp substrate inoculated with a fungus from the phylum Basidiomycota, whose fruiting bodies resembled the brackets produced by reishi.

The storage of these materials could have been a major challenge, because they had to be kept at approximately 4 °C to prevent the fungi from growing too quickly and fully consuming the substrate. Ideally, they should have been freight-shipped in refrigerated containers and if proper refrigerated storage was not available, immediately processed and packed into formworks. If kept unrefrigerated, the material would grow into a hardened mass in the bags within 3 to 4 days, making it labor- and time-intensive to break the hemp fibers apart. The structures grown at the UVA and K-State were grown in schools of architecture, which did not have access to large-scale refrigeration.

3. Results and Discussion: Two Serpentine Wall Prototypes

Each structure tested assembly strategies for growing large mycelium building units and assembling them into prototypes of efficiently formed serpentine wall prototypes. As a pair, they demonstrated the flexibility and facility of myco-materials to adapt to different approaches of fabrication based on the available tools, materials, and knowledge. La Parete Fungina demonstrated two undulating wall units standing side-by-side, each created from seventeen myco-welded slabs. L'Orso Fungino revisited the in-situ monolithic fabric forming of units that were repeated, stacked, and post-tensioned. While developing the two techniques, a major concern was to design the formwork systems to be re-usable. Consequently, the formal character and complexity of each structure were intimately related to their respective method of myco-fabrication. Both were assumed to be compression-bearing structures, even if they were not exhibited resisting external loads.

3.1. La Parete Fungina

For the structure grown by UVA students, pre-emptive planning took place two weeks before the workshop in January 2022. This included the development of the script described above in Section 2.1, hiring and coordinating with a research assistant at the UVA, and purchasing materials for the formwork. The five-day workshop was attended by undergraduate students from both engineering and design backgrounds during daily four-hour sessions. For the first three days of the workshop, the primary goals were fabricating the wooden formwork and creating a growing cart. Nine wooden formworks (Figure 9a) were hand-built from 17 mm unfinished whitewood boards, cut into 75 mm strips. Each slab was 1200 mm long end-to-end and had a common rectangular cross-section 150 mm wide and 70 mm thick. The underlying poly-line V-shape of this structure produced nine formworks whose forms lay between a V and a rectangle, required measuring several non-orthogonal cuts with varying angles. This was a minor technical challenge, but was time consuming and would have benefited from digital fabrication resources. Grow-space was limited too; no more than 3 m × 3 m under a staircase. A moveable growing cart (Figure 9b) was improvised using three heavy-duty wire shelves, plastic zip ties, and black plastic sheeting. The cart had five 1220 mm × 1370 mm shelves, resulting in approximately 8.3 square meters of growing surface.

The final two workdays during the workshop were used to form and start growing the slabs. First, the inoculated hemp substrate had to be broken up until the fibers were completely loose (Figure 9c). As it was being fiberized, 250 g of kitchen flour was mixed for each 5 kg bag of living substrate. The flour was recommended by the manufacturer as a nitrogen-rich nutrient to promote the rapid growth of fungal hyphae, but for this project, the recommended quantity was doubled. The intention was to have the flour act as a temporary binder while the mycelia formed their bonds between fibers. Water was added to the extent that when a handful of fibers were squeezed, only one drop of water was released. Once fully prepared, the loose fibers were compacted by hand into the formworks on the plastic lined shelves of the cart (Figure 9d), and the wood formwork could be carefully slipped off, and reused to form multiples of the same shape (Figure 9e). As a means of providing a clean and humid environment to each slab, they were covered in food-safe plastic film (Figure 9f). The cart was covered with a black plastic covering (Figure 9g) that kept the slabs in the dark while they grew, and, more importantly, provided a second means of keeping a clean and humid growing environment. Approximately one-third of the bags were used with two days of delivery to grow a first round of slabs. During the first week-long grow period, the students stored the remaining bags in a covered outdoor space, stacked on shelves and wrapped in black plastic sheeting. This was the best option due to the lack of access to large-scale cold storage. The bags encountered temperature swings between roughly −2 and 18 °C between day and night.

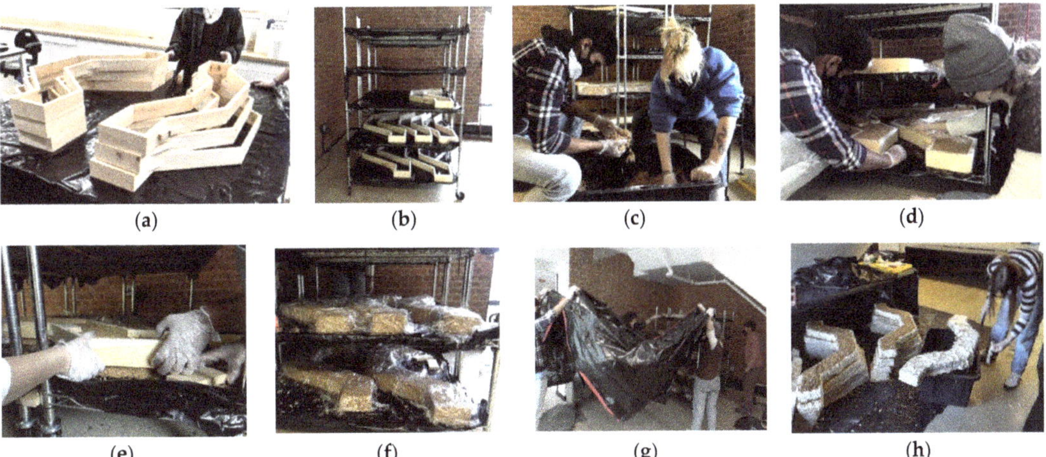

Figure 9. Preparation and process for forming and growing myco-material slabs: (**a**) re-usable wood formworks created in the first days of the workshop; (**b**) growing cart in its space under a staircase in the school of architecture; (**c**) fiberizing the living hemp substrate and mixing in additives prior to packing the formwork; (**d**) hand packing the wood formwork directly on the grow cart; (**e**) removing the mold for reuse; (**f**) slabs individually wrapped in food-safe plastic to keep the fibers humid and warm; (**g**) plastic "cloak" which covered the entire grow cart to keep growing specimens dark warm and humid; (**h**) myco-welding slabs with loose inoculated substrate as mortar. Photos by the author and Leila Ehtesham.

In the weeks that followed the workshop, the UVA research assistant and one of the participants continued to form and grow the remaining slabs while also myco-welding the slabs that were sufficiently cultivated. The living slabs were stacked into "chunks" between five and six layers thick, with loose substrate in between to level the assembly (Figure 9h). The assembly of living parts had to be completed in rigorously clean conditions, while keeping the assembly in the correct and intended configuration. While the slabs were being stacked, they were gelatinous and fragile and had to be handled with care by at least two people at a time. Furthermore, while slabs bonded and grew together, they needed to be kept in an appropriately clean, dark, warm, and humid environment. The wall chunks were grown into monolithic-like masses for roughly two weeks, after which they were passively dried until installation.

At the time of assembly, all the myco-welded chucks had at minimum one week of drying time. The on-site installation of La Parete Fungina was completed in approximately one hour. The six wall chunks were driven to the site in a small passenger van. Wooden anchoring stakes (40 mm diameter) were driven into the ground and the base chunk was friction-fitted in place (Figure 10a). The remaining chunks were dry stacked (Figure 10b) each with similar wooden stakes in between. The simple friction-fitting system was ideally suited for this application, because the exhibition was temporary and needed to be taken down after a period of three months with minimal impact or damage to the UVA grounds.

The two undulating units (Figure 10c) were approximately 1200 mm tall, configured in a manner such that the serpentine wall geometry produced a gap in the wall. The myco-welded objects were highly didactic due to their long growing time shown through artifacts such as changes in color to the formation of fruiting bodies. In the days following the installation, the structure was subjected to wind and snow, which did not cause a collapse, nor was the material compromised. The thick layer of overgrowth demonstrated its inherent resilience that would be well suited to its exhibition outdoors for roughly two months, after which it was taken down.

(a)   (b)   (c)

Figure 10. Installation of the prototype onsite: (**a**) friction fitting base chunk to the anchoring stakes in the ground; (**b**) friction fitting upper chunk of the myco-welded wall; (**c**) complete structure as it was exhibited at the biomaterials building exposition. Photos by the author and Leila Ehtesham.

3.2. L'Orso Fungino

For the structure grown at K-State, pre-emptive planning took place for two weeks following the week-long workshop at the UVA. With the change in strategy for growing the wall chunks, notable adjustments to the scale and geometry were determined. An important driver was that the structure needed to be shipped 1800 km from Manhattan, Kansas, to Charlottesville, Virginia. Within the budget, two pallets could be sent (1220 mm × 1016 mm in area for each). Sized according to the freight limitations, wall units were designed, each approximately 750 mm long and 750 mm tall. The prototypes intended to demonstrate the wall units was vertically stacked in twos. The formwork was designed to be quickly assembled, collapsible, and re-usable. The rigid portion of the apparatus (Figure 11a) was created with a combination of hand-cut nominal timber frames, CNC-cut plywood panels, and 3 mm plastic laminations for surfaces in direct contact with living materials. The hand-stretched fabric portions (Figure 11b,c) were composed of breathable synthetic geotextile. Several formworks were fabricated so that multiple units could be grown simultaneously.

The first fabric formwork apparatuses were packed roughly one month before the opening of the exposition. Within the time constraints, two units could be packed by six people. As a safeguard from potential failures, additional units were accounted for in the materials budget. Due to a limited supply of materials, it was decided that non-sterilized and non-inoculated hemp fibers would be mixed in with the inoculated substrate to increase the yield volume. Several previous experiments in the MycoMatters Laboratory successfully propagated Ecovative materials into ratios of up to one part inoculated to four parts non-inoculated and non-sterilized hemp fibers (1:4) by volume. The fabric formworks were packed with a 1:2 ratio to increase the volume with less risk. In addition to the hemp, 250 g of kitchen flour per 5 kg bag of living material and water was added such as above.

(a)   (b)   (c)

Figure 11. Fabric formwork apparatuses: (**a**) rigid elements of the apparatus composed of wood and plastic laminate sheet where there would be contact with living materials; (**b**) upholstered formwork with black synthetic geotextile fabric; (**c**) top of the formwork apparatus showing the geotextile upholstered into the rigid frame. Photos by the author.

Within a four-hour work period, two formworks were filled with inoculated substrate (Figure 12a). The formworks were packed monolithically by hand and with the help of tools to compress the material around the perforated cardboard tubes (Figure 12b). Each unit used between nine and ten bags of pre-inoculated material due to overpacking, which caused the fabric to stretch, ultimately requiring more material to fill the formwork (Figure 12c). The formwork was then covered with a black plastic covering that kept the material humid and dark while the mycelium grew (Figure 13a). After the first two units were packed, a third was packed a day later. During the growth period of the first three units, approximately fifteen bags could be stored in the lab refrigerator. The remaining twenty bags (approximately) had to be kept on a pallet in the lab. After only four days, mycelium from the first two units had already grown through the stretched fabric (Figure 13b). The formwork of the first two units was removed (Figure 13c), which meant they would have two full weeks to passively dry in the lab before being shipped. The third unit became contaminated (Figure 13d) deep in the monolithic colony, despite that surface mycelium managing to grow in many areas. This suggested that a contaminant from the non-sterilized hemp and a lack of air were probably contributing causes. The first two formwork apparatuses were re-assembled and re-used to grow two more units. Of those two, one more unit became contaminated. The remaining three fabric-formed monolithic units and a wooden formwork apparatus were palletized (Figure 14a) and shipped to the UVA.

The on-site installation of L'Orso Fungino was completed in approximately two hours. The assembly of the prototype began by stringing cables through cardboard conduits of the lower wall unit (Figure 14b). The steel cables were mechanically fastened to the plywood base compression plate. Next, the cables were strung through the top wall unit (Figure 14c), this time with more difficultly because one of the cardboard tubes bent during the packing process. None of the units was fully dry which was an advantage because the cable could be forcefully pushed through the spongy mycelial matrix. Cables were mechanically fastened to threaded eyebolts. The stack of two units (Figure 14d) had a top plate that was put in compression by the tightening of each eyebolt against a washer (Figure 15a). The assembly was allowed to compress by one centimeter from post-tensioning. L'Orso Fungino was exhibited in the gallery at the School of Architecture at the UVA, alongside the third wall unit and the formwork apparatus (Figure 15b,c). The post-tension system was successful as a technique for stabilizing myco-structures. The connection details are in need of future iterations. A larger open question was how such post-tensioned wall structures would support the load of a vault, a truss, or a beam. In its current state, the post-tensioning system was simple to disassemble at the end of the exposition.

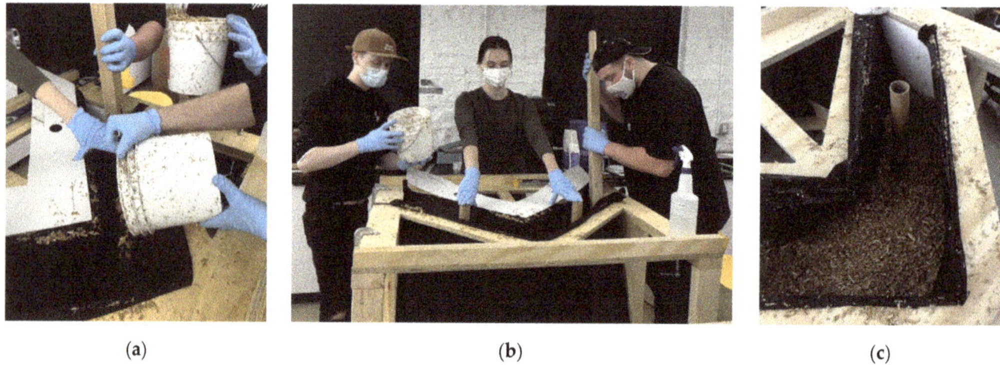

(a)　　　　　　　　　　　(b)　　　　　　　　　　　(c)

Figure 12. Packing the fabric formwork apparatus with living myco-materials: (**a**) loose inoculated substrate was poured into the apparatus; (**b**) material was compressed into the form while minding the cardboard conduits; (**c**) apparatus almost full of inoculated substrate. Photos by the author.

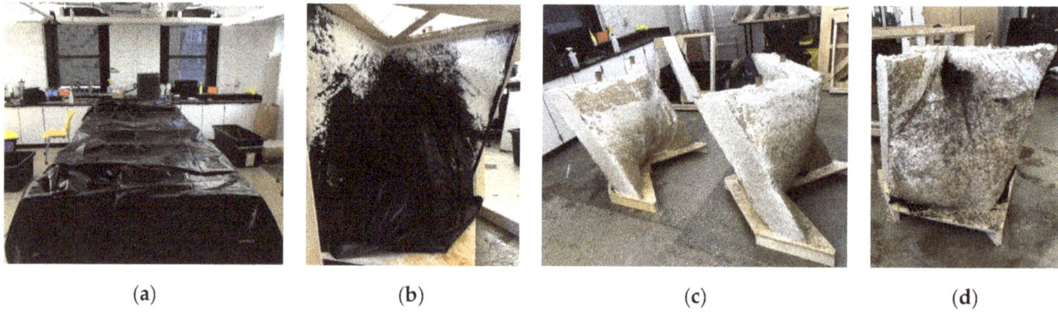

(a)　　　　　　　(b)　　　　　　　(c)　　　　　　　(d)

Figure 13. Growing fabric-formed monolithic wall units: (**a**) black plastic covered the formworks as the material grew in the MycoMatters Laboratory; (**b**) after four days of growth, healthy mycelium was found growing through the fabric; (**c**) first two fabric-formed monolithic units with all formworks removed after growing four days; (**d**) third unit found with a contamination compromising mycelial growth deep into the unit. Photos by the author.

(a)　　　　　　　(b)　　　　　　　(c)　　　　　　　(d)

Figure 14. Assembling the prototype: (**a**) palletized units and formwork apparatus before shipping to the UVA; (**b**) stringing cables from base plate through lower wall unit; (**c**) stacking two wall units while pulling tension cables through the assembly; (**d**) stacked undulating wall units before post-tensioning. Photos by the author and Leila Ehtesham.

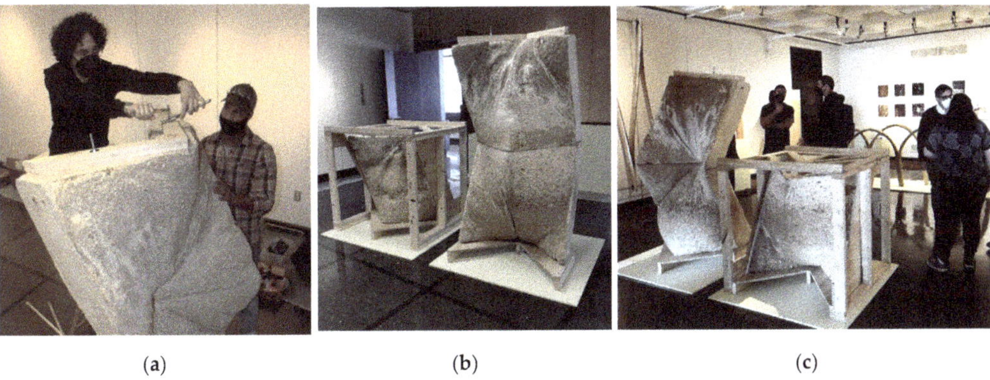

(a)   (b)   (c)

Figure 15. Post-tensioning and exhibiting the prototype: (**a**) tightening the internal cables to add compressive force to the assembly; (**b**) gallery installation with stacked units and wood formwork; (**c**) wall prototype during exhibition opening at the UVA. Photos by the author and Leila Ehtesham.

3.3. Discussion and Future Work

La Parete Fungina and L'Orso Fungino both served as demonstrations of wall assembly systems that challenged the status quo of myco-fabrication. Myco-welding and fabric-forming techniques were tested for their capacity to create complex yet efficiently formed wall structures from large building units grown in re-usable formwork systems. The structures initiated a new dialog of architecture-scale myco-fabrication techniques that were positioned between those which used building units the size of a brick and those which used units the size of a room. They demonstrated the flexibility and ease with which myco-materials could adapt based on available tools, materials, and knowledge.

Working in educational contexts, the strategies generated material knowledge directly through the production of physical artifacts. For novice student collaborators, this approach was productive toward fostering their appreciation and mastery of building material assemblies through the technical lens of myco-materials. Furthermore, students had the freedom to exercise their creativity and experience working at full-scale through experiments that left room for improvised adjustments. Insights on the craft of growing myco-structures were never assumed, were developed directly in collaboration with the students, and generated through creating. The students learned first-hand that building (with any material) is challenging, but also joyful and rewarding. Thus, the impacts of the methods presented below were both technological and educational. Myco-materials were challenging for students because they required greater care and attention compared to common, inert materials. For example, compared to handling live materials, personal protection practices, including wearing masks and gloves and thoroughly cleaning all working surfaces and formwork, are not an inherent protocol for a typical design or engineering student. Lessons learned through minor frictions, failures, and contaminations were vital to provoke important questions about the material ethics and the appropriateness of the methods.

Rather than demonstrate methods ready to grow buildings entirely from myco-materials the prototypes suggested new hybrid techniques that could contribute to the broader cannon of myco-fabrication; adding to well-established discrete element, 3D printing, and monolithic techniques. The prototypes detailed above operated between the scales of a brick and a monolithic pavilion. At the scale of a house (for example), a building structure grown from myco-materials would likely require multiple prefabricated units or "chunks" that would be assembled on-site. Without commercial-scale growing resources, there are practical and biological limits to the scales of colonies one can grow. The Monolito Micelio [47] was 2.5 m × 2.5 m × 2.5 m, and its scale presented several notable disadvantages, including the significant demand for time and labor, the risk of handling such significant quantities of living material in uncontrolled environments, the need for pliable internal reinforcing, and the challenge of drying large colonies to stop growth without large ovens. While there were few active energy inputs needed to grow the myco-materials, the near-ubiquitous plastic formworks in which they were grown presented an ethical dilemma. The plastic-lined plywood and woven nylon fabric formwork system that formed the Monolito Micelio were a waste byproduct that resulted in landfill disposal. However, if a sufficient volume of living myco-materials was available and there was enough funding to hire labor for such a project, most of the formwork could have been reused to create several pavilion-scale units that could be aggregated to form larger spaces. This suggested that the techniques demonstrated by the Monolito Micelio could be scaled to grow parts two to four times larger. The goal for myco-fabrication in architecture should be to develop the capacity to grow parts at the same scale as current glue-laminated and cross-laminated timber manufacturing.

Experimentation with myco-fabrication at architectural scale is still nascent as a field and faces massive challenges that need to be overcome to achieve results other than prototype structures and fanciful pavilions. Limitations for building-scale deployments of myco-materials are foremost caused by the challenge of supply and access to commercial quantities of material. Currently, the lack of myco-material production infrastructure makes it both energetically and financially costly to transport wet, living mycelium composites over long distances. Ecovative, which is based in Troy, NY, has pioneered the scaling-up of myco-material production by closely studying and collaborating with commercial mushroom farming industries, especially in the neighboring state of Pennsylvania, where over 60% of all mushrooms in the United States are grown [58]. As a result, they offer a biotechnology that grows quickly and reliably under favorable environmental conditions and with some resilience to contamination. The ability to grow mycelium entirely through the lignocellulosic substrate within just a few days is a major advantage that makes growing structures a relatively fast process. However, such speed and reliability come with costs that can be a barrier to using myco-materials at a large scale. Beyond the cost of the material itself, the need to hire expedited refrigerated transport and have access to large-scale refrigerated storage are other potential barriers.

Radical approaches to sustainable construction raise questions of material ethics when evaluating what and how much was wasted in different approaches to myco-fabrication. Although in many regional contexts myco-material technologies have the capacity to require less energy than petrochemical foams and plastics they seek to replace, there are critical ethical questions that must be considered. In addition to the near-ubiquitous use of plastic and issues of waste-producing formworks, transporting myco-materials long distances with petroleum-consuming vehicles should be scrutinized. For the prototypes presented here, live myco-materials were procured from Ecovative's spawn supplier in Pennsylvania and transported either to the UVA or K-State. For both, it was cost-prohibitive to hire a refrigerated truck to deliver from the spawn supplier directly to the university. Thus, more improvisatory and self-motivated methods were needed. For the UVA-grown structure, shipping delays were going to have the material arrive after the workshop was over. To keep the project on schedule, the author drove a 750 km round-trip to the spawn supplier. For the structure grown at K-State, creative logistical planning was required to organize a two-step relay delivery. For a pro-rated cost, the pallet was first "hitch-hiked" 1700 km with a scheduled refrigerated shipment of button mushrooms spawned on a farm in the neighboring state of Oklahoma. Research assistants in the MycoMatters Laboratory then drove an 800 km roundtrip to bring the pallet back from the farm. It would be ideal for the distribution of myco-materials to exist in a model where they are regionally produced as a way of reducing transportation distances. More immediately, the opportunity to overlay the myco-material demand with existing food supply chain maps from existing spawn suppliers presents an opportunity to cultivate deep bonds between agriculture and biomaterial industries.

For myco-materials to succeed in the future, they must remain in the current dialog, both in academic and professional contexts. Collaborations across the fields of design, engineering, biological sciences, agriculture, and economics (for example) are essential to developing new knowledge essential for the scaling of new material technologies in ways relevant to building construction. Importantly, knowledge about growing structures must be as commonly accessible such as knowledge needed for building with wood, steel, or concrete. The craft of growing myco-structures is clearly in need of time to mature, but will only do so through continued study and experimentation. Future work should seek cooperative logics between fungal growth, computational design, and digital fabrication to further discover constructive possibilities with myco-materials. For example, digital reference technologies, such as 3D-scanning and augmented reality, may be important components of future work on myco-welding. The ability to grow twine and other natural-fiber textiles into the material matrix suggests that through computational design and analysis, the strategic placement of such reinforcements could be deployed to selectively strengthen and enhance myco-materials. Such advancements could help fully integrate forming materials into the building components being grown. Pre-stressing and post-tensioning have probable futures in myco-fabrication for certain structural scenarios. Robotic fiber winding and CNC knitting are two technologies that have been widely demonstrated and could be immediately applied in the context of myco-fabrication. The raising popularity of "co-bots" also suggests a future in which machines could collaborate with craftsmen and carry out improvisational tactics with greater precision, reduced demand for labor, and potentially much safer and cleaner fabrication conditions.

4. Summary and Conclusions

There is an urgent need for low-energy and renewable building materials that divert building and demolition waste from landfills and lessen the impact of the construction industry on climate change. The ability to rapidly grow building structures from myco-materials, particularly for short-term or temporary functions, has the potential to greatly reduce building and demolition waste. This paper provided an extensive overview of the state-of-the-art in deploying myco-materials at the architectural scale, highlighted the numerous case studies of researchers in diverse global contexts growing building-scale structural parts and pavilions, and gave first-hand insight about the significant challenges and limitations associated with the application of myco-materials for architectural structures. The applied research presented here developed hybrid myco-fabrication techniques that overcame the challenge of cultivating large colonies of living myco-materials into precise forms, and demonstrated intuitive and re-usable formwork systems that reduced waste byproducts from growing and fabrication processes. The techniques were developed in academic environments that gave young designers and engineers the access, space, and resources for working with myco-materials. The two prototype wall structures demonstrated the ability to grow large and complex shapes outside of rigorously controlled biolab environments, and with fewer risks than monolithic structures grown in situ. The lightweight properties of mycelium composites were an advantage in this context, where large, complex building components could be pre-grown and pre-dried off-site in semi-controlled environments and assembled with less continuous on-site labor compared to the production of brick/block or monolithic mycelium structures.

Realistically, the greatest potential for these techniques is in applications that replace EPS and other varieties of foam and insulation materials for insulated concrete formworks, large-scale acoustical arrays, temporary self-supporting structures, interior furnishings, scenography or theatre stage projects, and others, leveraging the inherent absorptive, insulative, and fire-resistant properties of myco-materials. Whether these techniques are applicable to load-bearing building structures is still an open question that demands further research. Nonetheless, large-scale building and long-span myco-material structures continue to gain interest in trends of research and commercialization that seek to vastly offset the impact of the short lifespans of buildings in the modern era.

Funding: The 2022 biomaterial building exposition was hosted by the University of Virginia School of Architecture and funded by The Jefferson Trust and the Center for Global Inquiry & Innovation. Additional funding for student research assistants was provided through the Kansas State University Global Food Systems Seed Grant Program during the 2021–2022 funding period.

Acknowledgments: The work presented here would not have been possible without the research assistants Leila Ehtesham, Annabelle Woodcock, Holly Ellis, and Emmett Lockridge. Thanks to the students in the UVA workshop: Josh Cauthen, Abby Hassell, Mak Johansen, Jacob McLaughlin, and Ryan Naddoni. Additionally, thanks to the students in the MycoMatters seminar: Matthew Fuller, Eduardo Granillo, and Joey Jacobs. Additional thanks to Joe Hornung, Richard Thompson, Lisa Schubert, Katie Kingery-Page, and Dean Tim de Noble from the College of APDesign at Kansas State University for their belief in and support of this research.

Conflicts of Interest: The author declares no conflict of interest.

References

1. Liu, G.; Xu, K.; Chang, X.; Chang, G. Factors influencing the service lifespan of buildings: An improved hedonic model. *Habitat Int.* **2014**, *43*, 274–282. [CrossRef]
2. Wuyts, W.; Miatto, A.; Sedlitzky, R.; Tanikawa, H. Extending or ending the life of residential buildings in Japan: A social circular economy approach to the problem of short-lived constructions. *J. Clean. Prod.* **2019**, *231*, 660–670. [CrossRef]
3. Environmental Protection Agency. Sustainable Management of Construction and Demolition Waste. 2018. Available online: https://www.epa.gov/smm/sustainable-management-construction-and-demolition-materials (accessed on 15 June 2022).
4. Trubiano, F.; Onbargi, C.; Finaldi, A.; Whitlock, Z. *Fossil Fuels, The Building Industry, and Human Health*; Kleinman Center for Energy Policy: Philadelphia, PA, USA, 2019; Available online: https://kleinmanenergy.upenn.edu/research/publications/fossil-fuels-the-building-industry-and-human-health-evaluating-toxicity-in-architectural-plastics/ (accessed on 21 August 2022).
5. National Institute of Environmental Health Science. *Endocrine Disruptors*; National Institutes of Health: Washington, DC, USA, 2010. Available online: https://www.niehs.nih.gov/health/materials/endocrine_disruptors_508.pdf (accessed on 21 August 2022).
6. Stichnothe, H.; Azapagic, A. Life cycle assessment of recycling PVC window frames. *Resour. Conserv. Recycl.* **2013**, *71*, 40–47. [CrossRef]
7. Marten, B.; Hicks, A. Expanded Polystyrene Life Cycle Analysis Literature Review: An Analysis for Different Disposal Scenarios. *Sustainability* **2018**, *11*, 29–35. [CrossRef]
8. Environmental Protection Agency. Municipal Solid Waste Generation, Recycling, and Disposal in the United States. 2014. Available online: https://www.epa.gov/sites/default/files/2015-09/documents/2012_msw_dat_tbls.pdf (accessed on 21 August 2022).
9. Gray, A. Germany Recycles More Than Any Other Country. World Economic Forum. 2017. Available online: https://www.weforum.org/agenda/2017/12/germany-recycles-more-than-any-other-country/ (accessed on 21 August 2022).
10. Geyer, R.; Jambeck, J.; Law, K.L. Production, use, and fate of all plastics ever made. *Sci. Adv.* **2017**, *3*, e1700782. [CrossRef] [PubMed]
11. McDaniel, V. How the Pandemic Drove up The Cost of Wood Products. U.S. Department of Agriculture. 13 May 2022. Available online: https://www.fs.usda.gov/features/how-pandemic-drove-cost-wood-products (accessed on 21 August 2022).
12. Stamets, P. *Mycelium Running: How Mushrooms Can Help Save the World*; 10 Speed Press: Berkeley, CA, USA, 2005.
13. Ecovative Design. Available online: https://ecovativedesign.com/ (accessed on 15 June 2022).
14. Mycoworks. Available online: https://www.mycoworks.com/ (accessed on 15 June 2022).
15. Mogu.bio. Available online: https://mogu.bio/ (accessed on 15 June 2022).
16. Holt, G.A.; McIntyre, G.; Flagg, D.; Bayer, E.; Wanjura, J.D.; Pelletier, M.G. Fungal Mycelium and Cotton Plant Materials in the Manufacture of Biodegradable Molded Packaging Material: Evaluation Study of Select Blends of Cotton Byproducts. *J. Biobased Mater. Bioenergy* **2012**, *6*, 431–439. [CrossRef]
17. Mushroom Packaging. Available online: https://mushroompackaging.com/ (accessed on 15 June 2022).
18. Magical Mushroom Company. Available online: https://magicalmushroom.com/ (accessed on 15 June 2022).
19. Grown.Bio. Available online: https://www.grown.bio/shop/ (accessed on 15 June 2022).
20. Forager. Available online: https://forager.bio/ (accessed on 15 June 2022).
21. Bolt Threads. Available online: https://boltthreads.com/ (accessed on 15 June 2022).
22. Gandia, A.; van den Brandhoff, J.G.; Appels, F.V.W.; Jones, M.P. Flexible Fungal Materials: Shaping the Future. *Trends Biotechnol.* **2021**, *39*, 1321–1331. [CrossRef] [PubMed]
23. Bayer, E.; McIntyre, G. Method for Producing Grown Materials and Products Made Thereby. U.S. Patent 9485917B2, 8 November 2016. Available online: https://patents.google.com/patent/US9485917B2/en?oq=US9485917 (accessed on 15 June 2022).
24. Elsacker, E.; Vandelook, S.; Brancart, J.; Peeters, E.; De Laet, L. Mechanical, physical, and chemical characterization of mycelium-based composites with different types of lignocellulosic substrates. *PLoS ONE* **2019**, *14*, e0213954. [CrossRef] [PubMed]
25. Jones, M.; Mautner, A.; Luenco, S.; Bismark, A.; John, S. Engineered mycelium composite construction materials from fungal biorefineries: A critical review. *Mater. Des.* **2020**, *187*, 108397. [CrossRef]
26. Girometta, C.; Picco, A.M.; Baiguera, R.M.; Dondi, D.; Babbini, S.; Cartabia, M.; Pellegrini, M.; Savino, E. Physico-Mechanical and Thermodynamic Properties of Mycelium-Based Bio-composites: A Review. *Sustainability* **2019**, *11*, 281. [CrossRef]

27. Appels, F.V.W.; Camere, S.; Montaliti, M.; Karana, E.; Jansen, K.M.B.; Dijksterhuis, J.; Krijgsheld, P.; Wösten, H.A.B. Fabrication factors influencing mechanical, moisture- and water-related properties of mycelium-based composites. *Mater. Des.* **2019**, *161*, 64–71. [CrossRef]

28. Hsu, T.; Dessi-Olive, J. A design framework for absorption and diffusion panels with sustainable materials. In Proceedings of the 2021 Inter-Noise Conference, Washington, DC, USA, 1 August 2021.

29. Jones, M.; Bhat, T.; Kandare, E.; Thomas, A.; Joseph, P.; Dekiwadia, C.; Yuen, R.; John, S.; Ma, J.; Wang, C. Thermal Degradation and Fire Properties of Fungal Mycelium and Mycelium—Biomass Composite Materials. *Sci. Rep.* **2018**, *8*, 17583. [CrossRef] [PubMed]

30. Van Wylick, A.; Elsacker, E.; Yap, L.L.; Peeters, E.; de Laet, L. Mycelium Composites and their Biodegradability: An Exploration on the Disintegration of Mycelium-Based Materials in Soil. In *Construction Technologies and Architecture*, 4th ed.; International Conference on Bio-Based Building Materials; Trans Tech Publications Ltd.: Zurich, Switzerland, 2022.

31. Sydor, M.; Bonenberg, A.; Doczekalska, B.; Cofta, G. Mycelium-Based Composites in Art, Architecture, and Interior Design: A Review. *Polymers* **2022**, *14*, 145. [CrossRef] [PubMed]

32. Zeitoun, L. Shell Mycelium: Exploring Fungus Growth as a Possible Building Block. *Designboom*. 25 July 2017. Available online: https://www.designboom.com/architecture/shell-mycelium-degradation-movement-manifesto-07-25-2017/ (accessed on 15 June 2022).

33. The Growing Pavilion. Available online: https://thegrowingpavilion.com/ (accessed on 15 June 2022).

34. Rensselaer Polytechnic Institute Summer Architecture Studio, Mycelium Pavilion. 2019. Available online: https://www.arch.rpi.edu/2019/09/2019su-summerstudio/ (accessed on 15 June 2022).

35. Mushroom Tiny House. Available online: https://mushroomtinyhouse.com/ (accessed on 15 June 2022).

36. Mok, K. Mycotecture: Building with Mushrooms? Inventor Says Yes. *Treehugger*. 26 September 2012. Available online: https://www.treehugger.com/mycotecture-mushroom-bricks-philip-ross-4857225 (accessed on 15 June 2022).

37. Saporta, S.; Yang, F.; Clark, M. Design and delivery of structural material innovations. *Struct. Congr.* **2015**, *2015*, 1253–1265.

38. Heisel, F.; Lee, J.; Schlesier, K.; Rippmann, M.; Saeidi, N.; Javadian, A.; Nugroho, A.R.; Van Mele, T.; Block, P.; Hebel, D.E. Design, Cultivation and Application of Load-Bearing Mycelium Components: The MycoTree at the 2017 Seoul Biennale of Architecture and Urbanism. *Int. J. Sustain. Energy Dev. (IJSED)* **2018**, *6*, 296–303. [CrossRef]

39. Soh, E.; Chew, Z.Y.; Saeidi, N.; Javadian, A.; Hebel, D.; Le Ferrand, H. Development of an extrudable paste to build mycelium-bound composites. *Mater. Des.* **2020**, *195*, 109058. [CrossRef]

40. Goidea, A.; Floudas, D.; Andreen, D. Pulp Faction: 3d printed material assemblies through microbial biotransformation. In *Fabricate 2020*; UCL Press: London, UK, 2020.

41. Blast Studio. Available online: https://www.blast-studio.com/ (accessed on 15 June 2022).

42. Baricci, R.A. Structural Analysis and Form-Finding of Mycelium-Based Monolithic Domes. Master's Thesis, University of Miami, Coral Gables, FL, USA, 2016.

43. Pallister, J. 3D-Printed Mushroom Roots Could Be Used to Build Houses. *Dezeen*. 6 March 2014. Available online: https://www.dezeen.com/2014/03/06/movie-eric-klarenbeek-mushroom-roots-fungus-3d-printed-chair/ (accessed on 15 June 2022).

44. Mycelium Chair by Ecovative at Biodesign: From Inspiration to Integration, an Exhibition Curated in Collaboration with William Myers at the Road Island School of Design (RISD). Available online: https://naturelab.risd.edu/events/biodesign-from-inspiration-to-integration/ (accessed on 15 June 2022).

45. Piórecka, N. MYCOsella: Growing the Mycelium Chair. Bachelor's Thesis in Architecture Dissertation, Newcastle University, Newcastle upon Tyne, UK, 2019. Available online: https://issuu.com/nataliapiorecka/docs/dissertation_project_ba_architectur (accessed on 15 June 2022).

46. Kuta, S. Is Fungus the Answer to Climate Change? Student Who Grew a Mushroom Canoe Says Yes. Available online: https://www.nbcnews.com/news/us-news/fungus-answer-climate-change-student-who-grew-mushroom-canoe-says-n1185401 (accessed on 15 June 2022).

47. Dessi-Olive, J. Monolithic Mycelium: Growing vault structures. In Proceedings of the International Conference on Non-Conventional Materials and Technologies (NOCMAT), Nairobi, Kenya, 24–26 July 2019; pp. 2–15.

48. Adamatzky, A.; Ayres, P.; Belotti, G.; Wösten, H. Fungal Architecture. *Int. J. Unconv. Comput.* **2019**, *14*, 397–441.

49. Dessi-Olive, J. Craft and structural innovation of mycelium-structures in architectural education. In *Structures and Architecture A Viable Urban Perspective?* 1st ed.; Hvejsel, M.F., Cruz, P.J.S., Eds.; CRC Press: London, UK, 2022. [CrossRef]

50. Modanloo, B.; Ghazvinian, A.; Matini, M.; Andaroodi, E. Tilted Arch; Implementation of Additive Manufacturing and Bio-welding of Mycelium-Based Composites. *Biomimetics* **2021**, *6*, 68. [CrossRef] [PubMed]

51. Dahmen, J. Soft Futures: Mushrooms and Regenerative Design. *J. Archit. Educ.* **2017**, *71*, 57–64. [CrossRef]

52. Elsacker, E.; Søndergaard, A.; Van Wylick, A.; Peeters, E.; De Laet, L. Growing living and multifunctional mycelium composites for large-scale formwork applications using robotic abrasive wire cutting. *Constr. Build. Mater.* **2021**, *283*, 122732. [CrossRef]

53. Saporta, S.; Clark, M. "Bio-Welding" of Mycelium-based Materials. In Proceedings of the 2019 IASS Symposium, Barcelona, Spain, 7–10 October 2019.

54. Biomaterials Building Exposition, Curated by Katie MacDonald and Kyle Schumann. Available online: https://www.biomaterialbuilding.com/ (accessed on 15 June 2022).

55. President's Commision on Slavery and the University. University of Virginia. 2018. Available online: https://dei.virginia.edu/sites/g/files/jsddwu511/files/inline-files/PCSU%20Report%20FINAL_July%202018.pdf (accessed on 15 June 2022).

56. Trimble, B.E. Design of unique landscape walls and their use in building facades. In Proceedings of the 12th Canadian Masonry Symposium, Vancouver, BC, Canada, 2–5 June 2013.
57. Rhinoceros by McNeal. Available online: https://www.rhino3d.com/ (accessed on 15 June 2022).
58. Mushroom Production in Pennsylvania. Available online: https://extension.psu.edu/forage-and-food-crops/mushrooms (accessed on 15 June 2022).

 *biomimetics*

Article

Exploring the Binding Capacity of Mycelium and Wood-Based Composites for Use in Construction

Dana Saez [1,*], Denis Grizmann [1], Martin Trautz [1] and Anett Werner [2]

1   Chair of Structures and Structural Design (Trako), Faculty of Architecture, RWTH Aachen University, Schinkelstraße 1, 52062 Aachen, Germany; grizmann@trako.arch.rwth-aachen.de (D.G.); trautz@trako.arch.rwth-aachen.de (M.T.)
2   Group Enzyme Technology, Chair Bioprocess Engineering, Faculty of Mechanical Science and Engineering, Institute of Natural Materials Technology, Technische Universität Dresden, 01069 Dresden, Germany; anett.werner@tu-dresden.de
*   Correspondence: saez@trako.arch.rwth-aachen.de

Abstract: Existing research on mycelium-based materials recognizes the binding capacity of fungal hyphae. Fungal hyphae digest and bond to the surface of the substrate, form entangled networks, and enhance the mechanical strength of mycelium-based composites. This investigation was driven by the results of an ongoing project, where we attempt to provide basic concepts for a broad application of a mycelium and chipped wood composite for building components. Simultaneously, we further explore the binding capacity of mycelium and chipped wood composites with a series of experiments involving different mechanical interlocking patterns. Although the matrix material was analyzed on a micro-scale, the samples were developed on a meso-scale to enhance the bonding surface. The meso-scale allows exploring the potential of the bio-based material for use in novel construction systems. The outcome of this study provides a better understanding of the material and geometrical features of mycelium-based building elements.

Keywords: binding capacity; bio-adhesives; bio-composites; biomaterials; building biomaterials; fungal mycelium; mechanical performance

Citation: Saez, D.; Grizmann, D.; Trautz, M.; Werner, A. Exploring the Binding Capacity of Mycelium and Wood-Based Composites for Use in Construction. *Biomimetics* **2022**, *7*, 78. https://doi.org/10.3390/biomimetics7020078

Academic Editors: Andrew Adamatzky, Han A.B. Wösten, Phil Ayres and Stanislav N. Gorb

Received: 19 March 2022
Accepted: 9 June 2022
Published: 11 June 2022

Publisher's Note: MDPI stays neutral with regard to jurisdictional claims in published maps and institutional affiliations.

1. Introduction

According to the 2020 Global Status Report for Buildings and Construction published in 2019, the sector moved away from the Paris Agreement goals by causing the highest CO_2 emissions ever recorded: around 10 Gt CO_2, or 28% of total global energy-related CO_2 emissions [1,2]. The increase is mainly related to the carbon-intensive manufacturing processes of building construction materials; therefore, it is crucial to develop novel material strategies to mitigate carbon emissions during the lifecycle of buildings.

Although using renewable raw materials, e.g., wood, presents itself as a logical strategy to withstand CO_2 emissions, the main problem lies in their production, or, more precisely, their growing time. Looking for alternative renewable materials, recent research has suggested that fast-growing organisms such as Fungal mycelium can be engineered to produce novel construction materials [3–6]. Mycelium-based composite production is based on the use of lignocellulosic substrates in combination with the natural growth of the vegetative component of the mycelium of filamentous fungi. As filamentous fungi grow, they form hyphae, which result in a close-meshed network and give the resulting material a solid structure. Such composites have many advantages, such as good thermal insulation, low dry density, and sound absorption. These properties make them suitable for use as building materials (e.g., as insulating materials), but they represent a challenge in their load-bearing capacity.

Our team has been conducting research experiments by developing methods for influencing hyphal growth with the primary objective to provide a mycelium-based composite with particular stability and increased strength [7–11]. In this paper, we describe a manufacturing method of mycelium and wood-based composites where the binding capacity of mycelium plays a crucial role. The fabrication process leads to the fungal mycelium forming predominantly skeletal hyphae at the joint interface of the composite material, which, due to its morphology, leads to increased material strength (Figure 1). This increased strength opens up new application possibilities that go beyond the applications as insulation material or leather substitute [12].

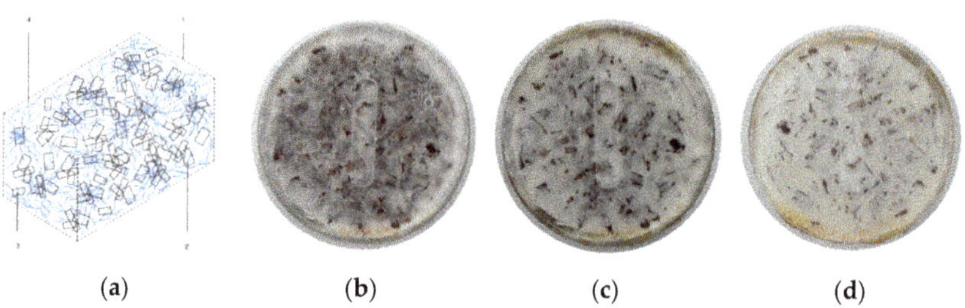

(a)   (b)   (c)   (d)

Figure 1. (**a**) The conceptual illustration describes the systematical formation of skeletal hyphae: the dashed line symbolizes the material interfaces (1); the rectangles, substrate consisting of coarse (2) and fine (3) wood chips; and the blue lines, the dense network of mycelial threads (4). The photos show the growth of mycelial cross-linked growth over time (**b**) after two weeks, (**c**) after three weeks, and (**d**) after four weeks.

2. Materials and Methods

2.1. Fungal–Substrate Composition

The findings of this study concern a specific fungal–substrate composition: the fungal mycelium is derived from *Ganoderma lucidum* (GL) and *Pycnoporus sanguineus* (PS), and the substrate is beechwood. Fungi belonging to the genus *Basidiomycetes*, such as GL, *Ganoderma applanatum*, *Trametes hirsuta*, *Trametes versicolor*, or *Fomes fomentarius*, are mostly found in forests and fulfill, among other things, the task of decomposing deadwood. Consequently, we chose timber as the lignocellulosic substrate. Wood consists of approx. 25–30 wt% lignin, 25–30 wt% pentosans (hemicellulose), 40–50 wt% cellulose, and other components such as resinous substances, terpenes fats, fatty acids, proteins, and minerals. Fungi can decompose lignin, hemicellulose, and cellulose into their subunits by releasing enzymes such as cellulases, laccases, amylases, proteases, or lipases into the immediate environment to degrade the substrate. Subsequently, the degradation products are absorbed by the hyphae and used to grow the fungus [10]. The selection of beechwood as the substrate is a consequence of this research team's previous investigations [2,4,10]. Although the methodology we are about to describe could be transferred to other fungal–substrate compositions, the test results may differ.

2.2. Binding-Specific Manufacturing

As mentioned before, we seek to provide a mycelium-based composite with a particularly stable and increased strength for use in construction. The manufacturing method proposed consists of the following steps:

(1) Selecting the lignocellulosic substrate.
(2) Inoculating the substrate with fungal spores and fungal mycelium.
(3) Mixing the inoculated substrate so that a homogeneous growth of the mycelium can be achieved.
(4) Incubating the obtained mixture from Step (3) in a first incubation phase for a time between 5 and 7 days, at a temperature ranging from 20 °C to 28 °C, and at humidity ranging from 80% to 95% to achieve the cross-linked growth of the mycelium around the substrate.

(5) Placing the obtained incubated mixture from Step (4) in shaping containers defining the shape of the base unit of the composite material, and incubating the mixture in a second incubation phase for a time between 3 and 10 days, at a temperature ranging from 20 °C to 30 °C, and at humidity ranging from 80% to 95% in order to obtain the cross-linked growth of the mycelium around the substrate.

(6) Obtaining at least two base units of the composite with at least one bonding interface.

(7) Joining at least two basic units through the binding interface and incubating for a time between 10 and 30 days, at a temperature in the range of 15 °C to 30 °C, and at a humidity in the range of 80% to 95% to promote the formation of search hyphae and skeletal hyphae between said basic units and to obtain mycelium and wood-based composite specimens (3rd incubation phase) (Figure 2).

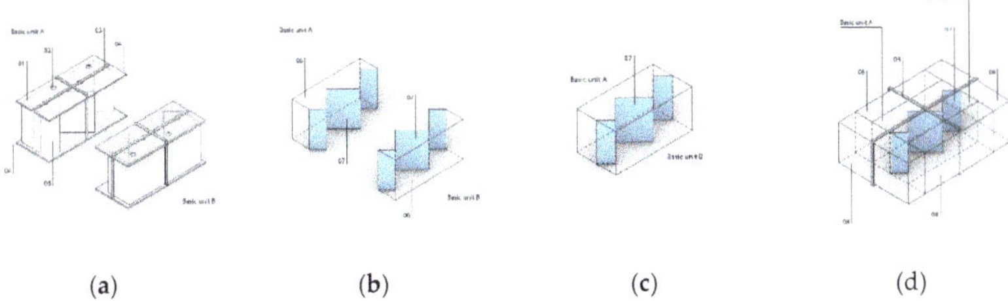

(**a**)   (**b**)   (**c**)   (**d**)

Figure 2. Conceptual illustrations describing the manufacturing process: (**a**) incubation phase II of basic units; (**b**) basic units with binding interface after incubation; (**c**) joining of two basic units; (**d**) incubation phase III of the composite test specimen.

(8) Denaturizing the specimen at a temperature range of 65 °C to 90 °C and obtaining a mycelium-based lignocellulosic composite with a residual moisture content of 10% to 12% by weight based on the total weight of the composite.

Steps (1)–(4) of the manufacturing process are similar to those described in the previous publications of this research team [10]. It is in Step (5) that upon completion of the second incubation phase, at least one base unit of the composite material with at least one binding interface is obtained. A "binding interface" in the present research context refers to the surface to which a different base unit of the composite material can be attached. In the subsequent Step (6), at least two base units of the composite material are provided and subsequently joined so that the binding interfaces of the subunits are joined, as well.

A proper joining configuration is achieved by placing at least two basic units against each other, even though more basic units are able to be attached to form the compound units. Subsequently, the joined basic units are incubated in a third incubation phase (Step (7)). The fungal mycelium grows within the binding interface and couples the two basic units of the composite material.

Finally, the composite material obtained in Step (8) is denaturized to bulk consistency. A mycelium and wood-based lignocellulose composite material is obtained with 10% to 15% residual moisture content.

All manufacturing process steps described above result in the fungal mycelium forming predominantly skeletal hyphae at the joint interface of the composite material, which, due to its morphology, leads to the increased strength of the material in the binding interface.

2.3. Planar and Non-Planar Binding Interface

The present research developed a method wherein the binding interface of at least two basic units has a non-planar surface and wherein a factor between 1.2 and 5 enlarges the binding interface compared to that of a planar surface. Whether the surface area of the bonding interfaces is increased by non-planar joint means that the surface of the connection interface has teeth-like interlocking. It could be arranged symmetrically or asymmetrically, which can be connected by the interface of at least two basic units with teeth-like interlocking in the longitudinal section.

The surface of the connecting interface of the base units is enlarged with teeth-like interlocking with jagged (see test specimens Sch2 and Sch3) or rounded protrusions (see test specimens Sch4 and Sch5). These protrusions may be continuous or have at least one short planar section between each jag or curve.

Six different shaping containers were developed to compare, on the one hand, a single test specimen as a single solid unit (Sch0), without binding interface, with at least two base units test samples (Sch1–5). On the other hand, different geometries of the binding interface were developed to compare the influence of the planar and non-planar binding interfaces.

Each container had a 134.4 cm^3 capacity. Figure 3 and Table 1 describe the exact shapes and sizes of the junction interfaces.

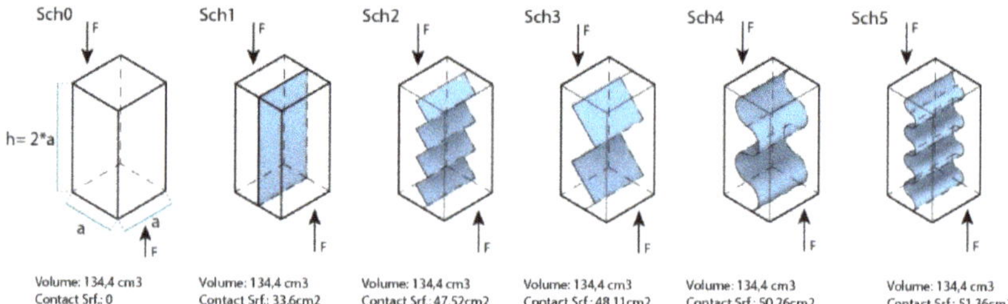

Figure 3. Meso-scale samples for shear tests.

Table 1. Surface and geometry of the binding interface of the test specimens described in Figure 1.

Test Specimen	Sch0	Sch1	Sch2	Sch3	Sch4	Sch5
Geometry of the binding interface	-	planar	jagged small	jagged big	rounded big	rounded small
Surface of the binding interface (cm^2)	0	33.6	47.52	48.11	50.26	51.36

2.4. Shear Tests

Test methods determined the shear strength of the test specimens according to DIN-EN 12090:2013 test standards as illustrated in Figure 4. The test specimens were manufactured as described in Section 2.2 with a consistent filling density of 0.5 g/cm^3 in a cube-like container (a $\times$ a $\times$ h). Due to the exploratory character of the testing, two different substrate–fungus combinations were tested: beechwood–GL and beechwood–PS. This variation led to a total amount of 24 Sch test specimens. The shear tests were conducted under laboratory conditions with the corresponding setup, using a Zwick Zmart.Pro testing machine with a testing speed of 3 mm/min. The tests were documented with the help of photographs, and the force deformation curve was digitally recorded with ZickRoell software. The recorded data were subsequently edited and graphically represented in strength diagrams.

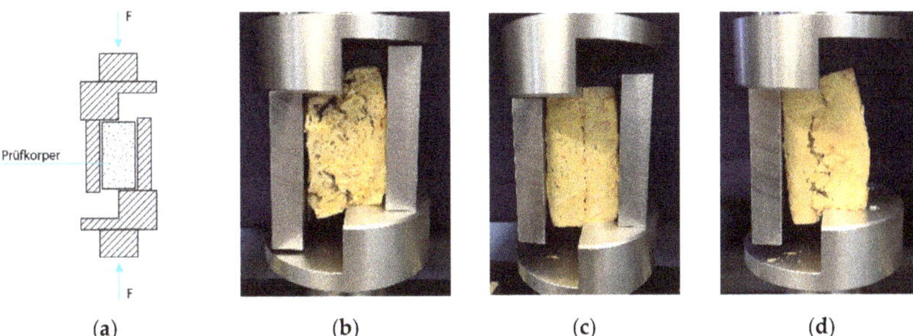

Figure 4. (**a**) Conceptual illustration describing the testing setup. Test specimens being tested under DIN-EN Standards: (**b**) Sch0, specimen with no binding interface; (**c**) Sch1, specimen with planar binding interface; and (**d**) Sch3, specimen with big jagged interface.

3. Results

The obtained material was examined to determine its shear strength. As mentioned before, the tests were carried out on Sch0 to Sch5 test specimens under DIN-EN 12090:2013. The force–displacement curve was recorded for each test, and the shear strength was derived from it (Figure 5).

The first set of analyses presented an apparent variation in fracture behavior. As shown in Figure 5, the blue curves reported significantly less resistance than the rest. Therefore, we can conclude an increase in the material strength of the test specimens developed under binding-specific manufacturing (described in Section 2.2, tests Sch1–5) against those produced under single manufacturing without a binding interface [10]. By repeating the shear tests, this observation was confirmed. These tests were conducted with two different substrate–fungi variations: My8, to beechwood–GL, and My9, to beechwood–PS. Compared with My8, the tests on My9 present significantly less shear resistance.

Non-homogeneous growth on the test specimens may have contributed to the increased variation in the curves in both My8 and My9. The growth variety may result from non-proper environmental conditions in one or more of the cultivation stages during manufacturing.

From the observation of the red curve (Sch1 = planar binding interface) in contrast to the rest of them (Sch0 = no binding interface; Sch2–5 = non-planar binding interface), an influence on the geometry of the inner binding interfaces can be deduced. Consequently, different interface geometries influence the material's behavior in terms of failure mode, stiffness, and shear strength. The force–displacement curve of specimen Sch1 (red curves) with a planar connection interface, for example, demonstrates an early start of material fracture, which is indicated by a sharp drop in the curve. However, due to the significant curve variation, it has not been possible to determine whether the surface area increase or geometry variation in the binding interface has increased the resistance. Additionally, it would be necessary to repeat the tests with a more significant number of test specimens to determine if the variation between jagged and rounded geometry plays a role in the binding capacity.

The results show that by using the binding ability of the mycelium, it is possible to influence the material properties of composites through the targeted arrangement of joining surfaces with higher stiffness. In this way, direction-dependent material behavior can also be generated. It is also interesting to note that the study of shear strengths reveals that both the interface arrangement and the shear strength are essentially derived from the interface configuration.

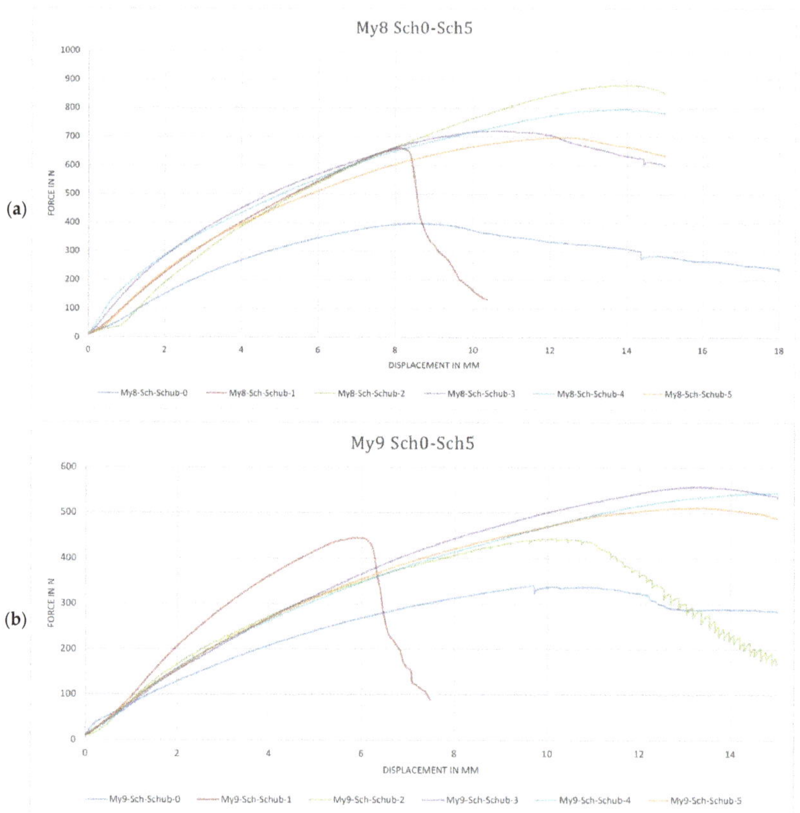

Figure 5. Diagrams showing the recorded force–displacement curves of conducted tests: (**a**) My8—specimen of beechwood–ganoderma lucidum; (**b**) My9—specimen of beechwood–pycnoporus sanguineus.

4. Discussion

The attractiveness of mycelium as a matrix material lies on the one hand in its biological origin, which due to its chemically untreated state, can be easily integrated into the biological material cycle for the circular economy [13]. Due to its manufacturing process, mycelium-based materials have a minimal CO_2 footprint compared to the vast majority of standard building materials. A wide variety of mycelium-based products have been developed, such as packaging [12] or insulation materials [14]. In general, mycelium-based materials have gained popularity by exploiting the rapid virtual growth of hyphae, which allows the production of conglomerate materials. On the contrary, the positive results on mycelium and wood-based material binding interfaces present this material as an ideal candidate for manufacturing laminar materials. According to the data recorded in Figure 5, the application of binding-specific manufacturing clearly increases the shear strength of the test specimens by at least 50% (average for Sch1 = 0.173 N/mm^2). Although non-planar binding interfaces present higher shear strength (up to 83%), the results obtained by planar binding interfaces have significant implications for mycelium-based materials. This process could allow the production of multi-laminar elements and could replace oil-based glue materials. Binding-specific manufacturing will doubtless be much scrutinized, but due to the increase in the shear strength, we can conclude that the production process presents advantageous conditions for conglomerates and laminar mycelium and wood-based materials.

With the investigations presented in this paper, it could be assumed that several basic units' laminar composition can positively influence the load-bearing behavior of mycelium-based materials. However, multiple basic unit composition is not the only factor to consider to achieve building standard requirements. Other factors that positively influence the strength and stiffness properties of mycelium-based materials are mycelium–substrate combinations which could lead to optimal growth or optimal mechanical qualities, and the increase in the material's density by compression [15]. Further studies that take the latter variables into account will need to be undertaken in combination with binding-specific manufacturing processes.

One of the most important findings of this research is that the geometry of the interfaces can influence fracture behavior. Figure 6 shows how the non-planar binding interfaces show an increased strength. These findings contribute in several ways to a deeper understanding of the binding capacity of mycelial hyphae in combination with wood substrate and provide a basis for further development of mycelium-based material properties.

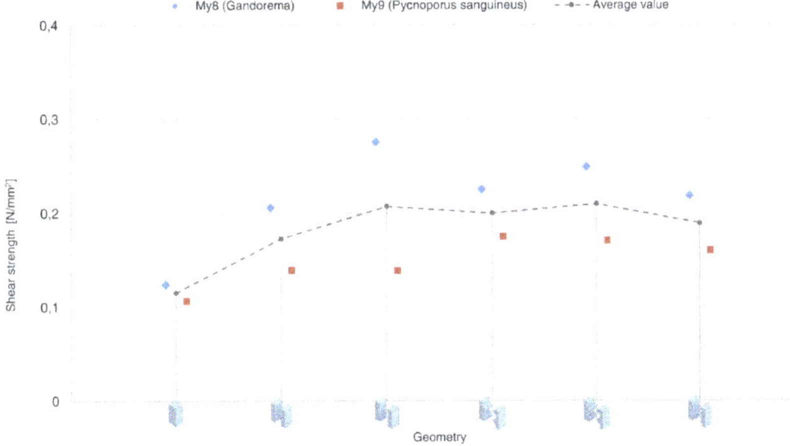

Figure 6. Shear strength to geometry diagram. Test specimen geometries from left to right: Sch0, Sch1, Sch2, Sch3, Sch4, and Sch5.

5. Patents

The work reported in this manuscript resulted in the Patent Application No. (DRN) 2021122113010200DE.

Author Contributions: Conceptualization, D.S., M.T. and D.G.; methodology, A.W., D.S. and D.G.; validation, M.T., A.W., D.S. and D.G.; investigation, D.S. and D.G.; data curation, D.S.; writing—original draft preparation, D.S.; writing—review and editing, D.S. and D.G.; visualization, D.S. and D.G.; photos in Figure 4, Ibac; supervision, M.T.; project administration, D.S. and D.G. Manufacturing steps (1) to (4) were developed and conducted by A.W. at AG Enzymtechnik, Dresden University of Technology. Manufacturing steps (5) to (7) were developed and conducted by Trako, led by M.T., Faculty of Architecture, RWTH Aachen University. All tests were conducted at the Institute and Chair of Building Materials Research (Ibac), RWTH Aachen University. Test analysis and interpretation by D.S. and D.G. All authors have read and agreed to the published version of the manuscript.

Funding: This research received no external funding.

Institutional Review Board Statement: Not applicable.

Informed Consent Statement: Not applicable.

Data Availability Statement: Not applicable.

Acknowledgments: This paper and the research behind it would not have been possible without the support of our colleagues at Ibac, RWTH Aachen University: Clarissa Glawe, Hendrik Morgenstern, and Cynthia Morales Cruz. The authors would like to extend special thanks to Trako's assistant team members: Raman Suliman, Steliyana Yancheva, Lea Scholz, Helena Krapp, and Hlib Novosolov.

Conflicts of Interest: The authors declare no conflict of interest. The funders had no role in the design of the study; in the collection, analyses, or interpretation of data; in the writing of the manuscript, or in the decision to publish the results.

References

1. United Nations Environment Programme. *Global Status Report for Buildings and Construction: Towards a Zero-Emission, Efficient and Resilient Buildings and Con-Struction Sector*; United Nations Environment Programme: Nairobi, Kenya, 2020.
2. Hettinger, P. Structural Engineers Can Make a Big Difference in Terms of Sustainability. *DGNB Blog Sustainable Building*. Available online: https://blog.dgnb.de/en/structural-engineers-sustainability/#:~{}:text=%E2%80%9CStructural%20engineers%20 can%20make%20a%20big%20difference%20in%20terms%20of%20sustainability%E2%80%9D&text=Buildings%20contain%20 a%20lot%20of,materials%20on%20a%20massive%20scale. (accessed on 4 May 2022).
3. Sun, W.; Tajvidi, M.; Hunt, C.G.; McIntyre, G.; Gardner, D.J. Fully Bio-Based Hybrid Composites Made of Wood, Fungal Mycelium and Cellulose Nanofibrils. *Sci. Rep.* **2019**, *9*, 3766. [CrossRef] [PubMed]
4. Meyer, V.; Basenko, E.Y.; Benz, J.P.; Braus, G.H.; Caddick, M.X.; Csukai, M.; de Vries, R.P.; Endy, D.; Frisvad, J.C.; Gun-de-Cimerman, N.; et al. Growing a circular economy with fungal biotechnology: A white paper. *Fungal Biol. Biotechnol.* **2020**, *7*, 5. [CrossRef] [PubMed]
5. Jones, M.; Mautner, A.; Luenco, S.; Bismarck, A.; John, S. Engineered mycelium composite construction materials from fungal biorefineries: A critical review. *Mater. Des.* **2020**, *187*, 108397. [CrossRef]
6. Elsacker, E.; Vandelook, S.; Van Wylick, A.; Ruytinx, J.; De Laet, L.; Peeters, E. A comprehensive framework for the production of mycelium-based lignocellulosic composites. *Sci. Total Environ.* **2020**, *725*, 138431. [CrossRef] [PubMed]
7. Moser, F.; Trautz, M.; Beger, A.-L.; Löwer, M.; Jacobs, G.; Hillringhaus, F.; Wormit, A.; Usadel, B.; Reimer, J. Fungal mycelium as a building material 7. In Proceedings of the IASS Annual Symposia, Hamburg, Germany, 25–28 September 2017; Volume 2017, pp. 1–7.
8. Löser, T. Mycelium-Based Hybrid Materials–Development of Test Specimens and Strength Test for Their Charac-Terization. Master's Thesis, Dresden University of Technology, Faculty of Mechanical Engi-neering, Institute of Natural Materials Engineering, Chair of Bioprocess Engineering, Dresden, Germany, 2020.
9. Saez, D.; Grizmann, D.; Trautz, M.; Werner, A. Developing sandwich planels with a mid-layer of fungal mycelium composite for a timber panel construction system. In Proceedings of the World Conference on Timber Engineering, Santiago, Chile, 9–12 August 2021.
10. Saez, D.; Grizmann, D.; Werner, A.; Trautz, M. Analyzing a fungal mycelium and chipped wood composite for use in construction. In Proceedings of the IASS Annual Symposium 2020/21, Guilford, UK, 23–27 August 2021.
11. Saez, D.; Grizmann, D.; Trautz, M.; Werner, A. Building materials from wood and fungal mycelium for load-bearing structures. In Proceedings of the 1. Fachkongress Konstruktiver Ingenieurbau: Kompetenz-Plattform für die Bautechnische Gesamtplanung, 10–11 May 2022; Lochner-Aldinger, I., Ed.; Technische Akademie Esslingen e.V.: Ostfildern, Germany, 2022; pp. 259–264.
12. Ecovative. Available online: Ecovativedesign.com (accessed on 4 May 2022).
13. Ellen Mac Arthur Foundation. The Butterfly Diagram: Visualising the Circular Economy. Available online: https://ellenmacarthurfoundation.org/circular-economy-diagram (accessed on 4 May 2022).
14. MOGU. M_My-34_Mogu Website. Available online: https://mogu.bio/ (accessed on 4 May 2022).
15. Trautz, M. Timber and mycelium-based composite material for building construction. In Proceedings of the 6th International School and Conference on Biological Materials Science: Bioinspired Materials, Kostenz, Germany, 21–24 March 2022.

biomimetics

Article

Growth and Mechanical Characterization of Mycelium-Based Composites towards Future Bioremediation and Food Production in the Material Manufacturing Cycle

Thibaut Houette [1,*], Christopher Maurer [2], Remik Niewiarowski [1] and Petra Gruber [3]

1   Department of Biology, The University of Akron, Akron, OH 44325, USA; rdn19@uakron.edu
2   Redhouse Studio, Cleveland, OH 44113, USA; chris@redhousestudio.net
3   Transarch Office for Biomimetics and Transdisciplinary Architecture, 3370 Ybbs an der Donau, Austria; peg@transarch.org
*   Correspondence: th153@uakron.edu

Abstract: Today's architectural and agricultural practices negatively impact the planet. Mycelium-based composites are widely researched with the aim of producing sustainable building materials by upcycling organic byproducts. To go further, this study analyzed the growth process and tested the mechanical behavior of composite materials grown from fungal species used in bioremediation. Agricultural waste containing high levels of fertilizers serves as the substrate for mycelium growth to reduce chemical dispersal in the environment. Compression and three-point bending tests were conducted to evaluate the effects of the following variables on the mechanical behavior of mycelium-based materials: substrate particle size (with or without micro-particles), fungal species (*Pleurotus ostreatus* and *Coprinus comatus*), and post-growth treatment (dried, baked, compacted then dried, and compacted then baked). Overall, the density of the material positively correlated with its Young's and elastic moduli, showing higher moduli for composites made from substrate with micro-particles and for compacted composites. Compacted then baked composites grown on the substrate with micro-particles provided the highest elastic moduli in compression and flexural testing. In conclusion, this study provides valuable insight into the selection of substrate particle size, fungal species, and post-growth treatment for various applications with a focus on material manufacturing, food production, and bioremediation.

Keywords: mycelium; fungal architecture; myceliated material; living material; sustainability; biotechnology; compression; bending; waste upcycling; mycoremediation

Citation: Houette, T.; Maurer, C.; Niewiarowski, R.; Gruber, P. Growth and Mechanical Characterization of Mycelium-Based Composites towards Future Bioremediation and Food Production in the Material Manufacturing Cycle. *Biomimetics* **2022**, *7*, 103. https://doi.org/10.3390/biomimetics7030103

Academic Editor: Stanislav N. Gorb

Received: 7 May 2022
Accepted: 23 July 2022
Published: 28 July 2022

Publisher's Note: MDPI stays neutral with regard to jurisdictional claims in published maps and institutional affiliations.

1. Introduction

1.1. Problem Statement of Current Building Materials

Today's architectural practices negatively impact the planet's ecosystems. The building industry and its energy-intensive material manufacturing processes were responsible for a total of 37% of global carbon emissions in 2020, and these processes contribute significantly to anthropogenic climate change and resulting extreme weather events [1]. The push for sustainability in architecture needs to encompass the entire life cycle of materials and buildings, from resource extraction to repurposing or disposal. Even simple materials used in the building industry are becoming scarce and limited. At the other end of the building process, 600 million tons of construction and demolition waste were generated in 2018, with 145 million tons sent to landfills [2]. As a result, landfilled hazardous materials, such as lead, can contaminate ground water [3]. Even after waste removal, sites contaminated with heavy metals and toxic chemicals from industry still need to be cleaned to limit run-off and spread to the environment. Therefore, the building industry urgently needs to focus on cleaner materials that can be repurposed, to reduce both material scarcity and waste generation.

1.2. Towards Living Materials

In response to the current limitations of traditional architectural materials and their manufacturing practices, living materials (i.e., materials integrating biological organisms) have emerged in recent decades. Specifically, Engineered Living Materials (ELMs) are defined as genetically or mechanically "engineered materials composed of living cells that form or assemble the material itself, or modulate the functional performance of the material in some manner" [4]. The implementation of living organisms in technological materials allows engineers to benefit from the qualities of biological growth [5]. The overall advantages of ELMs are self-production, clean chemistry, sustainability, adaptability, self-healing, and the potential for added functionality through genetic and mechanical engineering. The wide scope, potential, and limitations of ELMs have been discussed in multiple reviews [4,6]. Examples of ELMs include microbially manufactured polymer matrices, soft living robots, smart living surfaces, living carbon composites, bacteria-based self-healing concrete, bacterial cellulose, biologically fabricated bricks, and mycelium-based materials. The long-term goal of ELMs is to build large-scale hierarchical material systems from simple autonomous micro-entities in situ. However, more research needs to be conducted to keep organisms alive in the applied setting, scale up production of a laboratory environment, and predict organisms' behavior.

1.3. Mycelium-Based Materials

Fungi are used for ELMs because of their mycelium, which forms a 3D binding network, secretion of enzymes, diversity of properties between species, and wide range of material applications. Mycelium-based materials are among the most successful large-scale living materials [4,7]. Mycelium-based materials are produced by growing fungal mycelium on an organic substrate (e.g., often agricultural byproducts) in a mold. A variety of post-growth treatments are applied based on the desired application. The material properties are highly tunable based on the selected substrate type (i.e., chemically, and related to size), fungal species, growth environment, and post-growth treatments [5,7–13]. The ability to fine tune the properties of mycelium-based materials increases their range of application, including packaging, electronics, acoustic absorbers, footwear, insulators, fire protection, and self-healing materials [8,14–23]. At the end of their life cycle these composite materials are biodegradable and can even be used as a substrate for growing new iterations of materials. Therefore, mycelium-based materials promote organic waste upcycling, low-energy material manufacturing, and biodegradable materials, making them an alternative to current architectural materials. To showcase their architectural potential, temporary installations have been built with mycelium-based materials [5,24–29]. To address the limitations of ELMs, more studies need to be conducted before their implementation in permanent buildings by exploring new substrate/fungus combinations, evaluating the effect of various parameters (e.g., fungal species, substrate type and size, growth environment, and post-growth treatments) on material properties, predicting material behavior, ensuring homogeneous material properties, and characterizing the accumulation and decomposition of toxic chemicals [5].

1.4. State of the Art of the Production and Mechanical Properties of Mycelium-Based Materials

Throughout the production process of myceliated materials, various factors influence their acoustic, thermal, mechanical, and physical characteristics [12,13]. Different fungal species will feed on different substrates and grow at different rates [30]. The mycelium anatomy and structure differ between species, with three main categories found in basidiomycetes: monomitic (i.e., generative), dimitic (i.e., generative, and skeletal), and trimitic (i.e., generative, binding, and skeletal) [13]. For example, myceliated materials grown from trimitic species (e.g., *Trametes versicolor*) display a higher compressive, tensile, and flexural strength than those grown from monomitic species (e.g., *Pleurotus ostreatus*) [31,32]. The substrate forms the base of the composite materials as it is not completely decomposed by the fungus during the mycelial growth process. Therefore, the substrate composition and particle size affect the material properties of the end material as it serves as its backbone structure [13,33]. To successfully introduce the desired fungus, the substrate must be cleaned of other species. By removing competing species, this fungus can grow over the entire substrate and produce a coherent material. Different techniques are used to kill competing species: sterilization (e.g., autoclaving, bathing in hydrogen peroxide, bathing in a basic solution, baking) or pasteurization (e.g., steaming) [5,13]. These techniques affect the mycelial growth speed and removal of competing species differently.

During inoculation, the proportions of the ingredients (e.g., dry substrate, water, mycelium spawn, nutrients) for optimal mycelial growth vary between fungal species. As mycelial growth highly depends on the nutrient profile of the substrate, food with high nutritious content is often added to the agricultural byproducts. Since mechanical failure always occurs in the mycelium binder, mycelial growth and, therefore, the nutrient profile of the substrate, are especially important for mechanical performance [13]. However, the use of more-nutritious substrate makes the material less impactful in terms of sustainability, waste upcycling, and in general, resource efficiency. Percentages of ingredients used in previous research projects are shown in Appendix A.

The growth environment (i.e., temperature, relative humidity, access to oxygen, clean/ventilated air exchange, and lighting conditions) also impacts mycelium growth. The temperature should be kept at around 25–30 °C [7,13,34] and the relative humidity around 70–80% [7]. Based on the fungal species, substrate, growth environment, and level of growth desired, the growth time may vary from 6 days [35] to 20 days [7], and up to months [13,34]. In the literature, the growth of fruiting bodies in the production of mycelium-based materials is avoided as it can consume resources otherwise used for mycelium growth, modify material shape, increase composite heterogeneity, require maintenance (i.e., harvest required in an environment that should remain as clean and undisturbed as possible to avoid contamination), and release spores that could cause allergic reactions or infections [32,36–41]. Fruiting body formation can be inhibited by controlling both the lighting conditions and carbon dioxide concentration (i.e., dark with low CO_2) or by introducing GSK-3 inhibitors [36,37]. During the growth process and the drying period, myceliated materials shrink. Knowing the shrinkage of the material is important to estimate and target specific product dimensions. For cylindrical specimens, the shrinkage has been estimated to be around 17% in height (vertical shrinkage) and around 10% in diameter (horizontal shrinkage) [9]. For rectangular specimens, shrinkages of 5.56% horizontally and 2.78% vertically have also been observed [42].

After the growth process, various treatments (e.g., drying, baking, compacting, coating) can be applied to the myceliated materials to tune their properties [13]. Existing studies do not usually differentiate drying and baking of the specimens. Furthermore, there is no standard practice for drying or baking, as their success depends on specimens' dimensions. For example, various research teams used the following drying/baking techniques: 40 °C for 72 h plus 2 h at 100 °C [33], 60 °C for 24 h [8], 60 °C for 2 h [7], 70 °C for 5 to 10 h until the weight is stabilized [9], 80 °C up to a constant weight [21], 80 °C for 24 h [32], 100 °C for 4 h [35,43], and 100 °C for several hours [44]. The mycelial structure is believed to degrade with temperatures of approximatively 225 to 300 °C [7].

Various research articles studied the mechanical properties of mycelium-based materials, which have been described by a two-phase particulate model with the mycelium as the matrix and substrate as the dispersed phase [7–11,42]. The differences in the production, dimensions, description, and testing procedures of such bio-composite materials make data comparison between studies difficult, as they severely impact their mechanical behavior [42]. The mechanical behavior of mycelium-based materials highly depends on the anisotropic substrate matrix. A study recently looked at the effect of fiber orientation within the substrate on the compressive behavior of mycelium-based materials [42]. The study found that adding fibers oriented in the direction of loading increased Young's modulus. Conversely, fibers oriented perpendicular to the loading direction produced a decrease in Young's modulus and ultimate strength. A study found that myceliated materials made from loose substrate had lower compressive Young's moduli than those from chopped substrates [9]. The same study also found variability in compressive Young's moduli based on substrate used, despite their similar density (around $100 \ \text{kg/m}^3$): from 0.14 MPa for loose pine shavings samples up to around 1.25 MPa for pre-compacted hemp and flax samples. Myceliated samples grown with *Pleurotus ostreatus* on hemp mat had a compressive strength of 0.19 MPa compared to 0.26 MPa for *Trametes versicolor* [31]. Another study found a compressive modulus of 1.3 MPa for *Ganoderma lucidum* grown on macerated red oak wood chips (5–15 mm) and a nutrient solution with a final density of $318 \ \text{kg/m}^3$ [45]. Vidholdová et al., 2019, grew low-density mycelial boards having a density of $103 \ \text{kg/m}^3$, which resulted in a compressive resistance at 20% strain of 23.95 kPa [11]. Compressive performance of porous materials increases with increasing density, which may result from a variety of parameters including the substrate used, its particle size, the degree of compaction, and the amount of substrate digested by the fungus [13,46,47]. Islam et al., 2017, studied the correlation between density and uniaxial elastic modulus on mycelium boards from Ecovative Design LLC (Green Island, NY, USA) [48]. They found that densities in the range of around 150 to $160 \ \text{kg/m}^3$ lead to elastic moduli in the range of 0.5 MPa to 1.1 MPa, respectively. Yang et al., 2017 found various Young's moduli ranging from around 5 to 50 MPa for densities ranging from around 160 to $280 \ \text{kg/m}^3$ made with *Irpex lacteus* mycelium grown on a variety of substrates including wood pulp, millet grain, wheat bran, natural fiber, and calcium sulfate [8]. In terms of bending properties, a study looked at the mechanical properties of bioresin-infused mycelium-based sandwich composite materials under 3 pt bending, which led to an elastic modulus of 1.13 MPa for a density of $121.7 \ \text{kg/m}^3$ [49]. Appels et al., 2019, found an increase in flexural moduli from non-pressed (ranging from 1 to 9 MPa for densities from 100 to $170 \ \text{kg/m}^3$) to cold-pressed (ranging from 12 to 15 MPa for a mean density of $240 \ \text{kg/m}^3$) and hot-pressed (ranging from 34 to 80 MPa for densities from 350 to $390 \ \text{kg/m}^3$) in materials grown from *Trametes multicolor* or *Pleurotus ostreatus* on rapeseed straw, beech sawdust, or cotton [32]. Further, Appels et al., 2018, found that controlling the lighting conditions and carbon dioxide levels of the growth environment (i.e., light with CO_2 content), in addition to deleting the hydrophobin gene *sc3*, increased composite density and resulting Young's modulus [38].

Due to the novelty of the research field of mycelium-based composites, many research questions still need to be addressed before their implementation in permanent architectural projects [5]. For instance, more studies should explore particle shapes, composition, and distributions due to their significant influence on mechanical properties [42]. The substrate particle size for an optimal balance between mycelial growth and mechanical performances is still subject to research. Higher substrate density lowers air transmission, resulting in limited mycelial growth inside the substrate if the substrate is not artificially aerated [13]. However, Islam et al., 2018, found that the generic trends of the stress–strain curves from compressing mycelial composite made of different particle sizes were not sensitive to the particle size, suggesting that the myceliated materials' compressive response is independent of substrate particle size [43]. In their study, they compared five different substrate particle sizes with varying aspect ratios (2, 5, or 8), sizes (2.5, 5, or 10 mm), and diameters (0.5, 1, or 2 mm). In another study, materials made from medium particles (0.75–3.0 mm) led to higher density, Young's modulus, and ultimate strength than others made from smaller (0.5–1.0 mm), larger (4.0–12.0 mm), or more diverse (0.5–12.0 mm) particles [42]. This study also showed that higher density did not increase compressive performance in all cases. To the author's knowledge, no study has questioned the effect of the very fine micro-particles on mycelial growth speed and mechanical performance. Moreover, the substrate particle size will likely influence the effects of different post-growth treatments. To further extend the analysis of substrate size, more studies should evaluate the subsequent effects of post-growth treatments on composites grown with these various substrate sizes. As a result, the combined effects of mutually responsive variables need to be addressed in future research.

1.5. Underutilized Benefits of Fungal Mycelium

The variety of fungal species possess many qualities unexploited in current mycelium-based materials. Fungi are known for their nutritional, medicinal, and bioremediation benefits [50,51]. Fungi release enzymes to break down substances that they can feed upon. This behavior makes them very interesting for mycofiltration (i.e., use of fungi to filter water) and mycoremediation (i.e., use of fungi to decontaminate/depollute the environment). Whereas mycofiltration relates to filtering water, mycoremediation aims at decontaminating a substrate. Mycoremediation is a subset of bioremediation practices, in which fungi serve to uptake and break down pollutants. It has been referenced as the cheapest remediation technique for polycyclic aromatic hydrocarbons, at approximately 50 USD/ton [51]. Paul Stamets described a circular model in which fungi serve to produce food and medicine, remediate soil, and facilitate plant growth [51]. Thanks to recent mycelium-based materials research, the production of building materials can now be integrated in this model by growing mycelium materials on contaminated substrate and/or harvesting fruiting bodies during the growth process. Depending on the type of contamination and fungal species employed, toxins present in the substrate are decomposed by the fungus, stored in the mycelium, or accumulated in the fruiting bodies, which are harvested for treatment. In all cases, the myceliated substrate can serve for producing composite materials once their toxicity is evaluated. As stated in Section 1.4, the growth of fruiting bodies is currently avoided in the production of composite materials for various reasons. In terms of mechanical performance, the authors did not find any study comparing the mechanical behavior of mycelium-based materials grown with and without fruiting bodies. Therefore, a study should be conducted to validate the hypothesis that fruiting body formation lowers mechanical performance of these materials. Depending on the targeted application, the authors believe that the benefits of producing fruiting bodies for food or medicinal applications outweigh the potential reduction in mechanical performance. For instance, fruiting bodies could be harvested during the mycelium growth of composite materials. Specific frames should be used to allow harvest without disturbing the growth environment and introducing contaminants. Another solution is to upcycle the fruiting block serving for mushroom production after it is spent, by using it as myceliated substrate for material production [52]. In conclusion, more studies should be performed to evaluate the potential integration of material production in the circular model of current utilizations of fungi.

1.6. Overall Goal of the Project

A potential for this updated model is to grow bioremediating species on contaminated organic substrate from sources including the agriculture and building industries. In addition to decontaminating the substrate, food (i.e., fruiting bodies) and materials (i.e., mycelium-based composites) can be produced after toxicity evaluation. Depending on the chemicals used to accumulate or decompose and the desired application, various bioremediating fungal species can be employed. For instance, *Pleurotus ostreatus* (commonly named oyster mushroom) is a recommended species to decontaminate petroleum products [51,53]. Another species often seen in polluted soils, *Coprinus comatus* (commonly named shaggy mane), is a bio-accumulator of heavy metals and a species recommended for decontamination of substrates with nitrates and phosphorus-bound toxins [51]. Therefore, specific fungal species can be used depending on the chemicals present in the substrate. During mycelial growth over the enriched substrate, fruiting bodies are harvested for chemical treatment or food based on their toxicity. Upon full mycelial coverage, the composite material receives various treatments to tune its properties. Finally, the mycelium-based material and fruiting bodies produced would be chemically tested to ensure that they are safe for the desired application.

This article presents a first step towards the implementation of this model in a case-study. Two bioremediating species (*Coprinus comatus* and *Pleurotus ostreatus*) were grown on an agricultural byproduct (straw) to produce composite materials. The mycelium-based materials produced were mechanically characterized through compression and bending testing to assess their potential for architectural purposes. These tests sought to evaluate the effects of the following variables on composite mechanical behavior: substrate size (with or without micro-particles), fungal species (*Coprinus comatus* or *Pleurotus ostreatus*), and post-growth treatment (dried, baked, compacted then dried, compacted then baked).

To complete the model validation, chemical tests would need to be performed in a further study. These tests were not generated in this part of the project due to a cut in resources (expertise and money) and delays emerging from COVID-19 regulations. However, all mycelium-based materials and samples of both substrate mixtures were stored to be used in future chemical analysis.

2. Materials and Methods

The main steps of the project and variables studied are illustrated in Figure 1.

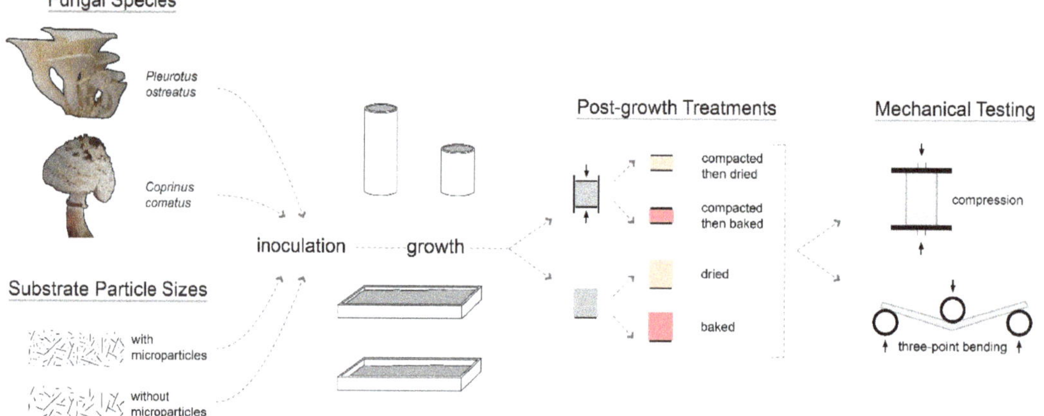

Figure 1. This figure shows the overall process of the project, including the variables evaluated (i.e., fungal species, substrate particle size, and post-growth treatment) for their effect on the mechanical properties (i.e., under compression and bending) of the composite material.

2.1. Substrate and Fungal Species

The substrate used for mycelium growth was straw, collected in bales from Dussel Farm, Ohio, USA. Different batches of mycelium-based materials were produced with two different fungal species (*Pleurotus ostreatus* and *Coprinus comatus*). Due to limited availability of resources and the scale of the project, the batches were not conducted at the same time. Mycelium-based materials grown with *Pleurotus ostreatus* were manufactured from January to June 2020, while those with *Coprinus comatus* were produced from June to September 2020. The same process was used for both batches: preparation of the straw substrate, inoculation, mycelium growth, application of post-growth treatments, and mechanical testing.

2.2. Substrate and Mold Preparation
2.2.1. Chipping

To reduce the size of the substrate particles, the straw was chipped (Done Right Chipper Shredder Premier 300, Generac Power Systems, Inc., Waukesha, WI, USA). Chipping straw increases substrate density, as a reduction in particle dimension increases packing density. The density of the straw increased from 121.69 kg/m^3 before chipping to 177.28 kg/m^3 after. The chipper produced particles with varying lengths, including microparticles. Due to the dispersal of particles into the environment, 17.93% and 21.62% of the straw weight was lost during the chipping process for the first and second batch, respectively. Both batches were chipped at different times of the year (January and June). The difference in straw weight loss likely results from the divergent environmental conditions (i.e., mostly humidity and wind) during chipping.

2.2.2. Sieving

Since the chipper produced particles of variable dimensions, the chipped particles were sieved to control the particle size. Two different particle size sets were produced for this research project to compare their effect on mycelial growth speed and mechanical performance. The first set of particles (S for small) was chipped and sieved through a 5.7 mm sieve to exclude large particles. The second set of particles (L for large) was similarly chipped and sieved through a 5.7 mm sieve. In addition, the particles were then sieved through a 1.5 mm sieve; only the material stuck on the sieve was kept and fine particles were removed. In conclusion, fine particles (passing through a 1.5 mm sieve) were kept in mixture S, while they were removed from mixture L. The two mixtures of particles (L and S) served as substrate to assess the effect of the fine particles on the mycelial growth and mechanical performance. The meshes of the 5.7 mm sieve (0.22-inch opening) and 1.5 mm sieve (0.06-inch opening) are respectively closest to standard US meshes No. 30 and No. 14. Additional details on sieving non-spherical objects and the sieving procedure are given in Appendix B.

For the first batch, the sieving process only resulted in 1.05% weight loss for the 5.7 mm sieve, and 0.51% for the 1.5 mm sieve. After the chipping and sieving processes of the first batch, 80.57% of the unchipped straw weight (100%) remained as the small mixture. Due to the removal of fine particles, only 42.73% of the unchipped straw weight remained as the large mixture. For the second batch, 66.13% and 38.52% of the unchipped straw weight resulted in the small and large mixtures, respectively. Both batches were sieved by a different individual, which may be the cause of the different percentages observed despite following the same procedure.

2.2.3. Quantifying Particle Sizes

As stated in Section 2.2.2., the dimensions of the particles passing through a sieve may be larger than the opening size of the sieve. Therefore, a technique was developed to quantify the mean size of the particles of mixtures S and L. The same chipping and sieving processes were performed on a known quantity of straw to produce both mixtures. Particles from each sieving step were then measured by placing them on a light table; a picture of them was taken and their 2D dimensions were extracted using an image analysis algorithm. This process is described in detail in Appendix C. For each set of particle sizes, this entire process was repeated five times with new particles from the same mixture.

2.2.4. Sterilization and Pasteurization

Due to inaccessibility of specific resources (e.g., pasteurization and inoculation equipment and workforce) under COVID-19 regulations, the first batch of substrate was sterilized and the second was pasteurized. For the first batch, the dry chipped and sieved substrate was sterilized in 10 autoclaving bags weighing around 2422 ± 109 g for mixture L and 1918 ± 152 g for mixture S. Each bag was sterilized for 30 min at 121 °C and 16 psi inside an autoclave at the Biology Department of The University of Akron. Autoclaved bags remained sealed until inoculation to reduce the entry of contaminants. For the second batch, both mixtures were pasteurized one after the other at Valley City Fungi, OH, USA. Mixture L was pasteurized first to avoid the transfer of fine particles from mixture S to L. The pasteurization process is described in Appendix D. Both mixtures were finally placed into sealed bags to avoid contamination in transport and following manipulations.

2.2.5. Mold Preparation

The molds into which the inoculated substrate was placed for mycelium growth were built based on the ASTM testing requirements. Due to the absence of ASTM standards specific to mycelium-based materials, various related standards had to be analyzed. Based on their similarity with mycelium-based physical properties and previous research studies, ASTM D2166/D2166M-13 was used for compression testing and ASTM D1037-12 for bending testing [8,9,54,55]. According to these standards, cylindrical molds (i.e., PVC pipes) were built for specimens to be tested in compression and rectangular molds (i.e., wooden frames) for those to be tested in bending.

For compression testing, the targeted dimensions were 7 cm in diameter and 14 cm in height (i.e., height-to-diameter ratio of 2:1) as described in ASTM D2166/D2166M-13 [55]. However, it was anticipated that specimens would shrink during the drying process. To anticipate shrinkage and test specimens following the dimensions specified by the ASTM standard, specimens were made 8% larger in diameter and 20% larger in height based on estimations from the literature and preliminary experiments. Furthermore, it was expected that half of the specimens would be compacted to half of their height after the growth period (post-growth treatment detailed in Section 2.4). In this regard, specimens were grown inside 3 in. × 10 ft. PVC tubes (Charlotte PVC 40 Plain-End DWV Pipes, Charlotte Pipe and Foundry, Charlotte, NC, USA), with a diameter of 7.7 cm and a height of 17 or 34 cm.

For the bending test in ASTM D1037-12, materials thicker than 0.6 cm should have a width of 7.6 ± 0.1 cm and a length of (5.1 cm + 24 × the nominal thickness) [54]. Bending tests were performed with an Instron 5567 electrochemical testing system (Instron, Norwood, MA, USA), which can fit specimens up to a length of around 35.56 cm. Therefore, the specimens' dimensions were selected to be as follows: a length of 32.66 cm, a width of 7.60 cm, and a thickness of 1.15 cm. The same shrinkage percentages as per compression testing specimens were used to produce specimens 8% larger in width and length and 20% thicker. To optimize space and material, each mold was built to contain six specimens side by side. Half of the specimens were also grown twice as thick to prepare for compaction to half of their thickness. These molds were made of $\frac{1}{2}$ in. medium-density fiberboard (MDF) boards (Home Depot, Atlanta, GA, USA) to grow a large mycelium-based specimen measuring 35.5 cm in width, 50 cm in length, and 1.4 or 2.8 cm in thickness. A layer of plastic made from polyethylene resin (Home Depot, USA) was laid inside the mold to keep the mycelium from growing on and sticking to the wooden mold. Additional rectangular molds measuring 49.2 cm in width, 55 cm in length, and 1.4 or 2.8 cm in thickness were produced to grow leftover material.

2.3. Inoculation

Using existing research as a baseline (see Appendix A) and discussions with mushroom farmer John Burmeister from Valley City Fungi, Ohio, USA, the team decided on a recipe consisting of 10 wt% *Pleurotus ostreatus* or *Coprinus comatus* grain spawn, 22.5 wt% dry substrate (chipped and sieved straw), and 67.5 wt% water for both batches. The first batch was inoculated in redhouse studio's warehouse in Cleveland, Ohio, USA, where all materials were grown. For the second batch, the inoculation of both mixtures was performed immediately after pasteurization in the same container at Valley City Fungi, OH, USA. The inoculation process of both batches and a summary of the quantity of specimens grown are detailed in Appendix E.

2.4. Mycelium Growth

To allow for comparison between fungal species, the growth environments and growth time were kept the same for both batches. The mycelium-based materials were grown for 6.5 weeks (i.e., 46 days) in a dark opaque tent. The temperature and relative humidity inside the tent were 25.7 ± 0.2 °C and $50.5 \pm 5.6\%$, respectively, during the growth period. However, each specimen was grown inside a closed mold and then placed inside the large tent. The relative humidity inside each closed mold was based on the water concentration of the mycelium mixture and was considered to have remained constant throughout the growth period, as each mold was sealed with non-woven housewrap. Batches were grown at a different time of the year: from 21 March to 6 May 2020 (first batch) and from 29 June to 14 August 2020 (second batch). Mycelial growth and presence of contaminants were visually assessed throughout the growth period through the clear plastic for both batches. After 6.5 weeks of growth, all the non-woven housewrap material covering the top of the molds was removed to start the drying process of the specimens. Additional details on the growth data collected and removal of contaminated parts are explained in Appendix F.

2.5. Application of Post-Growth Treatments

Grown materials received were either dried, baked, compacted then dried, or compacted then baked. Drying myceliated materials dehydrates them to make the fungus dormant and stop its growth. If the water content of the materials is increased after the drying process, the fungus is usually able to resume growth [5]. Myceliated materials were dried in an oven for a total of 12 h at 40 °C. Due to the restrictions of the COVID-19 regulations, they were dried in 2 sessions of 6 h each and kept in the oven between these sessions.

The baking process is supposed to kill the fungus. In comparison with the drying process, if the water content of the material is increased after the baking process, no fungal growth should be observed. They were placed in an oven for 6 h at 100 °C to bake.

Half of the specimens were compacted to half of their thickness with an industrial hydraulic press (20 TON shop press from Central Machinery, Moses Lake, WA, USA) between metal plates. Further details on the application of post-growth treatments are explained in Appendix G.

2.6. Mechanical Testing Procedure

2.6.1. Compression Testing

Due to the lack of testing standards to assess the uniaxial compressive properties of myceliated materials, various ASTM standards specific to soil (i.e., ASTM D2166/D2166M-13), wood-based panels (i.e., ASTM D3501 and ASTM D1037), and thermal insulations (i.e., ASTM C165-07) have been used by research teams [8,9,11,42,55–57]. Elsacker et al., 2019, followed ASTM D3501, whereas their use of cylindrical specimens followed ASTM D2166/D2166M-13 [9]. This latter standard specifies that the specimen should have a height:diameter ratio ranging from 2:1 to 2.5:1. In their study, Elsacker et al. opted for a 0.5:1 height:diameter ratio with specimens measuring 3.75 cm in height (or 10 cm in height for pre-compressed specimens) for 7.5 cm in diameter [9]. Yang et al., 2017, followed ASTM D2166/D2166M-13, while growing their specimens to a height of 6 cm and a diameter of 5 cm (i.e., 1.2:1 height:diameter ratio) [8]. To facilitate data comparison and reduce the effect emerging from corners in rectangular specimens, ASTM D2166/D2166M-13 was used [55]. Our cylindrical specimens were grown to attain a 2:1 height:diameter ratio after post-growth treatments, as specified by the standard. However, due to the large amount of contamination, the height had to be reduced to 11 cm before the application of post-growth treatments. Therefore, specimens were tested with a height of 11 cm for the uncompacted specimens and 5.5 cm for the compacted specimens. Since the diameter could not be reduced easily, the diameters remained the same (i.e., 7 cm), resulting in a 1.57:1 height:diameter ratio.

The uniaxial compressive tests were performed at room temperature at the University of Akron's Olson mechanical testing facility. The testing room had a temperature of 20.06 ± 0.39 °C and a relative humidity of 62.6 ± 1.36%. Due to the high humidity, the specimens were stored in the Biodesign lab, which had a temperature of 25.53 ± 0.20 °C and a relative humidity of 46.14 ± 0.64%. Previous research projects also performed mechanical tests on myceliated materials under ambient conditions: 23 °C [7], and 25°C for a 50% relative humidity [9]. A 5567 Instron with a loading cell of 10 kN was used for the compressive tests. Before starting the test, the loading cell was lowered onto the specimens until a load of 1 N was reached. ASTM 2166-13 requires a loading rate that will produce a strain of 0.5 to 2%/min. Yang et al., 2017 used a 2%/min loading rate and our team decided to follow the same rate [8]. The standard also specifies that the specimen should fail within 15 min, but our specimens did not fail. Elsacker et al., 2019 stopped their compressive tests when the strain produced a height deformation between 70 to 80% strain [9]. In this regard, our tests were stopped when the specimen experienced 75% deformation (i.e., strain of 0.75) or the maximum load of 10 kN was reached. The specimens' dimensions (i.e., height, diameter, and weight) were measured according to the ASTM C303-10 to determine specimens' density before the first test [58].

Since the mycelium-based specimens did not fail under the compressive load, the team decided to reload the specimens grown from *Pleurotus ostreatus* to study the hysteresis and dilatation of mycelium-based materials. Therefore, the specimens were first loaded up to a 75% height deformation (i.e., strain of 0.75) or a load of 10 kN. Once one of these limits was reached, the load was released. Specimens were left to dilate for 10 min. After 9 min of being released, the specimens' height and diameter were measured. After 10 min of being released, the specimen was loaded a second time with an updated loading rate. The same formula was used to calculate the rate of testing (i.e., 2%/min). The height of the specimen changed between the start of each test. Thus, the rate was recalculated with the updated height for the second test. The height of all specimens was taken before and after both tests, 1 h after the end of the second test, and after 1 week of testing to study the dilatation over time. The width of the specimens was also recorded but was not useable due to the pieces falling off from the sides. The weight of the specimens was not recorded after the beginning of the test as specimens broke into multiple pieces every time they were handled.

Different methods can be used to measure Young's modulus. The slope of the linear portion of the stress–strain curve is commonly used to calculate Young's modulus. Our materials did not fail, and the stress–strain curves did not possess linear portions. Elsacker et al., 2019, seemed to have used the entire stress–strain curve to calculate Youngs' modulus [9]. Since our specimens did not fail, the resistance to the load started an exponential trend towards the middle of the test (i.e., densification). Therefore, extracting Young's modulus from the entire deformation of the specimen seems to be incorrect. Two studies obtained Young's moduli of myceliated specimens at 20% height deformation (i.e., strain of 0.20) [10,11], so this technique was chosen. Furthermore, the top and bottom surfaces of the myceliated specimens were not entirely flat, but highly textured. Therefore, the data collected at the beginning of the tests (up to around a strain of 0.05) only showed the compressive resistance of parts of the specimens. Calculating Young's modulus at a strain of 0.2 (i.e., 20% height deformation) allowed the removal of all this variability in the data. To facilitate comparison between studies, Young's modulus was still calculated with three different techniques used in the literature. First, it was calculated from the slope of the stress–strain curve between strain values of 0.19 and 0.20. This technique shows Young's modulus at a strain of 0.20. Second, it was calculated from the slope from the start of the test up to a strain of 0.20. Third, it was calculated from the stress–strain slope of the entire specimen deformation. Additional information about the calculation of Young's modulus is presented in Appendix H.

2.6.2. Three-Point Bending Testing

The bending testing followed the procedure from ASTM D1037-12: Standard Test Methods for Evaluating Properties of Wood-base Fiber and Particle Panel Materials, Section 9. Static Bending [54]. Since each mold contained six specimens grown as one entity, each specimen was cut to the desired size (i.e., 76 mm) prior to testing.

The bending tests were performed with the same Instron that was used for compression tests. A 5567 Instron with a loading cell of 100 N was used for the bending tests. A specific testing apparatus was built according to the ASTM standard. Rounded supports had a span of 266 mm, which is equal to 24 times the average nominal thickness of the panels. Before starting the test, the upper point was lowered onto the specimens until a load of 0.01 N was reached. ASTM D1037-12 requires a uniform loading rate to achieve an outer fiber strain rate of 0.005 mm/mm/min. Since each specimen had a slightly different thickness, each specimen was tested with a different loading rate. As a reference, the average loading rate of all specimens tested in bending for the first batch was 5.043 mm/min.

Tests were stopped once the specimen failed, or the load applied reached less than 20% of the maximal load achieved during the test.

The specimens' dimensions were measured according to the ASTM C303-10 to determine specimens' density before testing [58]. Some specimens broke asymmetrically (i.e., they did not fracture at the location of the upper point pushing down onto the specimen or the middle of the specimen) and/or the fracture was not parallel to the apparatus supports. After the test, the fracture location and angle were measured to study asymmetric failure (Figure A1). Additional information about the bending test data analysis is provided in Appendix H.

3. Results

3.1. Quantifying Particle Sizes

A total of 4004 particles were measured for the large mixture and 7014 for the small mixture. In both mixtures, micro-particles (i.e., size less than 0.01 mm) were detected. Due to reorientation of the particles during the sieving process, a few particles longer than 20 mm were still present in both mixtures. However, there was a clear difference in the lengths and widths of the set of particles from both mixtures (Figure 2). For the large mixture, the mean length of particles was 5.89 ± 4.08 mm, with a length ranging from less than 0.01 to 51.78 mm. The mean width was 0.63 ± 0.69 mm, with values ranging from less than 0.01 to 8.29 mm. For the small mixture, the mean length of particles was 2.99 ± 3.57 mm, with lengths ranging from less than 0.01 to 72.95 mm. The mean width was 1.13 ± 0.80 mm with values ranging from less than 0.01 to 9.50 mm.

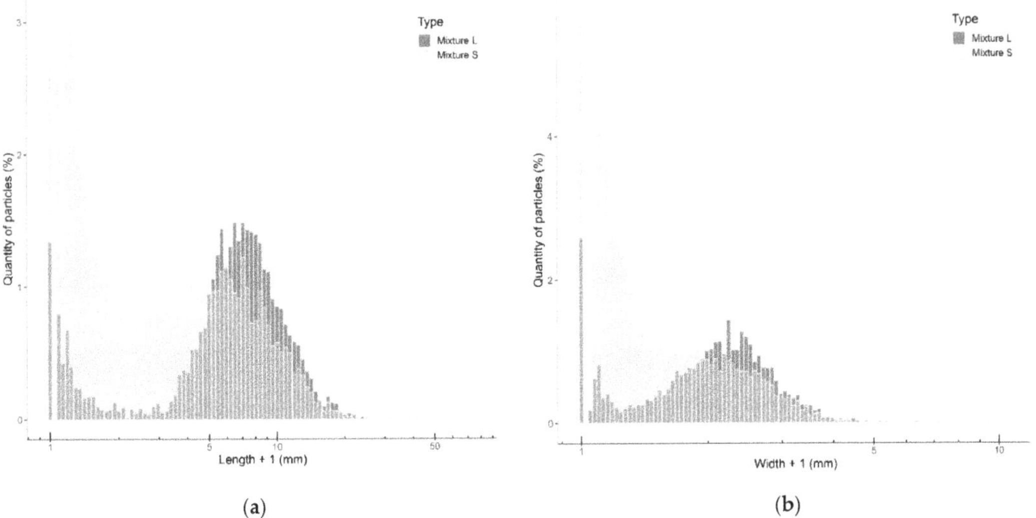

Figure 2. This figure shows the difference between the distribution particles' dimensions in both mixtures (L for large and S for small). The *x*-axis is displayed on a log-scale. Due to the presence of 0s in the data (emerging from micro-particles), which equal $-\infty$ on a log-scale, the data was offset by adding 1 to every value. Particles shown at 1 mm represent any micro-particles detected measuring less than 0.01 mm; (**a**) shows the distribution of particles' length + 1 in both mixtures; (**b**) shows the distribution of particles' width + 1 in both mixtures.

3.2. Growth of Mycelium-Based Materials with Pleurotus ostreatus

3.2.1. Growth Assessment

Appendix I contains information about the evolution of mycelium and contaminants' growth over time. All specimens grown for bending testing, and 65% of those for compression testing, exhibited some contamination by species believed to be *Trichoderma harzianum* or *Penicillium* sp. In addition, species visually resembling *Rhizopus* sp. (i.e., pin mold), and *Dactylium* sp. or *Hypomyces* sp. (i.e., cobweb mold), were also observed on some of the specimens. The contamination may have resulted from autoclaving large quantities of substrate or the mixing and placement in the molds (i.e., warehouse environment). The specimens grown in PVC pipes having a height of 34 cm had more contamination at the bottom of the specimen than the top, which may have resulted from higher humidity. Therefore, the growth of contaminants seems to correlate with high humidity. Throughout the growth and drying periods, fruiting bodies emerged from the specimens. These mushrooms had to be removed regularly as contaminants were growing on them. For the specimens grown for compression testing, the weight of contaminated material removed accounted for 59.16% of the initial weight from mixture L and 63.22% of that from mixture S. For the specimens grown for bending testing, the weight of material removed accounted for 40.99% of the initial weight from mixture L and 21.72% of that from mixture S. Therefore, one particle size did not seem to increase contamination in all cases (i.e., both compression and bending specimens).

In addition to the contamination, quantities of 805 and 697 g of fruiting bodies were removed from the specimens grown for compression and bending testing, respectively. Therefore, the weight of fruiting bodies collected equaled approximatively $3.07 \pm 3.48\%$ and $3.20 \pm 2.96\%$ of the initial specimen weight for both compression and bending specimens, respectively. Since the mixture was composed of 67.5wt% of water, the weight of fruiting bodies accounted for $9.45 \pm 10.70\%$ and $9.85 \pm 9.10\%$ of the initial dry weight of the specimens, respectively.

3.2.2. Shrinkage and Weight Reduction of Cylindrical Specimens Grown for Compression Testing

The shrinkage of the specimens was calculated for the entire manufacturing process. Detailed data about the shrinkage of the cylindrical specimens are shown in Appendix I. For these specimens, the mycelial growth process lowered the initial weight by 5% (Table A1). The specimens had to be slowly dried before applying the post-growth treatments. This step was necessary because post-growth treatments were applied in a non-controlled environment (i.e., warehouse) where specimens were exposed to various contaminating species. Furthermore, the materials were stuck inside their mold at the end of the growth process and could not be extracted without damaging them. Therefore, the cylindrical specimens were kept inside their mold during the drying period. This drying process reduced the weight of the specimens to 50% of their original weight. The compaction process did not significantly affect the weight of the specimens. The drying and baking treatments finally brought the weight of the materials down to around 22% of their initial weight at inoculation.

In terms of height, the growth process did not significantly modify the specimens' height. The drying period lowered their height by around 5%. After this drying period, the specimens were cut to a height of 11 cm. Therefore, the rest of the height evolution was considered to start with 11 cm equaling 100% at this stage. Compacted specimens were mechanically compacted to half of their height (i.e., 50% or 5.5 cm). Then, the baking and drying treatments reduced their height to 49.5% before testing. For the uncompacted specimens, the drying/baking process only reduced their height by around 0.3%. Therefore, there was a mean height shrinkage of 5.4% for the cylindrical specimens.

Shrinkage also happened horizontally (i.e., to the diameter of the cylindrical specimens). No shrinkage was observed during the growth period (Table A1). The drying period prior to post-growth treatments shrank the 17 cm high specimens (i.e., uncompacted) by an average of 3.24% An additional 0.28% diameter shrinkage was observed during the drying/baking process. The compacted specimens were compacted inside their PVC molds used during the growth period. Compaction treatment increased the specimens' width to their original width by pressing the material against the inner surface of the molds.

The specimens were removed from the molds for the baking/drying process. During this process, their diameter increased by around 5.31%. Therefore, only uncompacted specimens should be used to estimate the shrinkage observed in cylindrical specimens. The total shrinkage from inoculation to the beginning of mechanical testing would then average 3.52%.

3.2.3. Shrinkage and Weight Reduction of Panel Specimens Grown for Bending Testing

Compared to the cylindrical specimen analysis, the high level of contamination made the weight study inconclusive. The uncompacted specimens' height shrank by around 9.29% when considering all specimens (Table A2). The height of the compacted specimens was reduced by 62.68% during the entire process, including the compaction post-growth treatment. Knowing that the compaction post-growth treatment accounted for 50% of this height reduction, an average shrinkage of 12.68% was observed. In terms of horizontal shrinkage, the width of the uncompacted specimens was reduced by around 4.93%. The length shrinkage, however, could not be calculated because uncompacted specimens had to be trimmed due to contamination. Additional data about the shrinkage of the rectangular specimens are shown in Appendix I.

3.3. Growth of Mycelium-Based Materials with Coprinus comatus

In contrast to the first batch, mixtures L and S of the second batch contained different amounts of contamination. All specimens from mixture L were fully contaminated, and only displayed nil to low amounts of *Coprinus comatus* mycelium growth. In addition to the same species contaminating the first batch, materials grown from mixture L of the second batch also contained slime mold. All these specimens were documented and directly discarded to reduce the spread of contaminants to other specimens. In comparison, specimens grown from mixture S only displayed very low amounts of contamination across all specimens and were covered with mycelium. A share of 44% of the panels grown for bending, and 25% of the specimens grown for compression testing, exhibited some contamination. However, the levels of contamination were very low compared to the first batch, and no contamination was removed from the specimens. The difference in contamination levels is likely due to the pasteurization process not being successful for mixture L. The same pasteurization procedure was used for both mixtures, with the only difference being that mixture L was pasteurized before mixture S. In summary, all specimens grown from mixture L had to be thrown away due to the high levels of contamination and lack of *Coprinus comatus* mycelial growth. In contrast, all materials grown from mixture S were covered with mycelium and tested with no contamination removed.

3.4. Mechanical Testing

The mechanical testing served to compare the effects of three main variables on the biocomposite material's properties: fungal species (*Pleurotus ostreatus* and *Coprinus comatus*), substrate particle sizes (with or without micro-particles), and post-growth treatment (dried, baked, compacted then dried, and compacted then baked). The acronyms given to each sample are shown in Table 1.

3.4.1. Compression Testing

Of the 40 specimens grown with *Pleurotus ostreatus* (first batch) for compression tests, 26 were useable for testing due to the removal of contaminated specimens throughout all experimental groups. In this regard, the sample size of each experimental group was reduced from five to three or four depending on the group. By comparison, for the second batch, all mycelium-based materials grown from mixture S were usable (sample size of 10), but none from mixture L. Therefore, the comparison between both mixtures could not be performed for the *Coprinus comatus* species. Young's modulus of each sample is shown in Table 2. No specimen experienced sudden failure, as they were all compressed throughout the entirety of the test (i.e., until a strain of 0.75 or a load of 10 kN was reached).

The compressive Young's moduli were calculated following three methods. As stated in Section 2.6.1, the method considering the linear slope between a strain of 0.19 to 0.21 was selected for this analysis (bold in Table 2). The compressive Young's modulus ranged from 0.15 MPa (PO_LNCB) to 4.55 MPa (PO_SCB). A positive correlation between the density of specimens and their compressive Young's modulus was observed for all samples, except one (PO_SNCD). There was no significant difference between the compressive Young's modulus of the materials grown from *Pleurotus ostreatus* and *Coprinus comatus*.

Table 1. This table shows the names of the various samples produced in this study. The fungal species was added in front of these names: PO_ for *Pleurotus ostreatus* (e.g., PO_SNCD), and CC_ for *Coprinus comatus* (e.g., CC_SNCD).

Sample	Substrate Mixture	Compacted or Not	Dried or Baked
SNCD	Small (with fine particles)	Not compacted	Dried
SNCB	Small (with fine particles)	Not compacted	Baked
SCD	Small (with fine particles)	Compacted	Dried
SCB	Small (with fine particles)	Compacted	Baked
LNCD	Large (without fine particles)	Not compacted	Dried
LNCB	Large (without fine particles)	Not compacted	Baked
LCD	Large (without fine particles)	Compacted	Dried
LCB	Large (without fine particles)	Compacted	Baked

Table 2. This table shows relative humidity before testing, density, and Young's moduli of the samples for the three different calculation techniques. Acronyms for each sample are labeled in Table 1.

Sample	Sample Size	Relative Humidity (%)	Density (kg/m^3)	Compressive Young's Modulus (MPa)		
				Linear from 0.00 to 0.20 Strain	Linear from 0.19 to 0.21 Strain	Linear for Entire Test
PO_SNCD	3	9.23 ± 0.79	180.04 ± 8.54	0.63 ± 0.05	**0.29 ± 0.03**	2.56 ± 0.19
PO_SNCB	3	6.50 ± 0.32	186.13 ± 5.46	0.75 ± 0.13	**0.51 ± 0.12**	2.75 ± 0.52
PO_SCD	3	10.40 ± 1.10	258.93 ± 7.17	2.45 ± 0.34	**2.31 ± 0.13**	3.63 ± 0.14
PO_SCB	3	8.80 ± 1.09	283.07 ± 30.33	2.36 ± 0.79	**4.55 ± 2.29**	4.56 ± 0.73
PO_LNCD	3	7.78 ± 0.52	156.83 ± 5.45	0.44 ± 0.11	**0.36 ± 0.11**	2.08 ± 0.55
PO_LNCB	3	8.17 ± 1.12	136.22 ± 2.94	0.23 ± 0.01	**0.15 ± 0.03**	1.41 ± 0.16
PO_LCD	3	10.67 ± 0.34	188.88 ± 3.69	1.22 ± 0.30	**1.08 ± 0.28**	2.72 ± 0.29
PO_LCB	4	9.40 ± 0.90	217.46 ± 17.26	1.58 ± 0.24	**1.40 ± 0.22**	3.24 ± 0.21
CC_SNCD	10	6.51 ± 0.23	167.69 ± 3.83	0.57 ± 0.06	**0.61 ± 0.07**	3.17 ± 0.04
CC_SNCB	10	6.05 ± 0.23	175.79 ± 3.25	0.72 ± 0.07	**0.78 ± 0.12**	3.21 ± 0.05
CC_SCD	10	6.36 ± 0.45	264.55 ± 8.90	2.98 ± 0.30	**3.24 ± 0.25**	4.42 ± 0.17
CC_SCB	10	4.52 ± 0.97	270.21 ± 5.65	3.80 ± 0.17	**3.93 ± 0.30**	4.87 ± 0.17
CC_LNCD	0	/	/	/	/	/
CC_LNCB	0	/	/	/	/	/
CC_LCD	0	/	/	/	/	/
CC_LCB	0	/	/	/	/	/

In our first batch grown from *Pleurotus ostreatus*, the compressive Young's moduli of the samples ranged from 0.15 to 4.55 MPa. Specimens (PO_LNCB) made from mixture

L (substrate without fine particles) that were non-compacted and baked resulted in the lowest average Young's modulus at 0.15 MPa for a density of 136.22 kg/m^3. Specimens (PO_SCB) made from mixture S (substrate with fine particles) that were compacted and baked resulted in the highest average Young's modulus at 4.55 MPa. These specimens had a mean density of 283.07 kg/m^3. Overall, the average Young's moduli of the samples made from mixture S ranged from 0.29 to 4.55 MPa. The Young's moduli and stress–strain curves of the various samples grown from *Pleurotus ostreatus* show multiple trends (Figures 3 and 4): first, there was a positive correlation between the density of the samples and their compressive Young's moduli. Only one sample (PO_SNCD) of eight did not follow this correlation. Second, pre-compacting the myceliated specimens before the compression testing increased their Young's moduli as the four highest moduli were extracted from pre-compacted samples. Third, for the same post-growth treatments, specimens made out of mixture S (with fine particles) had higher Young's moduli than those from mixture L (without fine particles). Fourth, the baked materials had a higher Young's moduli than dried ones in three cases of four. The exception was the Young's modulus of LNCD being higher than that of LNCB. Finally, the mean relative humidity of the samples before testing was 8.82 ± 1.23%. The low variation in the relative humidity of the samples, especially due to the lack of correlation with the Young's moduli, did not likely affect the results of these compressive tests. The difference in relative humidity between samples is likely related to the post-growth treatment used. Baking resulted in a lower relative humidity than drying for 3 samples out of 4. However, the standard deviations show variations within each sample. Therefore, differences in relative humidity could also have emerged from variations in material heterogeneity (i.e., amount of mycelium growth, and distribution and orientation of particles). Since the weight of fruiting bodies harvested on each specimen was collected, the impact of fruiting body formation on mechanical performance of the composite material could be addressed. However, these results should be considered with care as they emerged from a study that was not designed for addressing such impact. Larger amounts of fruiting bodies harvested led to slightly lower Young's modulus in five of eight samples.

In the second batch grown from *Coprinus comatus*, the Young's moduli of specimens made from mixture S (substrate with fine particles) ranged from 0.58 to 3.69 MPa. Specimens (CC_SCB) that were compacted and baked had the highest average Young's modulus at 3.69 MPa with a mean density of 270.21 kg/m^3. Similar trends were observed in compression tests of materials grown from both species (Figures 3 and 4). There was a positive correlation between the samples' densities and their compressive Young's moduli. Compacted samples had higher Young's moduli than non-compacted ones. In both cases, baked materials showed a higher Young's modulus than dried ones. For non-compacted specimens, the difference between baking and drying was small. For compacted specimens, similar to the previous batch, baking resulted in higher Young's moduli than drying. However, the humidity of the baked samples was slightly lower than that of the dried samples. The mean relative humidity of the samples before testing was 5.86 ± 0.97%. The high sample size (i.e., 10) increases the validity of these results compared to those of the previous batch.

The study of materials' hysteresis showed that material height after test 2 was slightly lower than that after test 1 (Table 3 and Figure 5). Therefore, the material was still compressible after the first test (i.e., strain of 0.75 or 10 kN reached). In terms of dilatation, specimens did not dilate back to their initial height prior to testing (Table 3). During the first test, most uncompacted specimens reached the 0.75 strain limit first, while most of the pre-compacted ones reached the 10 kN limit first. For the second test, the maximal load of 10 kN was reached before 0.75 strain for all specimens. After the second test, most of the dilatation occurred within the first hour. Some dilatation was already observed 5 s after load removal (Figure 5). However, dilatation still occurred beyond one hour after compressive testing. A week after testing, pre-compacted specimens dilated back to a height closer to their initial height, in comparison to uncompacted specimens. In conclusion, both elastic and plastic deformations occurred in mycelium-based materials as all specimens experienced some dilatation after testing.

3.4.2. Bending Testing

For bending tests, 17 of 48 specimens grown with *Pleurotus ostreatus* (first batch) were useable for testing. Due to high levels of contamination throughout growth, extra specimens, grown from leftover mixtures in different molds, had to be cut down and used for bending testing. However, these extra specimens were not grown in the same environments nor processed along the same timeline. They were marked with the letter "I" to separate them from the regular bending specimens (e.g., PO_ILNCB and PO_ISCB). Therefore, the sample size of available specimens for testing ranged from 0 to 6 when considering both original and extra specimens separately (Table 4). For materials grown with *Coprinus comatus*, 52 of 54 mycelium-based materials grown from mixture S were usable, leading to sample sizes ranging from 11 to 15. Similar to specimens tested in compression, no specimens from mixture L were usable for bending testing. The comparison between both mixtures was again only possible for the *Pleurotus ostreatus* species. The elastic moduli of all samples are shown in Table 4 and Figure 6. No significant difference was observed between the elastic moduli of samples grown from the two fungal species. Each wooden mold contained six specimens, which were cut prior to testing. The edges of myceliated panels often exhibited more mycelial growth than their inside. Therefore, specimens on the edge of the panels were tracked. Yet, no significant difference in the elastic modulus was observed between edge and central specimens across all samples.

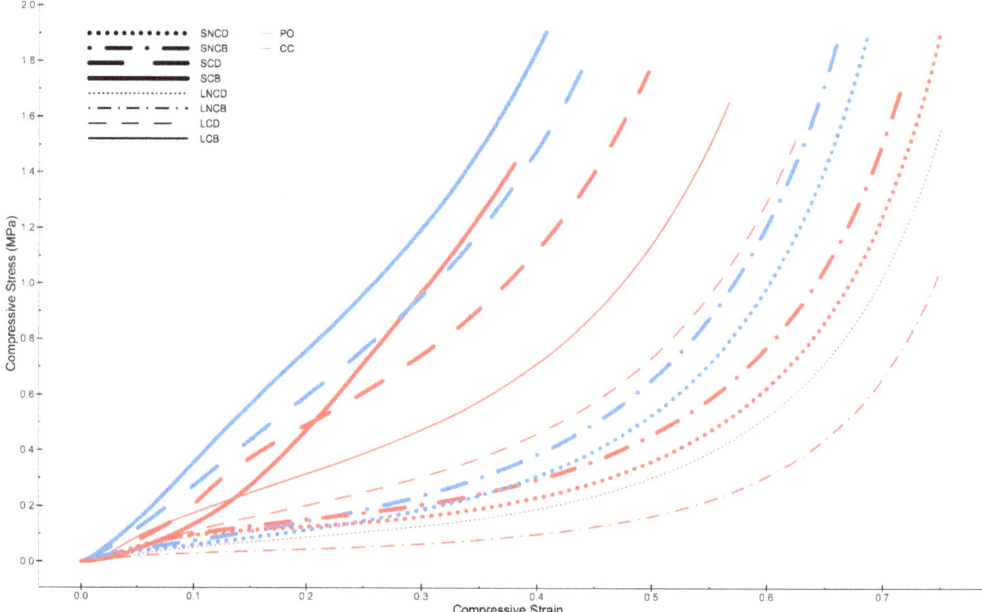

Figure 3. This figure shows the mean stress–strain curves for each sample of mycelium-based materials. The stress required to compress the material (i.e., increasing strain) increases linearly then exponentially. Due to the high variability in the stress–strain slope, the Young's modulus was calculated from 3 different strains as shown in Table 2. The Young's modulus used for comparison was calculated from 0.19 to 0.21 strain, to represent the slope of the stress/strain curve at 0.20 strain. "PO" stands for samples grown from *Pleurotus ostreatus*, and "CC" for *Coprinus comatus*. Acronyms for each sample are labeled in Table 1.

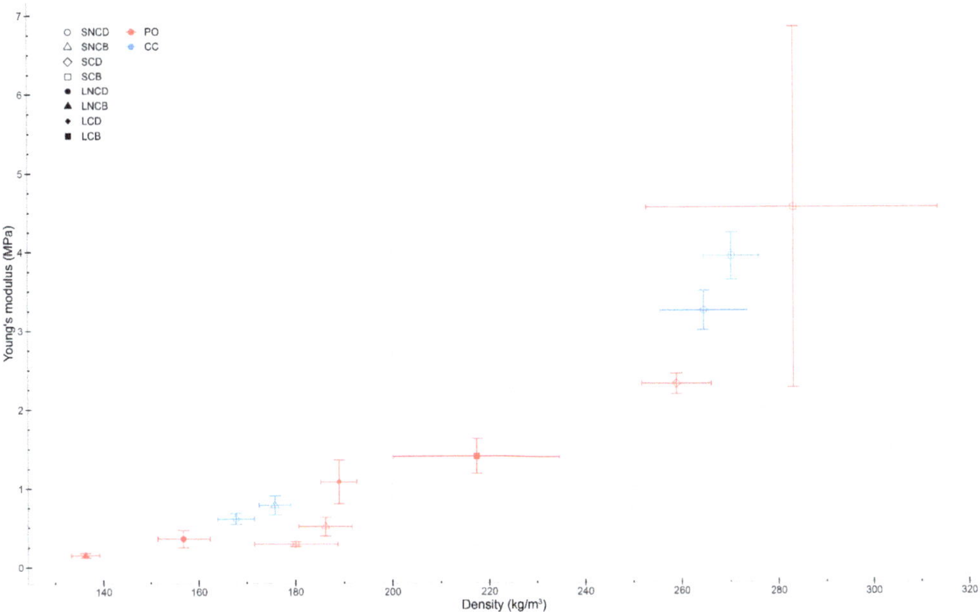

Figure 4. This figure shows the mean Young's moduli of the samples in relation to their mean density before compression testing. Young's moduli represented in this figure were calculated from 0.19 to 0.21 strain as it is closest to the traditional calculation of the Young's modulus. "PO" stands for samples grown from *Pleurotus ostreatus*, and "CC" for *Coprinus comatus*. Acronyms for each sample are labeled in Table 1.

Table 3. This table shows height deformation or strain of samples over time throughout the mechanical testing process (i.e., before the first test, immediately after the first test, before the second test (which equaled 9 min after the end of the first test), immediately after the second test, one hour after the second test, and one week after the test). Acronyms for each sample are labeled in Table 1.

Sample	Height Deformation (%)					
	Before Test 1	**After Test 1**	**Before Test 2**	**After Test 2**	**1 h after Test 2**	**1 Week after Test 2**
PO_SNCD	0.00	75.00	50.44	77.87	51.50	47.88
PO_SNCB	0.00	73.81	48.20	76.61	47.35	42.32
PO_SCD	0.00	50.39	21.01	51.94	21.95	18.06
PO_SCB	0.00	47.17	20.31	48.57	20.47	16.04
PO_LNCD	0.00	75.02	/	/	45.51	38.44
PO_LNCB	0.00	75.00	48.03	83.13	55.40	52.10
PO_LCD	0.00	61.15	/	/	29.80	23.91
PO_LCB	0.00	59.25	29.68	60.71	30.52	27.68

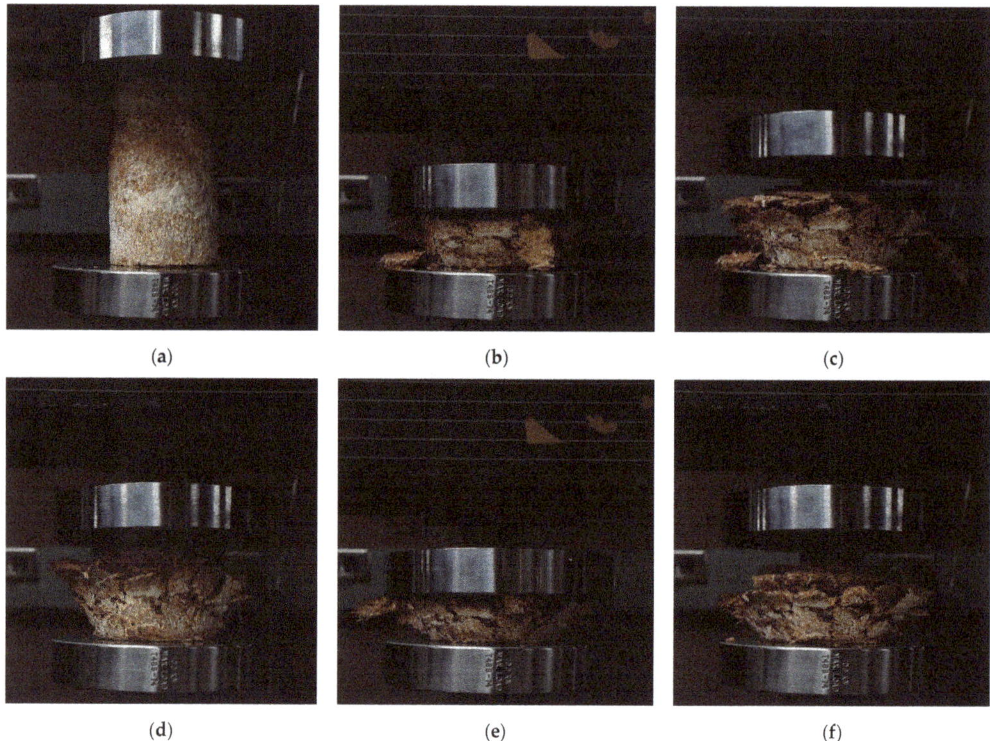

Figure 5. This figure shows the deformation of a specimen throughout the compression test: (**a**) beginning of the test; (**b**) end of the first test where a compressive strain of 0.75 was reached; (**c**) 5 s after removing the load from the first test; (**d**) 10 min after removing the load from the first test which equals the beginning of the second test; (**e**) end of the second test where a compressive strain of 0.62 and a load of 10 kN were reached; (**f**) 5 s after removing the load from the second test.

Table 4. This table shows relative humidity before testing, density, and the modulus of elasticity calculated by the Instron and from the ASTM D1037 formula (bold). It also presents the maximum load sustained by all samples, the distance between the center of the specimen and the fracture location (i.e., fracture offset from center), and the angle of that fracture in comparison to the axis of the bending supports. The letter "I" in the sample name means that the corresponding specimens were grown as an extra under different growth environments, mold sizes, and timelines. Acronyms for each sample are labeled in Table 1.

Sample	Sample Size	Relative Humidity (%)	Density (kg/m^3)	Elastic Modulus		Maximum Load (N)	Fracture Offset from Center (cm)	Fracture Angle (°)
				From Instron (MPa)	From ASTM D1037 (MPa)			
PO_SNCD	0	/	/	/	/	/	/	/
PO_SNCB	1	7.65	140.06	/	**4.88**	1.28	1.03	2.19
PO_SCD	6	6.03 ± 0.63	259.24 ± 17.86	31.66 ± 8.59	**22.39 ± 8.41**	3.62 ± 1.92	3.03 ± 1.86	3.60 ± 2.61
PO_SCB	0	/	/	/	/	/	/	/
PO_LNCD	2	5.68 ± 0.23	131.33 ± 4.36	6.37 ± 2.64	**5.64 ± 3.07**	0.64 ± 0.25	1.24 ± 0.34	13.35 ± 2.09
PO_LNCB	2	5.73 ± 0.13	123.63 ± 0.24	3.11 ± 0.29	**2.46 ± 0.23**	0.32 ± 0.08	0.81 ± 0.06	3.00 ± 2.24
PO_LCD	6	5.93 ± 0.56	235.32 ± 19.29	20.12 ± 17.44	**18.46 ± 16.95**	1.84 ± 1.48	4.29 ± 2.18	4.46 ± 3.23

Table 4. *Cont.*

Sample	Sample Size	Relative Humidity (%)	Density (kg/m³)	Elastic Modulus		Maximum Load (N)	Fracture Offset from Center (cm)	Fracture Angle (°)
				From Instron (MPa)	From ASTM D1037 (MPa)			
PO_LCB	0	/	/	/	/	/	/	/
PO_ISNCD	5	5.73 ± 0.48	154.04 ± 4.73	8.91 ± 2.13	**7.17 ± 1.60**	0.90 ± 0.33	1.18 ± 0.58	4.27 ± 5.21
PO_ISNCB	3	5.65 ± 0.24	145.78 ± 1.77	6.08 ± 0.39	**5.74 ± 0.58**	0.50 ± 0.12	2.86 ± 1.51	6.84 ± 5.59
PO_ISCD	0	/	/	/	/	/	/	/
PO_ISCB	1	5.35	160.60	0.70	**0.40**	0.24	0.53	8.13
PO_ILNCD	3	5.40 ± 0.11	127.17 ± 11.70	3.10 ± 1.50	**2.27 ± 1.16**	0.24 ± 0.14	1.87 ± 0.99	7.88 ± 3.97
PO_ILNCB	2	5.70 ± 0.15	148.56 ± 1.37	5.45 ± 0.78	**4.82 ± 0.68**	0.49 ± 0.01	0.86 ± 0.56	8.00 ± 4.99
PO_ILCD	0	/	/	/	/	/	/	/
PO_ILCB	4	5.76 ± 0.51	286.67 ± 6.84	110.42 ± 40.73	**106.08 ± 40.65**	8.85 ± 2.58	0.94 ± 1.32	6.31 ± 5.19
CC_SNCD	12	5.62 ± 0.21	133.02 ± 4.83	1.58 ± 0.70	**1.44 ± 0.72**	0.19 ± 0.08	1.90 ± 1.49	0.51 ± 0.28
CC_SNCB	11	5.50 ± 0.31	132.49 ± 6.66	1.35 ± 0.55	**1.16 ± 0.54**	0.20 ± 0.13	2.46 ± 1.60	0.60 ± 0.52
CC_SCD	14	7.85 ± 0.72	239.52 ± 27.84	25.84 ± 12.69	**22.34 ± 11.38**	2.85 ± 1.25	1.16 ± 1.06	0.58 ± 0.51
CC_SCB	15	6.70 ± 0.30	291.66 ± 25.11	53.06 ± 25.14	**48.39 ± 23.37**	5.77 ± 2.26	1.27 ± 0.81	0.67 ± 0.56
CC_LNCD	0	/	/	/	/	/	/	/
CC_LNCB	0	/	/	/	/	/	/	/
CC_LCD	0	/	/	/	/	/	/	/
CC_LCB	0	/	/	/	/	/	/	/

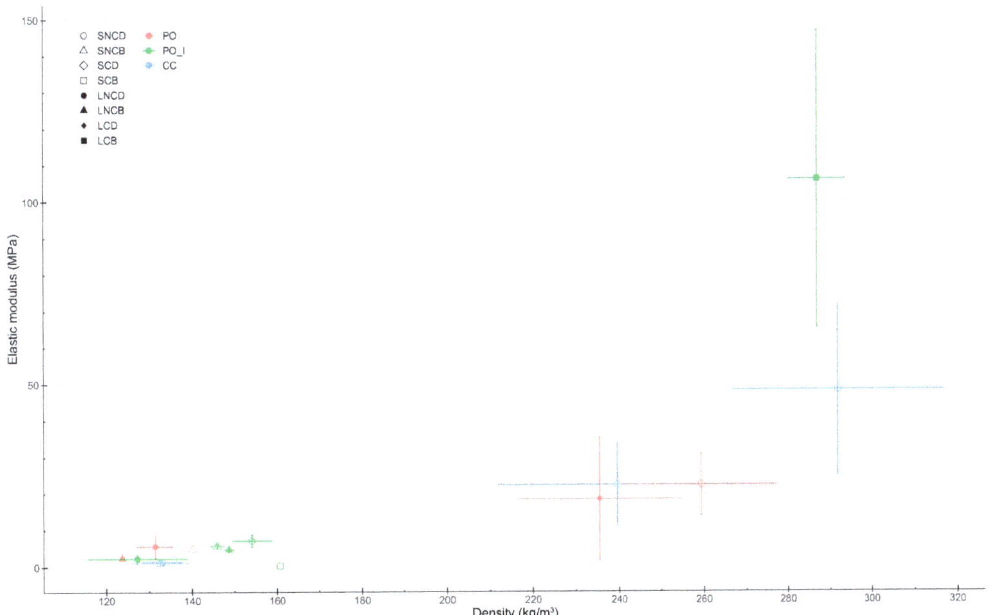

Figure 6. This figure shows the modulus of elasticity of samples in relation to their density prior to testing. "PO" stands for samples grown from *Pleurotus ostreatus* and the added "_I" means that corresponding samples were grown as extras under different growth environments, mold sizes, and timelines. "CC" refers to *Coprinus comatus* samples. Acronyms for each sample are labeled in Table 1.

For specimens from the first batch (grown with *Pleurotus ostreatus*) tested in bending, contradicting results were observed. However, specimens grown as extras were used due to high levels of contamination and resulting low sample size. Since they were grown in molds of a different size and processed on a delayed timeline, results should be analyzed with care. Therefore, data comparison between original specimens is more reliable than comparison between specimens grown as extras. Comparisons between original and extra specimens are only shown when the variable addressed cannot be analyzed between original or extra specimens. For instance, specimens (PO_ILCB) made from mixture L (substrate without fine particles), which were compacted and baked, resulted in the highest average elastic modulus at 106.08 $\pm$ 40.65 MPa for a density of 286.67 kg/m^3. In comparison, the rest of the samples ranged from 0.40 MPa (PO_ISCB) to 22.39 MPa (PO_SCD). One specimen (PO_ISCB) made from mixture S (substrate with fine particles), which was compacted and baked, resulted in the lowest average elastic modulus at 0.40 MPa for a density of 160.60 kg/m^3. This result is in contradiction with the trends observed across other samples and may be due to the different process used since this panel was grown as an extra from leftover mixture. In terms of variables, mixture S resulted in higher elastic moduli than mixture L in two samples out of two (PO_NCB and PO_CD) for the original specimens; and three samples out of four (PO_INCB, PO_INCD, PO_ICD and PO_ICB) for panels grown as extra. Drying resulted in higher moduli of elasticity than baking in one sample out of one (PO_LNC) for original specimens; one sample out of two (PO_ILNC and PO_ISNC) for panels grown as extra; and one sample out of two (PO_LC/PO_ILC and PO_SC/PO_ISC) when comparing original and extra specimens. Compacting led to higher moduli of elasticity in one sample out of one (PO_LD) for original specimens; one sample out of two (PO_ISB and PO_ILB) for panels grown as extra; and one sample out of one (PO_SD/PO_ISD) when comparing original and extra specimens. Such results should be interpreted with care since sample sizes ranged from one to six specimens. In conclusion, variables that led to higher elastic moduli were mixture S and compaction. The difference between the effects of drying or baking was less pronounced. For each sample tested under bending, specimens (i.e., that received the same treatments) were grown in the same mold. The quantity of fruiting bodies harvested was recorded for each mold. Therefore, no data analysis on fruiting bodies' effect on bending properties could be performed.

In the second batch, elastic moduli of mycelium-based samples made of *Coprinus comatus* grown on mixture S (with fine particles) ranged from 1.16 to 48.39 MPa. Specimens (CC_SNCB) that were not compacted and baked resulted in the lowest average elastic modulus at 1.16 $\pm$ 0.54 MPa for an average density of 132.49 kg/m^3. Specimens (CC_SCB) that were compacted and baked resulted in the highest average elastic modulus at 48.39 $\pm$ 23.37 MPa for an average density of 291.66 kg/m^3. In both cases, compacted specimens had substantially higher moduli of elasticity than non-compacted equivalents. Baking resulted in a higher modulus of elasticity than drying only for compacted specimens. This increase in modulus due to baking can partially be explained by an increase in density for these baked specimens. For non-compacted specimens, the process of baking or drying did not have a significant effect on their elastic modulus.

For all specimens, the location of failure (i.e., fracture) under the bending load was recorded (Table 4 and Figure A1). In a three-point bending test, failure should occur in the middle of the specimen. This fracture should also be parallel to the supports bending the specimen, so the angle between the axis of the fracture and the axis of the supports (i.e., its width as the supports were parallel to the width of the specimens) was calculated (Figure A1). Fracture location varied throughout all samples grown from both species (Figure 7), with no consistent trend observed (Table 4). In specimens grown from *Pleurotus ostreatus*, fracture angle heavily varied in comparison to the ideal failure scenario, ranging from 2.19° (PO_SNCB) to 13.35° (PO_LNCD) (Table 4). By comparison, specimens grown from *Coprinus comatus* failed at an angle close to the ideal scenario as fracture angle ranged from 0.51° (CC_SNCD) to 0.67° (CC_SCB). Within compacted specimens grown from *Coprinus comatus*, there was a decrease in modulus of elasticity as fracture happened further away from the specimen's center, especially in baked specimens (CC_SCB) (Figure 8). Therefore, location and angle of failure can be indicators of decreasing resistance under bending.

Figure 7. This figure shows the deformation of a PO_LNCB specimen throughout the bending test: (**a**) beginning of the test; (**b**) after 6 min and 25 s; (**c**) after 13 min (just before the end of the test). (**d**) shows the asymmetric deformation of a different specimen from the PO_LCD sample after 3 min.

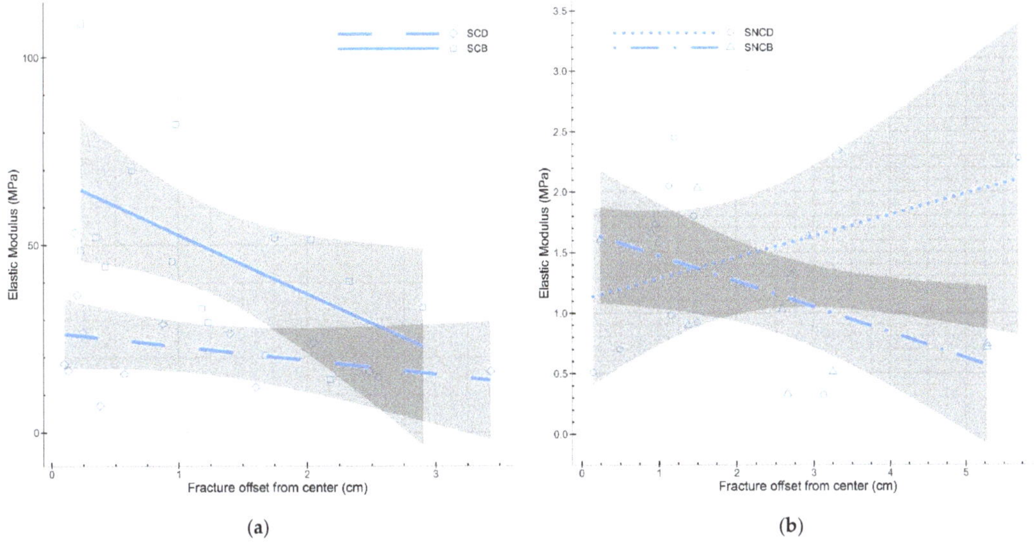

Figure 8. This figure shows elastic moduli calculated from the ASTM D1037 formula in relation to the distance between the fracture and the center of the specimen, grown from *Coprinus comatus*: (**a**) compacted samples; (**b**) uncompacted samples. For the compacted and baked non-compacted samples, elastic modulus decreases as fracture happens further away from the ideal scenario. Acronyms for each sample are labeled in Table 1.

4. Discussion

4.1. Mycelial Growth

4.1.1. Reducing Contamination and Promote Mycelial Growth

The mycelium-based materials recipe and processes impact the growth of contamination and mycelium, which impacts the performance of the materials. For the two batches grown in this study, the substrate was either sterilized (batch 1) or pasteurized (batch 2). Pasteurization is known to kill pests and competitors while minimizing the loss of beneficial micro-organisms [59]. In this regard, pasteurization resulted in lower contamination levels during mycelium growth over the substrate than sterilization [5]. In our study, medium levels of contamination were observed throughout all specimens during mycelium growth over sterilized substrate (i.e., both mixtures of batch 1). By comparison, specimens grown with pasteurized substrate (i.e., batch 2) were either fully contaminated (i.e., those grown with mixture L) or only displayed nil to very low amounts of contamination (i.e., those grown with mixture S containing fine particles). Since no significant difference in contamination was observed between specimens grown with both mixtures in batch 1, and mixtures with fine particles are harder to sterilize or pasteurize than those without, it seems unlikely that the mixture was the cause of the contamination levels. The variation in contamination levels between mixtures of batch 2 was most probably due to the success of the pasteurization process. Even if both sterilization/pasteurization processes were conducted on two different fungal species, pasteurization seems more effective as it can inhibit all contamination if successful [5]. The absence of contamination increases the process' potential for replication. In conclusion, future research should prioritize pasteurization of the substrate.

Furthermore, the growth of contaminants competes with the growth of the inoculated fungal mycelium species. Existing research projects have shown variations in growth time ranging from 6 days up to months [7,13,34,35]. In our case, access to the growth lab was limited due to COVID-19 regulations. Twenty-one days after inoculation, a thin mycelium mat was observed over most of the specimens' visible surfaces. The growth time was set to 46 days for both batches to ensure complete growth over all specimens. The addition of more-nutritious substrates or higher concentrations of mycelium spawn enhances mycelial growth and reduces contamination, resulting in higher mechanical properties [13]. However, this addition (not used in this project) decreases the sustainability of these materials as the substrate is composed of products other than waste or byproducts.

4.1.2. Food Production

During the growth and drying phases, fruiting bodies were collected on the mycelium-based materials grown with *Pleurotus ostreatus*, but not on those grown with *Coprinus comatus*. Yet, materials were grown in the same dark environment with the same temperature and relative humidity. However, they were not grown during the same season (i.e., March–May for *Pleurotus ostreatus* and June–August for *Coprinus comatus*). The formation of *Pleurotus ostreatus* fruiting bodies is most abundant in spring in temperatures ranging from 4 to 24 °C [51], which is the season in which our materials were grown. *Coprinus comatus* fruiting bodies are most prolific under spring and fall temperatures ranging from 4 to 16 °C [51], whereas our materials were grown in the summer. Therefore, the variation in mushroom yield may result from the difference in the fungal species and/or growth season.

The weight of *Pleurotus ostreatus* fruiting bodies harvested from compression and bending specimens, respectively, equaled 9.45 ± 10.70% and 9.85 ± 9.10% of the initial specimens' weight. Therefore, the biological efficiency (i.e., the effectiveness of the process for mushroom production) equaled 9.45% for the compression specimens and 9.85% for the bending specimens. A wide variation of yield was observed among our specimens. The biological efficiency ranged from 0.00 to 35.94% for the compression specimens and from 0.00 to 20.51% for the bending specimens. As a reference, the biological efficiency of various substrates grown with the same fungal species, *Pleurotus ostreatus*, ranged from none for elephant grass to 61.04% for composted sawdust [60]. However, our process mainly focused on producing mycelium-based materials and was not optimized for mushroom yield. Materials were grown inside a PVC tube or a flat rectangular mold enclosed with housewrap in a dark environment. Fruiting bodies were able to grow through the housewrap material that sealed the PVC tubes by enlarging the small holes already present in this material. Therefore, they could be harvested without removing the housewrap and exposing the mycelium-based material to potential contaminants. Mushrooms were not harvested in multiple flushes and their growth was stopped by drying the materials. Therefore, the biological efficiency is presented as a reference and only shows the potential of combined material and food production.

4.1.3. Material Shrinkage

Cylindrical specimens grown for compression testing shrank, on average, by 5.4% of their original height. These specimens measured 17 or 34 cm in height and were grown and dried inside PVC tubes. The specimens grown were taller than those produced in previous research projects and were dried inside their mold (i.e., PVC pipes). This set-up may have resulted in friction between the specimen and the internal walls of their mold, thus limiting shrinkage. In terms of horizontal shrinkage (i.e., in diameter), the shrinkage of uncompacted specimens from inoculation to the beginning of the mechanical testing was approximatively 3.52%.

For the rectangular specimens grown for bending testing, compacted materials shrank by an average of 12.68% in addition to the 50% compaction. Uncompacted specimens shrank by 9.29%. The specimens' width was reduced by around 4.93% for the uncompacted specimens. The lower shrinkage in width compared to height likely resulted from the friction caused by the bottom of the molds.

4.2. Compression Testing

In our analysis, the fungal species did not have a distinct effect on the compressive Young's modulus of the mycelium-based materials. In the literature, fungal species have been shown to affect the mechanical properties of mycelium-based materials. For instance, *Ganoderma* spp. generally have a higher strength than *Pleurotus* spp. [12], but Haneef et al. found contradictory results [7]. The effect of fungal species seems to depend on the hyphal types (i.e., monomitic, dimitic, and trimitic), which can be generative, binding, and/or skeletal [30,31].

Compressive Young's modulus was positively correlated with sample density. Density is known to have a significant effect on the stress–strain curve response of mycelium materials [48]. The correlation between porosity/density and compressive modulus was previously discussed in [13]. The density of the final mycelium-based materials can be controlled in two main steps of the process: during substrate processing and application of post-growth treatments. For example, sawdust is known to be denser than straw. However, as mycelium growth is based on oxygen access, it is usually reduced in the center of dense substrates [5,13]. By comparison, compaction after mycelial growth will partly break the mycelium network. Therefore, a trade-off between both options could be further studied. When comparing materials grown from *Pleurotus ostreatus*, materials made from mixture S (substrate with fine particles) showed an increase in density by 30% and resulting increase in compressive Young's modulus by 156% compared to those made from mixture L (without fine particles). In comparison, Rigobello and Ayres found that materials made from substrate particles having sizes ranging from 0.5 to 12.0 mm had a slightly higher density but lower Young's modulus than those made from particles having sizes ranging from 4.0 to 12.0 mm [42]. In their case, the presence of smaller particles reduced Young's modulus. In terms of post-growth treatments, material compaction increased density by 48% when considering all specimens but increased compressive Young's modulus by 511%. For structural applications, the material should be compacted after the growth of mycelium to increase its compressive modulus while avoiding mycelium growth inhibition emerging from a lack of oxygen access. When averaging all samples, baking increased material density by 4% but increased compressive Young's modulus by 43%. Baking (at 100 °C) is supposed to kill the fungus that may damage the mycelial network throughout the specimen. By comparison, drying (at 40 °C) is only supposed to put the fungus in a dormant stage. In this regard, the choice of drying or baking the materials depends on the desired application. For instance, keeping the fungus alive can potentially induce self-healing properties, but may not be desired in environments where humans may potentially be exposed to large quantities of fungal spores. The slight humidity difference between samples may be another factor influencing the variation in Young's modulus, since the strength of a material is roughly inversely proportional to its moisture content [45]. However, no correlation was observed between those values.

The substrate preparation, growth conditions, and post-growth processes used in this research resulted in similar or slightly higher compressive Young's moduli than those recorded in previous research projects, as described in Section 1. Introduction [8,9,11,31,45,48]. For instance, materials grown from *Pleurotus ostreatus* on hemp mat reached a compressive strength of 0.19 MPa [31]. In comparison, our mycelium-based samples made of *Pleurotus ostreatus* grown on chipped and sieved straw resulted in compressive Young's moduli ranging from 0.15 to 4.55 MPa. To the authors' knowledge, no studies have tested the mechanical properties of materials grown with *Coprinus comatus*. Current mycelium research explores the use of mycelium-based materials to replace traditional foam materials. In this case, the goal is to obtain high mechanical strength with low density. However, difficulties in manufacturing lead to inconsistencies in mechanical behavior of composite materials. For instance, materials made from the same substrate still show large standard deviations due to varying particle distribution and orientation [42].

Studying the hysteresis and dilatation of mycelium-based materials is important to predict their final dimensions and behavior. Since the height of all specimens increased after the compressive tests, materials experienced some elastic deformation. In addition, since no specimen dilated back to their initial height, plastic deformation also occurred in all materials tested. Drying the grown materials seemed to increase dilatation after compression of materials made from the large mixture. For specimens grown with the small mixture (i.e., containing fine particles), baking resulted in increased dilatation after compression testing. As expected, specimens that experienced higher strain (i.e., uncompacted specimens) were less likely to dilate back to a height closer to their initial height. Depending on the application, various substrate mixtures and post-growth treatments can be employed to target desired material behavior under compression, including maximum load, elasticity, hysteresis, and dilatation.

4.3. Bending Testing

Similar to compression testing, the elastic modulus of the mycelium-based materials was not thoroughly affected by the difference in fungal species, but the modulus was strongly correlated with material density. The results from the *Pleurotus ostreatus* materials were less consistent than those grown from *Coprinus comatus*. The main difference in behavior between the two fungal species was observed in the failure angle. Even if the specimens of each sample were produced with the same process, the variation in elastic moduli remained high, especially for materials grown from *Pleurotus ostreatus*. The variations in elastic moduli and failure angle among samples may partially be explained by the heterogeneous properties of the material due to irregular substrate particle distribution and orientation, along with heterogeneous mycelium growth. Furthermore, results obtained from the bending materials grown with *Pleurotus ostreatus* should be interpreted with care due to the small sample size and testing of materials initially grown as extra.

On average, the presence of fine particles in the substrate mixture resulted in a higher density, yielding higher elastic moduli. The difference between drying and baking did not have a significant effect on the material's elastic modulus, whereas compaction substantially increased the elastic modulus. For instance, the compaction of materials grown from *Coprinus comatus* resulted in a density increase of 100% and an increase in the elastic modulus of 2611%. In comparison to Appels et al., 2019, our compacted then baked specimens achieved flexural moduli in the same range as those of heat-pressed specimens having higher densities [32]. Therefore, the substrate preparation and post-growth treatments can be used to tune the behavior of the materials under bending for the application of interest.

4.4. Emerging Synergies between Bioremediation, and the Production of Food, Materials, and Medicine

This project is a first step towards a model in which fungi, especially their mycelium, can serve to decontaminate substrates, while producing fruiting bodies, medicinal drugs, fertilizers, and materials (e.g., for buildings or packaging). It stems from a circular model described by Paul Stamets, in which fungi are used for food, medicine, bioremediation, and fertilizer [51]. The updated model can be implemented in the two following examples.

The first example aims at reducing the environmental dispersal of chemicals that negatively impact the environment. The run-off of nitrates and phosphates, coming from agricultural fertilizers, is the main cause of harmful algae blooms in the Great Lakes [61,62]. The bloom of harmful algae and cyanobacteria lowers water oxygen levels, kills marine life, and negatively affects human health and agriculture [63]. Furthermore, the dispersal in the environment of these limited resources contributes to their increasing scarcity. As a response, hyper-accumulators of nitrates and phosphates, in the form of prairie grasses, could be planted along the edge of over-fertilized agricultural fields to limit environmental dispersal of these chemicals. The prairie grasses could then be harvested according to the growing seasons. They would then serve as a substrate for fungal growth to filter and uptake the substances, thus limiting their return in the environment. A variety of fungal species could be grown on this substrate to provide various functions, such as harvesting scarce substances in the fruiting bodies, producing mycelium-based materials, supplying food through edible fruiting bodies, and manufacturing of medicinal drugs. Chemical testing would be conducted throughout the process to track the substances and ensure safe levels depending on the targeted application. At the end of their life cycle, mycelium-based materials are biodegradable and can be reused as compost for fertilizers or as substrate for new generations of mycelium growth. Furthermore, a wider variety of organic substrates could be used to increase the impact of this upcycling process. For instance, algae can be harvested using algal turf scrubbers to produce living materials and various substances [4,64–68].

Similarly, organic waste from building and demolition sites (e.g., wood) can serve as substrate for mycelium growth to produce food and materials while decontaminating it. A large quantity of waste placed in landfills comes from construction and demolition sites, and its toxicity can contaminate the environment, including ground water [2,3]. Various studies have shown that children from various countries, such as USA, Nigeria, and France, are being poisoned with lead from buildings, especially through paint degrading into dust particles [69–74]. As a response, various fungal species can serve to substantially reduce lead concentration in aqueous substrates [75]. Fungi secrete enzymes known to break down aromatic compounds, including Polycyclic Aromatic Hydrocarbons (PAHs), present in waste from construction materials [76]. *Coprinus comatus*, a bio-accumulator of heavy metals, is a species recommended for decontamination of substrates with nitrates and phosphorus-bound toxins [51]. Research shows that using biochar and mycoremediation can lock away contaminants, such as lead, from the water cycle and bioavailability [75,77,78]. Chemical analysis should be performed based on the substrate and fungal species used to quantify the accumulation and breakdown of the various chemicals contaminating the substrate. Such analysis, called Toxicity Characteristic Leaching Procedures (TLCPs), would evaluate the effect of mycelium growth on bioremediation while ensuring the edibility of the fruiting bodies.

5. Conclusions

The diversity of fungal species and their functions has the capability to serve multiple functions in our industrialized society, from decontamination to production of food, materials, and medicine. This study shows the potential of two fungal species commonly used in bioremediation practices to produce building materials. These mycelium-based materials showed Young's and elastic moduli comparable to or slightly higher than those from other mycelium-based materials, depending on the variables studied. The effect of the selected variables (e.g., fungal species, substrate particle size, and post-growth treatments) tested in this study enhances knowledge about which parameters serve to tune mechanical properties of the final material. Density of composite materials was positively correlated with mechanical performance. The presence of micro-particles and compaction of grown material (as a post-growth treatment) increased density and the resulting Young's modulus, in addition to the elastic modulus. Differences between baking or drying grown material were less pronounced, with baking leading to slightly higher compressive moduli in most cases. The standard deviations observed within samples exemplify the need to better understand variables affecting material properties and to develop more accurate manufacturing procedures. Heterogeneity of produced materials was particularly apparent when analyzing the fracture location's eccentricity of materials under bending. The study of shrinkage during growth and drying periods provides valuable insight into predicting the materials' final dimensions. Both fungal species led to similar mechanical behavior. In addition to *Pleurotus ostreatus*, *Coprinus comatus*, a species known for bioremediation and food production, can also serve for material production and, potentially, a combination of all three. Integration of material production is suggested as an enhancement of the circular model described by Paul Stamets [51]. Further studies still need to be conducted to validate the combined use of these species' mycelium as a binder for building materials while decontaminating substrate and producing edible fruiting bodies.

Author Contributions: Conceptualization, C.M., P.G. and T.H.; methodology, C.M., P.G. and T.H.; software, T.H.; validation, T.H. and P.G.; formal analysis, T.H.; investigation, T.H. and R.N.; resources, C.M. and P.G.; data curation, T.H., P.G. and R.N.; writing—original draft preparation, T.H.; writing— review and editing, T.H., P.G., C.M. and R.N..; visualization, T.H.; supervision, P.G. and C.M.; project administration, P.G.; funding acquisition, P.G. and C.M. All authors have read and agreed to the published version of the manuscript.

Funding: This research was partially funded by USDA—Rural Business Development Grant (RBDG) under the Project Title: "Develop prototype panels for customer evaluation" processed by the University of Akron Research Foundation (UARF) in 2019 and 2020.

Institutional Review Board Statement: Not applicable.

Informed Consent Statement: Not applicable.

Data Availability Statement: Data is available upon request to the authors.

Acknowledgments: The authors would like to thank: John Burmeister from Valley City Fungi for providing knowledge and help with pasteurizing substrates; Henry Astley for helping with writing python code to extract particle size dimensions; Hunter King for his suggestions to measure particle dimensions of sieved materials; Jiansheng Feng for his support with mechanical testing; David McVaney for his ideas on mechanical testing procedures; Nicholas Mazzocca, Derek Jurestovsky and Jessica Tingle for their help with basics of making plots in R; Randy Mitchell and Richard Einsporn for their advice with data analysis; and Elena Stachew for her comments and suggestions throughout the project.

Conflicts of Interest: The authors declare no conflict of interest.

Appendix A

The percentages of each ingredient used in inoculation recipes from the literature are presented here. The GrAB project team used the following averaged percentages: 10 wt% of *Pleurotus ostreatus* spawn, 14 wt% of substrate, and 76 wt% of water [79]. Jones et al., 2018, soaked the substrate in Type 1 Milli-Q® water and sterilized the mixture before adding 25 wt% of *Trametes versicolor* spawn, while explaining that lower percentages of spawn slowed mycelial growth and increased risks of contamination [18]. Elsacker et al., 2019, used the following recipe in weight percentages: 10 wt% of *Trametes versicolor* spawn, 20 wt% of substrate (chipped, sieved and sterilize), and 70 wt% of sterile demineralized water [9]. The LIWAS project used the following recipe: 6 wt% of *Hypsizygus ulmarius* mycelium spawn, 34.5 wt% of substrate, and 59.5 wt% of water [5]. In comparison with other projects, these last percentages produced mixed mycelium growth results, likely due to the low amounts of mycelium spawn provided [18].

Appendix B

Details about sieving non-spherical objects and the sieving procedure conducted are described here. Due to the high aspect ratio between the length and width and/or thickness of the chipped material, the use of sieves is not as straightforward as with more spherical objects. The amount of material passing through the sieve depends on the amount of material placed on the sieve, the movement produced for sieving, and the sieving time. If too much material is placed on the sieve, particles are more likely to be arranged in a position perpendicular to the sieve holes (vertically). The horizontal sieving technique was chosen to limit the vertical repositioning of the particles after being positioned on a sieve. The speed of the circular sieve movement can also cause the particles to reposition vertically. Therefore, these three variables were kept constant throughout the sieving of the chipped straw.

The sieving process for the 5.7 mm sieve consisted of placing 140 g of chipped material on the 5.7 mm sieve and making 10 horizontal circles with the sieve for 8 s in a clockwise direction, then in an anti-clockwise direction. The sieving process for the 1.5 mm sieve consisted of placing 70 g of chipped material on the 1.5 mm sieve and making 5 horizontal circles with the sieve for 4 s in a clockwise direction, then in an anti-clockwise direction.

Appendix C

The details of the procedure performed to quantify the particle sizes of both mixture are presented here. First, particles were positioned on a light table while making sure that they did not touch each other. If particles were touching each other, the algorithm would recognize them as one, and resulting dimensions would be false. A picture of these particles was taken from a top view with a Sony alpha 7R II camera (Sony, New York City, NY, USA) mounted on a tripod. The following settings were used: f/4, 1/100 s, ISO-100 at a focal length of 35 mm. The 35 mm focal length was used to reduce lens distortion. The picture was then opened in Adobe Lightroom (version 6.10.1, Adobe™, San Jose, CA, USA) to crop, increase contrast, and correct the lens distortion. The adjusted pictures were then imported into a Python script (version 3.7, Python Software Foundation, Wilmington, DE, USA). The script turned the image into a binary image (e.g., only composed of black or white pixels), labeled each individual item, and calculated the number of objects and the area, perimeter, eccentricity, major axis length, and minor axis length of each item. A ruler integrated in the light table was used to scale the image by converting the number of pixels from the picture to mm.

Appendix D

The pasteurization process can be divided into the following steps: first, the substrate mixture (S or L) is placed in a large steamer with a rotating arm in its center. Then, the required amount of water is added to the mixture based on the necessary percentages from the inoculation recipe (detailed in Section 2.3). The steamer is then closed and both mixing and steaming are started. The steamer is kept on for one hour at a temperature of 64–65 °C to allow proper pasteurization. A thermal blanket is placed around the mixer to retain the heat. The mixture is then allowed to cool for one hour until it reaches a temperature of 27 °C. To speed up the cooling, the mixer is kept on and the fan of the steamer blows filtered outdoor air through the mixture. To enhance the pasteurization process, water is introduced before pasteurizing the substrate. However, the pasteurization process also brings moisture in. To quantify the quantity of water added to the mixture during the pasteurization process, the team ran the steamer without any substrate. The steamer produced approximatively 5.90 L of water, which represented 13.40% of the water in our inoculation recipe.

Appendix E

The inoculation procedure of both batches and summary of the quantity of materials grown are described here. Mixture L of the first batch was inoculated as follows: first, 9303.3 g of sterilized dry substrate was placed inside a sterilized drum mixer. With the drum mixer on (i.e., rotating), 28 L of distilled water was slowly poured inside the mixer. The mixer was left running to mix the substrate and water for 10 min. Then, 4134.8 g of *Pleurotus ostreatus* grain spawn was added. The whole mixture was mixed for an additional 30 min. The molds were sterilized with 70% isopropyl alcohol. After 30 min of mixing, the mixture was placed and compacted into the sterilized molds. The mixer was kept running while the team filled the molds to prevent the mixture from sitting still for too long. An acrylic disk was used for pushing by hand as hard as possible to compact the mixture multiple times during the filling process. For the 17 cm tall PVC tubes, the mixture was compacted twice: when the mixture filled half of the tube height and when it filled the entire height. For the 34 cm tall PVC tubes, the mixture was compacted four times (i.e., at each quarter of the height). Once the molds were filled, specimens for compression testing were covered with a layer of non-woven housewrap (Everbilt, Home Depot, USA) to maintain constant humidity, reduce the exposure to contaminants, and limit mycelium growth on the outer surfaces. In mycelium materials, the surfaces exposed to ambient air display increased mycelial growth. This phenomenon impacts mechanical behavior as shown in a study about jacketing the materials [44]. The bias emerging from increased mycelium growth on external surface was reduced in our study by compacting mixtures in the molds and growing all specimens in PVC tubes with top and bottom surfaces covered with housewrap. Specimens for bending testing were placed inside a ridge-like tent. The ridge-like tent was made of non-woven housewrap and clear plastic, to ensure mycelium growth was visible throughout the growth period. All covered molds were finally placed inside a sterilized large tent in order to lower the risks of contamination. An opaque tarp covered the large tents containing the specimens during the growth process to block sunlight, which is a stimulus for the formation of fruiting bodies. Filtered air flowed into this tent throughout the mycelial growth period. The same process was performed for mixture S with the following quantities: 11,841.3 g of sterilized dry substrate, 35.5 L of distilled water, and 5262.8 g of *Pleurotus ostreatus* grain spawn. The leftover inoculated mixtures, S and L, filled four additional rectangular molds each. The filled molds were then sealed with a layer of non-woven housewrap.

For the second batch, water was added before the pasteurization process and the mycelium spawn after it. Once the mixture had cooled to 27 °C, the mycelium spawn was added. After 5 min of mixing, the inoculated mixture was placed inside opaque trash bags to be transported to the growth facility (redhouse studio's warehouse in Cleveland, OH, USA). The temperature of the inoculated mixture should remain constant to avoid early growth and contaminants. Therefore, bags were only half filled with an average of 13.4 kg of inoculated mixture. These bags were then spread on a grid to reduce the heat produced by this biomass. Due to the time needed to fill all the molds, mixture S was placed in the molds on the day following inoculation, whereas mixture L was placed two days after inoculation. After being sterilized with 70% ethanol, the molds were filled with the inoculated mixture and pressed to achieve the desired heights akin to the previous batch. Finally, the filled molds were weighed and placed in their respective growth environments. The growth environments used for the first batch were reused for the second batch. In the first batch, the closed PVC tubes were set on a planar surface to maintain a flat surface on the specimens' bottom sides. However, this configuration did not let the specimens fully breathe during the growth period resulting in high humidity of the lower parts of the specimens and the presence of higher rates of contamination. Therefore, for the second batch, the filled PVC molds were placed on a metal grid to let them breathe. The leftover inoculated mixtures, S and L, filled eight additional rectangular molds each. Similar to the first batch, a layer of non-woven housewrap was used to cover these filled molds.

A total of 120 specimens were grown to evaluate the effects of the following variables on the compressive properties of mycelium-based composites: fungal species (*Pleurotus ostreatus* and *Coprinus comatus*), substrate particle sizes (with or without micro-particles), and post-growth treatments (dried, baked, compacted then dried, and compacted then baked). A total of 40 specimens were grown for the first batch with *Pleurotus ostreatus* to achieve a sample size of five. Due to increased availability of materials, space, time, and levels of contaminations observed in the first batch, the sample size was increased to 10 (80 specimens total) for the second batch grown with *Coprinus comatus*. For bending testing, the effects of the same variables were evaluated with a total of 148 specimens. The sample sizes used were 6 for the first batch (48 specimens) and 12 or 15 for the second batch (100 specimens), depending on the number of molds and material available.

Appendix F

Additional details about the growth data collected and removal of contaminated parts of the materials are discussed here. At the end of the growth process, each specimen was weighed and photographed, and its dimensions were measured. The contamination and fruiting bodies were removed throughout the drying process. In addition, pictures were taken before and after the removal of both fruiting bodies and contaminated material. Fruiting bodies and contamination were weighed as they were removed. Contaminants were seen growing on fruiting bodies. Since all specimens were kept in the same tent, keeping contaminated specimens would increase the risk of contamination of the non-contaminated specimens. However, as mycelial growth requires oxygen, mycelium growth was less abundant inside the specimens than on their surface. Therefore, cutting off contaminated parts exposed the interior of the specimens which contained substrate mixture with lower mycelial growth. This exposure promoted contaminant growth on drying panels. In this regard, the specimens were either left untouched, partly cut off, or fully disposed of, based on the amount of contamination.

Appendix G

Additional information about the application of post-growth treatments is presented here. Post-growth treatments (drying, baking, compacting then drying, or compacting then baking) were applied to the fully grown specimens in two locations: the Biodesign lab at the University of Akron and redhouse studio's warehouse where the specimens were grown. The relative humidity and presence of contaminants was lower in the Biodesign lab than at the warehouse. For instance, no growth of contaminants was observed on the panels brought to the Biodesign lab, whereas some contaminants started to grow on the panels kept at redhouse studio's warehouse throughout the post-growth treatment procedure. Therefore, non-compacted panels were baked or dried at the University of Akron to reduce the exposure to humid air and contaminants from the warehouse. The rest of the specimens were compacted, and dried or baked at redhouse studio's facility due to equipment availability. With the post-growth treatments applied, materials were moved to the Biodesign lab at the University of Akron for storage and mechanical testing.

To compact materials from the first batch, layers of aluminum foil were placed between the specimen and the metal plates to prevent the myceliated specimens from sticking to the metal plates. After the baking or drying process, some of the aluminum foil remained stuck to the myceliated specimens and metal plates. Separating the specimens from the aluminum foil and metal plates damaged some specimens. For the second batch, the aluminum foil was replaced with a polyester film, called mylar or biaxially-oriented polyethylene terephthalate (BoPET). In addition to its resistance to temperatures up to 100 °C, this material did not stick to the myceliated specimens. Once the compaction load was removed, the specimen recovered thickness. Therefore, the panel specimens (i.e., grown for bending testing) were kept compacted with metallic clamps. These clamps were only removed after the full drying or baking process. To avoid lateral expansion upon compaction, cylindrical specimens (i.e., grown for compression testing) were compacted inside of their respective growth environments (i.e., PVC pipes). However, clamps could not be placed until the specimens were taken out of the tubes. The load was kept on the specimen for a minimum of 5 min. Then, the specimens were removed from the pipes and kept compacted with clamps to maintain the desired thickness.

Appendix H

Additional details about the calculation of Young's modulus and the analysis of the bending test data are described here. Throughout compression testing, the cross-sectional area of the specimen increases. To calculate the compressive stress, ASTM D2166/D2166M-13 states that the cross-sectional area of the specimen used in the equation should account for specimen deformation throughout the test [55]. The ASTM standard contains a formula to estimate the cross-sectional area of a specimen for a given load. This formula was used for one of the specimens that had an initial cross-sectional area of 42.19 cm^2 before the test. The formula estimated a cross-sectional of 168.76 cm^2 after 75% deformation. However, the measured cross-sectional area was 59.58 cm^2 after 75% deformation. Therefore, this formula was not used in the calculations of compressive stresses.

In addition to the load-deflection data, the modulus of elasticity, maximum load, and flexure extension at maximum load were collected from the Instron. However, ASTM D1037-12 contains a specific formula to calculate the modulus of elasticity. The moduli from the Instron reading and the ASTM calculation were slightly different. The modulus calculated by the Instron was higher in most cases. Since the calculation used to find the modulus by the Instron is not known by the authors, the modulus of elasticity calculated from the ASTM standard was kept for data comparison. Following the standard, the slope of the straight-line portion of the load–deflection curve was calculated with a linear regression of the load–deflection curve between 10% and 40% of the maximum load.

The figure below presents how the fracture offset from the center and fracture angle were measured for the analysis of the specimens after the bending tests.

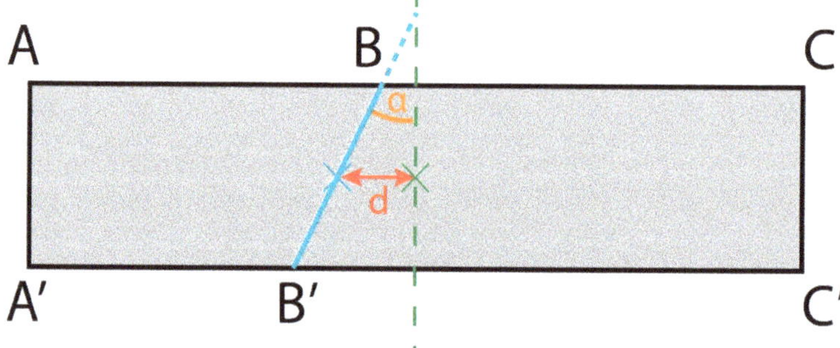

Figure A1. This abstract diagram shows the measurements used to compare the fracture observed in each specimen under bending to the ideal scenario. A bending specimen is displayed from the top view. The failure of the specimen (i.e., fracture) is shown in blue, whereas the ideal failure scenario is shown as a green dashed line. The two broken halves of the specimens are measured (A, B, C, A′, B′, and C′) to find the location of the specimen's failure. The fracture offset from center (d) represents the distance between the middle of the ideal failure scenario (green point) and the middle of the actual failure location (blue point). The fracture angle is measured by comparing the axes of the ideal and actual failures (α).

Appendix I

The evolution of mycelium and contaminant growth in the first batch is briefly described here. During the growth process, the state of the panels placed was visually assessed in a see-through enclosure. Environments were kept closed to limit contamination. Eleven days after inoculation, small amounts of mycelial growth were observed on the specimens. However, grey and green spots, likely contaminants, were seen. Twenty-one days after inoculation, mycelium formed a thin mat over most of the visible specimens' surfaces. The grey and green spots were still present. Forty-six days after inoculation, the mycelium formed a thick mat over the specimens, and the growth environments were opened for thorough assessment.

Further information about the height shrinkage and weight reduction of both cylindrical and rectangular specimen types is presented here. The shrinkage of the specimens grown for compression testing (i.e., in PVC pipes) was calculated by comparing the weight and dimensions of all specimens at the following stages: (1) inoculation, (2) beginning of drying (i.e., when growth environments were opened at the end of the growth period), (3) before cutting (i.e., specimens had to be cut to the sizes required for testing), (4) before the application of post-growth treatments, (5) before baking or drying, and (6) before mechanical testing (Table A1). To limit the growth of the contaminants after the mycelium growth period, contaminated parts of the specimens were removed from the grown materials. These parts do not possess mycelial growth. As such, they should not be used for the mechanical tests since they would generate invalid data. Removing these pieces jeopardized the comparison of the height and weight measurements. The width was not affected as the contaminated pieces were mostly slices at the top and bottom of the specimens. Therefore, upper or lower portions of the specimens were removed. In this regard, the weight evolution was analyzed for the specimens, with less than 10 g of contamination removed throughout the entire process.

Table A1. This table shows the evolution of the weight, height, and diameter from inoculation to the beginning of compression testing. Specimens were cut between steps 3 and 4. The weight was therefore kept between these two steps to ensure the comparison was continuous. The height was reset to 100% once the specimens were cut.

Step of the Process	Weight Compared to Initial (%)	Height Compared to Initial (%)		Diameter Compared to Initial (%)
	All Samples	Uncompacted	Compacted	All Samples
1. Inoculation	100.00	100.00		100.00
2. Beginning of drying	95.37	100.00		100.00
3. Before cutting	50.95	95.00		/
4. Before post-growth treat.	50.95	100.00	100.00	96.76
5. Before baking or drying	51.90	/	50.14	/
6. Before mechanical test	21.91	99.56	49.54	96.48

The weight and dimensions of all specimens grown for bending testing were measured at the following steps: (1) inoculation, (2) beginning of drying, (3) before the application of post-growth treatments, (5) before baking or drying, and (6) before cutting to required sizes for testing (Table A2). In contrast to cylindrical specimens, the post-growth treatments were applied to the specimens before cutting them to fit testing requirements for practicability. Each grown specimen was cut into six test specimens directly before the bending testing. All specimens grown for bending contained some contaminants. The weight of contamination manually removed during step 2 accounted for 28.84% of the initial weight of the specimens at inoculation. However, the uncontaminated parts of the specimens had to be trimmed further to fit the size requirements for testing. In total, 42.78% of the initial weight was removed and could not be used for testing. Therefore, tracking the evolution of these measurements was more challenging. The data were divided into two sets to facilitate the analysis: all specimens, and only those on which less than 10% of contamination was removed throughout the entire process.

Table A2. This table shows the evolution of the weight, height, width, and length for the specimens grown for bending. Uncompacted (NC) and compacted (C) specimens were analyzed separately since compacted specimens were constrained in metal frames used for compaction.

Step of the Process	Weight Compared to Initial (%)		Height Compared to Initial (%)		Width Compared to Initial (%)		Length Compared to Initial (%)	
	NC	C	NC	C	NC	C	NC	C
Specimens with less than 10% contamination removed								
1. Inoculation	100.00	100.00	100.00	100.00	100.00	100.00	100.00	100.00
2. Beginning of drying	78.82	85.21	111.90	99.93	97.65	98.40	98.50	99.13
3. Before post-growth treat.	17.00	50.13	91.07	91.07	94.93	/	96.50	/
4. Before baking or drying	/	/	/	/	/	/	/	/
5. Before cutting	/	15.17	87.14	37.32	/	92.35	/	93.05
All specimens								
1. Inoculation	100.00	100.00	100.00	100.00	100.00	100.00	100.00	100.00
2. Beginning of drying	81.17	84.27	102.32	99.93	98.25	98.40	99.00	99.13
3. Before post-growth treat.	37.65	50.13	91.96	91.07	94.93	/	96.50	/
4. Before baking or drying	/	/	/	/	/	/	/	/
5. Before cutting	/	15.17	90.71	37.32	95.07	92.35	/	93.05

References

1. United Nations Environment Programme. *2021 Global Status Report for Buildings and Construction: Towards a Zero-Emission, Efficient and Resilient Buildings and Construction Sector*; United Nations Environment Programme: Nairobi, Kenya, 2021.
2. United States Environmental Protection Agency. *Advancing Sustainable Materials Management: 2018 Fact Sheet*; United States Environmental Protection Agency: Washington, DC, USA, 2020.
3. Wadanambi, L.; Dubey, B.; Townsend, T. The Leaching of Lead from Lead-Based Paint in Landfill Environments. *J. Hazard. Mater.* **2008**, *157*, 194–200. [CrossRef] [PubMed]
4. Nguyen, P.Q.; Courchesne, N.-M.D.; Duraj-Thatte, A.; Praveschotinunt, P.; Joshi, N.S. Engineered Living Materials: Prospects and Challenges for Using Biological Systems to Direct the Assembly of Smart Materials. *Adv. Mater.* **2018**, *30*, 1704847. [CrossRef]
5. Houette, T.; Foresi, B.; Maurer, C.; Gruber, P. Growing Myceliated Facades-Manufacturing and Exposing Experimental Panels in a Facade Setting. In Proceedings of the Facade Tectonics World Congress 2020, 5–27 August 2020.
6. Srubar, W.V. Engineered Living Materials: Taxonomies and Emerging Trends. *Trends Biotechnol.* **2020**, *39*, 574–583. [CrossRef]
7. Haneef, M.; Ceseracciu, L.; Canale, C.; Bayer, I.S.; Heredia-Guerrero, J.A.; Athanassiou, A. Advanced Materials From Fungal Mycelium: Fabrication and Tuning of Physical Properties. *Sci. Rep.* **2017**, *7*, 41292. [CrossRef]
8. Yang, Z.J.; Zhang, F.; Still, B.; White, M.; Amstislavski, P. Physical and Mechanical Properties of Fungal Mycelium-Based Biofoam. *J. Mater. Civ. Eng.* **2017**, *29*. [CrossRef]
9. Elsacker, E.; Vandelook, S.; Brancart, J.; Peeters, E.; De Laet, L. Mechanical, Physical and Chemical Characterisation of Mycelium-Based Composites with Different Types of Lignocellulosic Substrates. *PLoS ONE* **2019**, *14*, e0213954. [CrossRef] [PubMed]
10. Ghazvinian, A.; Farrokhsiar, P.; Vieira, F.; Pecchia, J.; Gursoy, B. Mycelium-Based Bio-Composites for Architecture: Assessing the Effects of Cultivation Factors on Compressive Strength. In Proceedings of the International Conference on Education and Research in Computer Aided Architectural Design in Europe, Porto, Portugal, 11–13 September 2019; Volume 2, pp. 505–514. [CrossRef]
11. Vidholdová, Z.; Kormúthová, D.; Ždinský, J.I.; Lagaňa, R. Compressive Resistance of the Mycelium Composite. *Ann. WULS For. Wood Technol.* **2019**, *107*, 31–36. [CrossRef]
12. Girometta, C.; Picco, A.M.; Baiguera, R.M.; Dondi, D.; Babbini, S.; Cartabia, M.; Pellegrini, M.; Savino, E. Physico-Mechanical and Thermodynamic Properties of Mycelium-Based Biocomposites: A Review. *Sustainability* **2019**, *11*, 281. [CrossRef]
13. Jones, M.; Mautner, A.; Luenco, S.; Bismarck, A.; John, S. Engineered Mycelium Composite Construction Materials from Fungal Biorefineries: A Critical Review. *Mater. Des.* **2020**, *187*, 108397. [CrossRef]

14. Lazaro Vasquez, E.S.; Vega, K. From Plastic to Biomaterials: Prototyping DIY Electronics with Mycelium. In Proceedings of the UbiComp/ISWC 2019-Adjunct Proceedings of the 2019 ACM International Joint Conference on Pervasive and Ubiquitous Computing and Proceedings of the 2019 ACM International Symposium on Wearable Computers, ISWC, London, UK, 9 September 2019; Association for Computing Machinery: New York, NY, USA; pp. 308–311. [CrossRef]
15. Lazaro Vasquez, E.S.; Vega, K. Myco-Accessories: Sustainable Wearables with Biodegradable Materials. In Proceedings of the 23rd International Symposium on Wearable Computers, ISWC, London, UK, 9 September 2019; Association for Computing Machinery: New York, NY, USA; pp. 306–311. [CrossRef]
16. Silverman, J. *Development and Testing of Mycelium-Based Composite Materials for Shoe Sole Applications*; University of Delaware: Newark, DE, USA, 2018.
17. Jones, M.; Bhat, T.; Huynh, T.; Kandare, E.; Yuen, R.; Wang, C.H.; John, S. Waste-Derived Low-Cost Mycelium Composite Construction Materials with Improved Fire Safety. *Fire Mater.* **2018**, *42*, 816–825. [CrossRef]
18. Jones, M.; Bhat, T.; Kandare, E.; Thomas, A.; Joseph, P.; Dekiwadia, C.; Yuen, R.; John, S.; Ma, J.; Wang, C.-H. Thermal Degradation and Fire Properties of Fungal Mycelium and Mycelium-Biomass Composite Materials. *Sci. Rep.* **2018**, *8*, 17583. [CrossRef]
19. Karana, E.; Blauwhoff, D.; Hultink, E.-J.; Camere, S. When the Material Grows: A Case Study on Designing (with) Mycelium-Based Materials. *Int. J. Des.* **2018**, *12*, 119–136.
20. Luo, J.; Chen, X.; Crump, J.; Zhou, H.; Davies, D.G.; Zhou, G.; Zhang, N.; Jin, C. Interactions of Fungi with Concrete: Significant Importance for Bio-Based Self-Healing Concrete. *Constr. Build. Mater.* **2018**, *164*, 275–285. [CrossRef]
21. Abhijith, R.; Ashok, A.; Rejeesh, C.R. Sustainable Packaging Applications from Mycelium to Substitute Polystyrene: A Review. *Mater. Today Proc.* **2018**, *5*, 2139–2145. [CrossRef]
22. Pelletier, M.G.; Holt, G.A.; Wanjura, J.D.; Bayer, E.; McIntyre, G. An Evaluation Study of Mycelium Based Acoustic Absorbers Grown on Agricultural By-Product Substrates. *Ind. Crops Prod.* **2013**, *51*, 480–485. [CrossRef]
23. Atila, F. Effect of Different Substrate Disinfection Methods on the Production of Pleurotus Ostreatus. *J. Agric. Stud.* **2016**, *4*, 52–64. [CrossRef]
24. Inhabitat Phillip Ross Molds Fast-Growing Fungi Into Mushroom Building Bricks That Are Stronger Than Concrete. Available online: https://inhabitat.com/phillip-ross-molds-fast-growing-fungi-into-mushroom-building-bricks-that-are-stronger-than-concrete/ (accessed on 17 February 2019).
25. Stott, R. Hy-Fi, the Organic Mushroom-Brick Tower Opens at MoMA's PS1 Courtyard. Available online: https://www.archdaily.com/521266/hy-fi-the-organic-mushroom-brick-tower-opens-at-moma-s-ps1-courtyard (accessed on 17 February 2019).
26. Heisel, F.; Schlesier, K.; Lee, J.; Rippmann, M.; Saeidi, N.; Javadian, A.; Nugroho Adi, R.; Hebel, D.; Block, P. Design of a Load-Bearing Mycelium Structure through Informed Structural Engineering. In Proceedings of the World Congress on Sustainable Technologies (WCST), Cambridge, UK, 11–14 December 2017; pp. 1–5.
27. Archdaily This Pavillion Lives and Dies through Its Sustainable Agenda. Available online: https://www.archdaily.com/878519/this-pavillion-lives-and-dies-through-its-sustainable-agenda (accessed on 18 February 2019).
28. Dessi-Olive, J. Monolithic Mycelium: Growing Vault Structures. In Proceedings of the 18th International Conference on Non-Conventional Materials and Technologies "Construction Materials & Technologies for Sustainability", Nairobi, Kenya, 24–26 July 2019.
29. Company New Heroes. The Growing Pavilion. Available online: https://thegrowingpavilion.com/ (accessed on 20 October 2020).
30. Jones, M.; Huynh, T.T.; John, S. Inherent Species Characteristic Influence and Growth Performance Assessment for Mycelium Composite Applications Development of Microstrip Patch Antenna Strain Sensors for Wireless Structural Health Monitoring View Project Industrial Crops View Project. *Adv. Mater. Lett.* **2018**, *9*, 71–80. [CrossRef]
31. Lelivelt, R.J.J. *The Mechanical Possibilities of Mycelium Materials*; Eidhoven University of Technology: Eidhoven, The Netherlands, 2015.
32. Appels, F.V.W.; Camere, S.; Montalti, M.; Karana, E.; Jansen, K.M.B.; Dijksterhuis, J.; Krijgsheld, P.; Wösten, H.A.B. Fabrication Factors Influencing Mechanical, Moisture- and Water-Related Properties of Mycelium-Based Composites. *Mater. Des.* **2019**, *161*, 64–71. [CrossRef]
33. Attias, N.; Danai, O.; Abitbol, T.; Tarazi, E.; Ezov, N.; Pereman, I.; Grobman, Y.J. Mycelium Bio-Composites in Industrial Design and Architecture: Comparative Review and Experimental Analysis. *J. Clean. Prod.* **2020**, *246*, 119037. [CrossRef]
34. Jones, M.; Huynh, T.; Dekiwadia, C.; Daver, F.; John, S. Mycelium Composites: A Review of Engineering Characteristics and Growth Kinetics. *J. Bionanoscience* **2017**, *11*, 241–257. [CrossRef]
35. Ecovative Design. Mycocomposite. Available online: https://ecovativedesign.com/ (accessed on 27 November 2018).
36. Chang, J.; Chan, P.L.; Xie, Y.; Ma, K.L.; Cheung, M.K.; Kwan, H.S. Modified Recipe to Inhibit Fruiting Body Formation for Living Fungal Biomaterial Manufacture. *PLoS ONE* **2019**, *14*, e0209812. [CrossRef] [PubMed]
37. Lelivelt, R.J.J.; Lindner, G.; Teuffel, P.; Lamers, H. The Production Process and Compressive Strength of Mycelium-Based Materials. In Proceedings of the First International Conference on Bio-based Building Materials, Clermont-Ferrand, France, 22–25 June 2015; pp. 1–6.
38. Appels, F.V.W.; Dijksterhuis, J.; Lukasiewicz, C.E.; Jansen, K.M.B.; Wösten, H.A.B.; Krijgsheld, P. Hydrophobin Gene Deletion and Environmental Growth Conditions Impact Mechanical Properties of Mycelium by Affecting the Density of the Material. *Sci. Rep.* **2018**, *8*, 4703. [CrossRef] [PubMed]

39. Elsacker, E.; Vandelook, S.; Van Wylick, A.; Ruytinx, J.; De Laet, L.; Peeters, E. A Comprehensive Framework for the Production of Mycelium-Based Lignocellulosic Composites. *Sci. Total Environ.* **2020**, *725*, 138431. [CrossRef] [PubMed]

40. Elsacker, E.; Søndergaard, A.; Van Wylick, A.; Peeters, E.; De Laet, L. Growing Living and Multifunctional Mycelium Composites for Large-Scale Formwork Applications Using Robotic Abrasive Wire-Cutting. *Constr. Build. Mater.* **2021**, *283*, 122732. [CrossRef]

41. Manan, S.; Ullah, M.W.; Ul-Islam, M.; Atta, O.M.; Yang, G. Synthesis and Applications of Fungal Mycelium-Based Advanced Functional Materials. *J. Bioresour. Bioprod.* **2021**, *6*, 1–10. [CrossRef]

42. Rigobello, A.; Ayres, P. Compressive Behaviour of Anisotropic Mycelium-Based Composites. *Sci. Rep.* **2022**, *12*, 6846. [CrossRef] [PubMed]

43. Islam, M.R.; Tudryn, G.; Bucinell, R.; Schadler, L.; Picu, R.C. Mechanical Behavior of Mycelium-Based Particulate Composites. *J. Mater. Sci.* **2018**, *53*, 16371–16382. [CrossRef]

44. Tudryn, G.J.; Smith, L.C.; Freitag, J.; Bucinell, R.; Schadler, L.S. Processing and Morphology Impacts on Mechanical Properties of Fungal Based Biopolymer Composites. *J. Polym. Environ.* **2018**, *26*, 1473–1483. [CrossRef]

45. Travaglini, S.; Noble, J.; Ross, P.G.; Dharan, C.K.H. Mycology Matrix Composites. In Proceedings of the 28th Annual Technical Conference of the American Society for Composites 2013, State College, PA, USA, 9–11 September 2013; Curran Associates, Inc.: Red Hook, NY, USA, 2013; pp. 517–535.

46. Ashby, M.F.; Shercliff, H.; Cebon, D. *Materials: Engineering, Science, Processing and Design*, 4th ed.; Butterworth-Heinemann: Oxford, UK, 2018; ISBN 9780081023778.

47. Xia, X.C.; Chen, X.W.; Zhang, Z.; Chen, X.; Zhao, W.M.; Liao, B.; Hur, B. Effects of Porosity and Pore Size on the Compressive Properties of Closed-Cell Mg Alloy Foam. *J. Magnes. Alloy.* **2013**, *1*, 330–335. [CrossRef]

48. Islam, M.R.; Tudryn, G.; Bucinell, R.; Schadler, L.; Picu, R.C. Morphology and Mechanics of Fungal Mycelium. *Sci. Rep.* **2017**, *7*. [CrossRef] [PubMed]

49. Jiang, L.; Walczyk, D.; McIntyre, G.; Bucinell, R.B. A New Approach to Manufacturing Biocomposite Sandwich Structures: Mycelium-Based Cores. In Proceedings of the ASME 2016 11th International Manufacturing Science and Engineering Conference, Volume 1: Processing, Blacksburg, VA, USA, 27 June–1 July 2016. [CrossRef]

50. Stamets, P. *Growing Gourmet and Medicinal Mushrooms*; Ten Speed Press: Berkeley, CA, USA, 1993; ISBN 1580081754.

51. Stamets, P. *Mycelium Running: How Mushrooms Can Help Save the World*, 1st ed.; Ten Speed Press: Berkeley, CA, USA, 2005.

52. Grimm, D.; Wösten, H.A.B. Mushroom Cultivation in the Circular Economy. *Appl. Microbiol. Biotechnol.* **2018**, *102*, 7795–7803. [CrossRef] [PubMed]

53. Durr, L. Mycoremediation Project: Using Mycelium to Clean Up Diesel-Contaminated Soil in Orleans, California. 2016. Available online: https://static1.squarespace.com/static/61ad3dbc1635812fe1f11cae/t/61b975f2ef9ebd32bb073a9b/163954431928 3/MycoremediationReport_FungaiaFarm_2016.pdf/ (accessed on 5 August 2020).

54. *ASTM D1037-12*; Standard Test Methods for Evaluating Properties of Wood-Base Fiber and Particle Panel Materials. American Society for Testing and Materials: West Conshohocken, PA, USA, 2012. [CrossRef]

55. *ASTM D2166/D2166M-13*; Standard Test Method for Unconfined Compressive Strength of Cohesive Soil. American Society for Testing and Materials: West Conshohocken, PA, USA, 2013. [CrossRef]

56. *ASTM D3501–94*; Standard Test Methods for Wood-Based Structural Panels in Compression. American Society for Testing and Materials: West Conshohocken, PA, USA, 2018. [CrossRef]

57. *ASTM C165-07*; Standard Test Method for Measuring Compressive Properties of Thermal Insulations. American Society for Testing and Materials: West Conshohocken, PA, USA, 2017. [CrossRef]

58. *ASTM C303-10*; Standard Test Method for Dimensions and Density of Preformed Block and Board–Type Thermal Insulation. American Society for Testing and Materials: West Conshohocken, PA, USA, 2002. [CrossRef]

59. Watkinson, S.; Boddy, L.; Money, N. *The Fungi*, 3rd ed.; Academic Press: Waltham, MA, USA, 2015; ISBN 978-0-12-382034-1.

60. Obodai, M.; Cleland-Okine, J.; Vowotor, K.A. Comparative Study on the Growth and Yield of Pleurotus Ostreatus Mushroom on Different Lignocellulosic By-Products. *J. Ind. Microbiol. Biotechnol.* **2003**, *30*, 146–149. [CrossRef] [PubMed]

61. Fried, S.; Mackie, B.; Nothwehr, E. Nitrate and Phosphate Levels Positively Affect the Growth of Algae Species Found in Perry Pond. *Tillers* **2012**, *4*, 21–24.

62. Smith, D.R.; King, K.W.; Williams, M.R. What Is Causing the Harmful Algal Blooms in Lake Erie? *J. Soil Water Conserv.* **2015**, *70*, 27A–29A. [CrossRef]

63. Carmichael, W.W.; Boyer, G.L. Health Impacts from Cyanobacteria Harmful Algae Blooms: Implications for the North American Great Lakes. *Harmful Algae* **2016**, *54*, 194–212. [CrossRef]

64. Adey, W.H.; Kangas, P.C.; Mulbry, W. Algal Turf Scrubbing: Cleaning Surface Waters with Solar Energy While Producing a Biofuel. *Bioscience* **2011**, *61*, 434–441. [CrossRef]

65. Mulbry, W.; Kangas, P.; Kondrad, S. Toward Scrubbing the Bay: Nutrient Removal Using Small Algal Turf Scrubbers on Chesapeake Bay Tributaries. *Ecol. Eng.* **2010**, *36*, 536–541. [CrossRef]

66. Camere, S.; Karana, E. Growing Materials for Product Design. In Proceedings of the EKSIG2017-International Conference on Experiential Knowledge and Emerging Materials, Delft, The Netherlands, 19–20 June 2017; pp. 101–115.

67. Du, Z.Y.; Zienkiewicz, K.; Pol, N.V.; Ostrom, N.E.; Benning, C.; Bonito, G.M. Algal-Fungal Symbiosis Leads to Photosynthetic Mycelium. *Elife* **2019**, *8*, e47815. [CrossRef]

68. Liber, J.A.; Bryson, A.E.; Bonito, G.; Du, Z.Y. Harvesting Microalgae for Food and Energy Products. *Small Methods* **2020**, *4*, 2000349. [CrossRef]
69. Ginot, L.; Peyr, C.; Fontaine, A.; Cheymol, J.; Buisson, B.; Bellia, G.; Da Cruz, F.; Buisson, J. Dépistage du saturnisme infantile à partir de la recherche de plomb dans l'habitat: Une étude en région parisienne [Screening for Lead Poisoning in Children by Measuring Lead Levels in Housing: A Study of the Paris Region]. *Rev. Epidemiol. Sante Publique* **1995**, *43*, 477–484. [PubMed]
70. Nduka, J.; Orisakwe, O.; Maduawguna, C. Lead Levels in Paint Flakes from Buildings in Nigeria: A Preliminary Study. *Toxicol. Ind. Health* **2008**, *24*, 539–542. [CrossRef] [PubMed]
71. Tehranifar, P.; Leighton, J.; Auchincloss, A.H.; Faciano, A.; Alper, H.; Paykin, A.; Wu, S. Immigration and Risk of Childhood Lead Poisoning: Findings from a Case-Control Study of New York City Children. *Am. J. Public Health* **2008**, *98*, 92–97. [CrossRef] [PubMed]
72. Reyes, N.L.; Wong, L.Y.; MacRoy, P.M.; Curtis, G.; Meyer, P.A.; Evens, A.; Brown, M.J. Identifying Housing That Poisons: A Critical Step in Eliminating Childhood Lead Poisoning. *J. Public Health Manag. Pract.* **2006**, *12*, 563–569. [CrossRef]
73. Mattheck, C.; Sörensen, J.; Bethge, K. A Graphic Way for Notch Shape Optimization. *WIT Trans. Ecol. Environ.* **2006**, *87*, 13–21. [CrossRef]
74. Sorensen, L.C.; Fox, A.M.; Jung, H.; Martin, E.G. Lead Exposure and Academic Achievement: Evidence from Childhood Lead Poisoning Prevention Efforts. *J. Popul. Econ.* **2019**, *32*, 179–218. [CrossRef]
75. Kumar, A.; Yadav, A.N.; Mondal, R.; Kour, D.; Subrahmanyam, G.; Shabnam, A.A.; Khan, S.A.; Yadav, K.K.; Sharma, G.K.; Cabral-Pinto, M.; et al. Myco-Remediation: A Mechanistic Understanding of Contaminants Alleviation from Natural Environment and Future Prospect. *Chemosphere* **2021**, *284*, 131325. [CrossRef]
76. Gadd, G.M. *Fungi in Bioremediation*; Gadd, G.M., Ed.; Cambridge University Press: Cambridge, UK, 2001.
77. Kumar, V.; Dwivedi, S.K. Mycoremediation of Heavy Metals: Processes, Mechanisms, and Affecting Factors. *Environ. Sci. Pollut. Res.* **2021**, *28*, 10375–10412. [CrossRef]
78. Wang, J.; Xiong, Z.; Kuzyakov, Y. Biochar Stability in Soil: Meta-Analysis of Decomposition and Priming Effects. *GCB Bioenergy* **2016**, *8*, 512–523. [CrossRef]
79. Imhof, B.; Gruber, P. *Built to Grow: Blending Architecture and Biology*; Birkhäuser: Basel, Switzerland, 2015; ISBN 9783035609202.

 *biomimetics*

Article

Multi-Organism Composites: Combined Growth Potential of Mycelium and Bacterial Cellulose

Aileen Hoenerloh *, Dilan Ozkan and Jane Scott

Hub for Biotechnology in the Built Environment, School of Architecture, Planning and Landscape, Newcastle University, Newcastle upon Tyne NE1 7RU, UK; d.ozkan2@ncl.ac.uk (D.O.); jane.scott@newcastle.ac.uk (J.S.)
* Correspondence: a.t.hoenerloh2@ncl.ac.uk

Abstract: The demand for sustainable materials derived from renewable resources has led to significant research exploring the performance and functionality of biomaterials such as mycelium and bacterial cellulose. Whilst the growing conditions and performance of individual biomaterials are understood, to achieve additional new and enhanced functionality, an understanding of how biomaterials can be used together as composites and hybrids is required. This paper investigates the compatibility of mycelium and bacterial cellulose as two biomaterials with different qualities for the development of a large-scale biohybrid structure, the *BioKnit* prototype. Their compatibility was tested through preliminary design experiments and a material tinkering approach. The findings demonstrate that under optimal conditions mycelium and bacterial cellulose can grow in each other's presence and create composites with an extensive array of functions. However, there is a need to develop further fabrication settings that help to maintain optimal growing conditions and nutrition levels, whilst eliminating problems such as contamination and competition during growth.

Keywords: mycelium; bacterial cellulose; biocompatibility; knitted fabric; material tinkering

Citation: Hoenerloh, A.; Ozkan, D.; Scott, J. Multi-Organism Composites: Combined Growth Potential of Mycelium and Bacterial Cellulose. *Biomimetics* **2022**, 7, 55. https://doi.org/10.3390/biomimetics7020055

Academic Editors: Andrew Adamatzky, Han A.B. Wösten and Phil Ayres

Received: 15 March 2022
Accepted: 29 April 2022
Published: 3 May 2022

Publisher's Note: MDPI stays neutral with regard to jurisdictional claims in published maps and institutional affiliations.

1. Introduction

The growing interest in the use of living materials in the design field is motivated by the desire to create a renewable resource of sustainable materials that is biodegradable and grown using waste materials. This approach offers the potential to produce a low-cost alternative to commercial synthetic materials whilst developing a new spectrum of functional properties [1]. For instance, while fungi as mycelium composites can be used as a structural bulk material as in the MycoTree, MycoCreate-2, and El Monolito Micelio projects [2–5], bacterial cellulose shares many properties with plant cellulose and is already successfully used in the medical field [6] and in fashion design projects [7,8].

Whilst current research has focused on optimising the growth and performance of individual microorganisms for design applications, our research considers the potential to combine different microbial systems in order to achieve new functional possibilities. For example, can multi-microbial systems produce stronger, more durable biomaterials whilst transforming the look and feel of materials for architecture? Using biological materials at different stages of their lifecycle for different purposes requires developing new methods of fabrication that allow the designers to reveal their various features. *BioKnit* is a prototype designed and built in the Hub for Biotechnology in the Built Environment (HBBE) under the Living Construction theme [9]. It focuses on bringing mycelium and bacterial cellulose together with textiles using knitting technologies. Knitted fabric, the first component of the prototype, acts as a scaffold and guides, enhances, and restricts the organism as it grows or attaches to it. Mycelium in a composite form, the second component, acts as a bulk material that gives sufficient compressive strength to the knitted structure to enable the production of a 1.8 m high, free-standing vault. Bacterial cellulose (BC) is the third component and adds a layer of complexity to the system by adding a new optical and tactile quality. BC is used as a surface treatment to coat the knit/mycelium composite and as a tactile skin, self-adhered to the mycelium/knit composite. Mycelium and BC, as two commonly used biomaterials in the design field, were chosen as organisms since cellulose can be used to cultivate mycelium [9]. It proves that they can establish a relationship, and one of the organisms could be used to support the other.

Fungi mainly consist of two parts: the fruiting body/mushroom and the mycelium/roots. The mycelium is of general interest as a biomaterial due to its ability to rapidly grow on various forms of waste as a composite, creating a bulk building material [5]. There are numerous studies on the properties of mycelium composite materials grown on agricultural waste, such as the works of Appels et al. [10] and Jones et al. [5]. Another approach was taken by Elsacker et al. by growing BC to be utilized as a nutritious substrate additive with the focus being on improving the mycelium growth and incorporating the bio-organism in a dried form [11].

BC is a biopolymer that can be grown as a pure culture from a single bacterial strain in sterile lab conditions or with a symbiotic culture of bacteria and yeast (SCOBY), which is commonly used to create fermented kombucha tea in household kitchens. Properties such as fine crystalline structures, biodegradability and biocompatibility, good water-holding capacity, and chemical stability can explain the growing interest in BC as a biomaterial [6]. Due to the large-scale nature of the *BioKnit* prototype, the more resilient kombucha method was chosen for these experiments.

This paper introduces the initial design experiments that were conducted during the prototyping process and asks the following questions: (1) how two organisms grow together and how they influence each other, (2) what is the potential of a knit scaffold as a technique to assemble multiple living materials through growth, and (3) how can the challenge of con-tamination due to varying growth requirements be addressed? These questions are tested to develop protocols for multi-kingdom textile composites through a methodology based on explorative experimentation using material tinkering [12] and based within established biomaterial protocols rather than a biomimetic investigation translating functional models from nature. Each organism's behavior was observed in an iterative array of experiments.

2. Materials and Methods

2.1. Mycelium Composite Preparation

Mycelium grown in the experiments was inoculated with a mixed substrate (10 g of strawbale, 10 g of wood shavings, 10 g of coffee grounds), which was sterilized in an autoclave at 121 °C for 15 min. The sterile mixture was seeded with 10 g of oyster mush-room spawn from GroCycle, UK and kept in sealed plastic boxes ($100 \times 100 \times 30$ mm), in the dark, at ambient temperature. After three weeks, the samples were taken out of the boxes and kept in three different forms: oven-dried, air-dried, and living. The first set was air-dried for 8 days and then oven-dried (for 2 h under 60 °C). The second set was air-dried for 2 weeks at room temperature. Finally, the third set of tiles were kept alive in a closed container. These three sets of tiles were then integrated in the bacterial cellulose experiments.

2.2. Bacterial Cellulose Preparation

All BC was grown using the kombucha technique with a tea-based medium and a piece of SCOBY (symbiotic culture of bacteria and yeast) purchased from "Happy Kombucha". A new SCOBY was used for each experiment. The medium was prepared in 3-litre batches with tap water, pasteurized apple cider vinegar (62.5 mL/L), white refined sugar (75 g/L), and black tea (2 g/L) (Tetley Black Tea bags). To prepare the medium, the water was brought to boiling point before adding the tea bags to infuse for 15 min. The vinegar and sugar were added, then stirred until all the sugar had dissolved. The addition of vinegar lowers the pH from 4.9 to 3.5 ($\pm$0.1). The medium was cooled down to <30 °C before being used. The BC in experiments 1–3 was grown in a plastic box ($110 \times 190 \times 330$ mm) with 2.2 L medium, covered with a cotton cloth. Experiment 4 was grown in a (360 × 26 × 140 mm) box with the medium quantity adjusted throughout.

2.3. The Experimental Setting

The associations of BC and mycelium in different metabolic stages were tested in four experiments. Mycelium composites were shaped as tiles, and BC was oriented flat due to the necessity in its growth conditions.

The experiments were documented using photography (iPhone SE and Fuji film X-T2 with 80 mm lens) and microscopy (Dino-Lite digital microscope at 70× magnification). The images from photography and microscopy captured the details, smaller features, and non-measurable characteristics such as the surface texture and BC attachment. They generated qualitative data for this study.

In experiments 1, 2, and 3, the compatibility of BC (secondary organism) growing around already-established mycelium (primary organism) was explored. In experiment 4, the compatibility of mycelium (secondary organism) growing around already-established BC (primary organism) was explored. In each experiment, the success of the growth of the secondary organism was monitored alongside the contamination that occurred during the growth stage.

2.4. Experiment 1 (Oven-Dried Mycelium with BC)

After preparing the kombucha culture, a fresh SCOBY was placed in the medium, as seen in Figure 1. Then, the first set of oven-dried mycelium tiles were positioned on the surface. It was not necessary to add support and keep the mycelium at the liquid surface, as the hydrophobicity of it kept the tiles afloat. The SCOBY was positioned in between the tiles towards the center of the container. This setup was left in a static condition to ferment for 14 days.

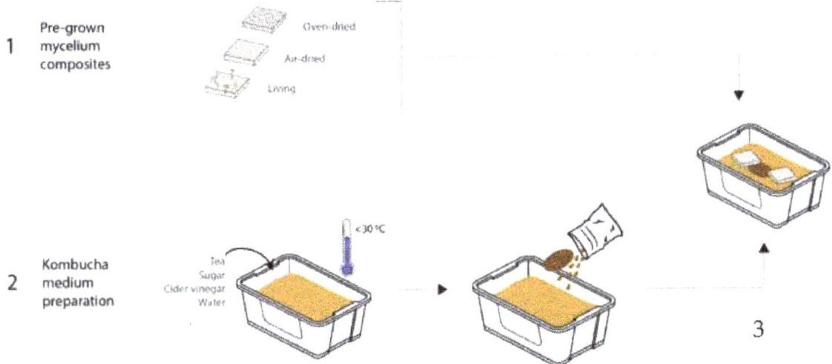

Figure 1. The setup for experiments 1, 2, and 3 (image credit: authors).

To harvest the composite material of BC and mycelium of experiments 1–3, the growth was removed from the liquid medium and washed with antibacterial dish soap and cold water after 14 days of static fermenting. While the BC was thoroughly kneaded with the soap, care was taken to not wet the mycelium or introduce stress to the joint locations of the BC to the tiles.

2.5. Experiment 2–3 (Air-Dried Mycelium with BC and Living Mycelium with BC)

For experiments 2 and 3, the hydrophobicity of the air-dried and living mycelium was reduced and not sufficient to keep the tiles afloat; therefore, the placement of the tiles in the liquid medium was adjusted. Two small glass jars were used to support each tile: one upside down and submerged in the liquid to place the tile on (see Figure 2) and the other one placed on top to weigh down the otherwise floating tile. The jar was placed with the opening facing down to allow ventilation and avoid the growth of mold underneath.

The quantity of medium was adjusted to cover the bottom half of the tiles, and the SCOBY was placed in between the tiles in the center of the box.

(a)   (b)

Figure 2. (**a**) Setup of experiments 2 and 3 with (from right to left) air-dried mycelium, living mycelium, and a control culture. (**b**) Closeup of the positioning of the tiles within the liquid using jars (image credit: authors).

The harvesting process for these experiments was the same as for experiment 1.

2.6. Experiment 4 (BC with Living Mycelium inside Soft Scaffold)

In experiments 1–3, the compatibility of BC growing around already-established mycelium was explored; however, it was also important to understand whether the order of growth could be reversed, with mycelium growing as a secondary organism alongside already-established BC. This was tested in experiment 4. For the mycelium to grow next to or inside wet bacterial cellulose, a support that can hold and contain the substrate is needed. In line with the *BioKnit* background of this research, a knitted pocket (linen, circular plain, 8 gg Dubied) was used to contain the mycelium substrate. To avoid killing the bacteria in the BC, the mycelium substrate was autoclaved and pre-inoculated for 5–10 days. Simultaneously, the BC was grown on the soft scaffold. To stop the BC from merging both sides of the pocket together, one side was placed on top of an acrylic scaffold inside the kombucha medium to be held on the air–liquid interface while the other side was pulled up on the corners with acrylic string (see Figure 3a). Fresh medium was added underneath the growing pellicle every 3 days to compensate the evaporated medium and keep the air–liquid interface stable.

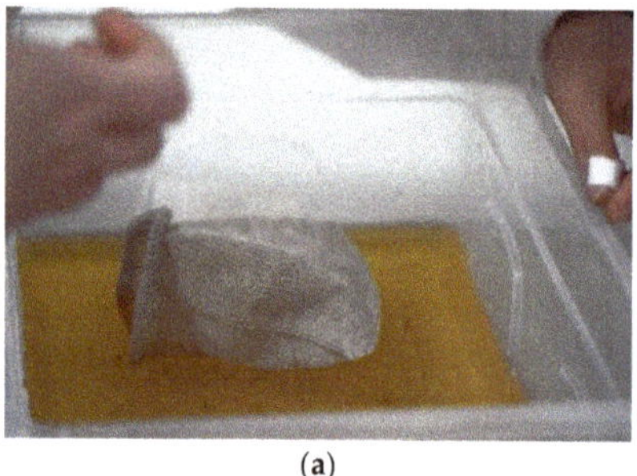
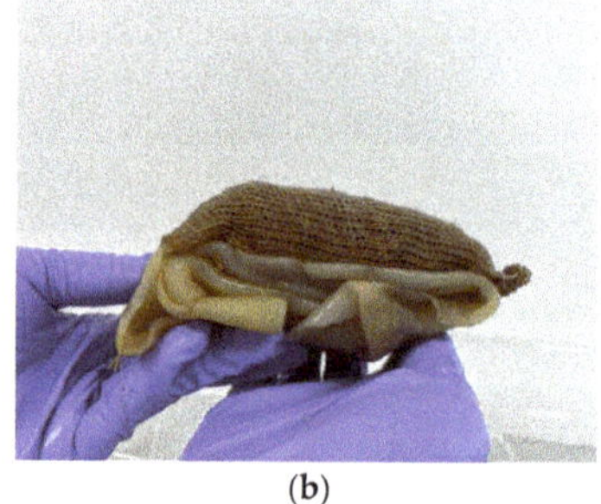

(a)   (b)

Figure 3. (**a**) The knitted pocket being placed inside the kombucha culture on top of the acrylic scaffold with the top half being lifted. (**b**) Pocket filled with pre-inoculated mycelium substrate and BC folded in 9 layers underneath (image credit: authors).

The second step of the experiment involved harvesting the BC knit composite and filling the pocket with the pre-inoculated mycelium substrate (see Figures 3b and 4). The pocket was sewn shut using cotton string, piercing through the layer of BC and the knit. To allow the mycelium to breathe, the excess BC was folded underneath the pocket on the side which had the BC growth on it (total of 9 BC layers), leaving one knitted side exposed (Figure 3b). This composite was placed inside a plastic box with lid, BC side facing down, and left to grow for 12 days.

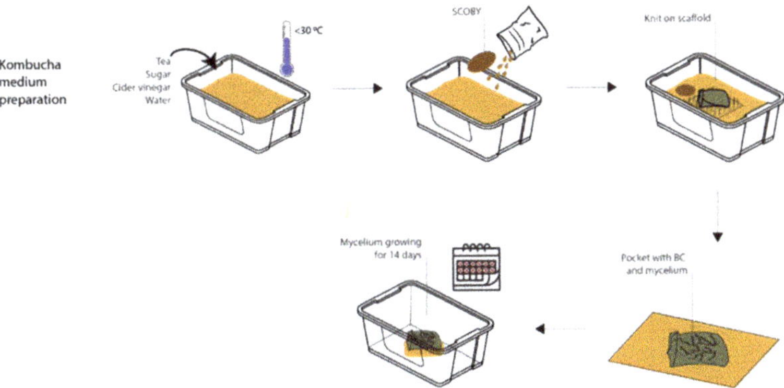

Figure 4. The setup for experiment 4 (image credit: authors).

Experiment 4 required two separate harvesting processes. First, the BC was harvested after 14 days of static growth and washed as described above. The top side of the knitted pocket was also cleaned with dish soap to remove any media that had been drawn up through the fibers. The fabric was thoroughly rinsed to eliminate the risk of soap residue hindering the mycelium growth. The second harvest stage was after 12 days of mycelium growth, where the composite material was removed from the plastic box and dried. Because contamination occurred, the composite was dried in an industrial oven at 50 °C for 8 h.

Throughout the 14 days of BC growth, the formation of air bubbles was handled as in the previous experiments and pushed out to the sides.

2.7. Drying Process

The drying process for experiment 4 is described above. For experiments 1–2, the BC was washed, padded dry using paper towels, and placed on a clean bamboo cutting to dry uncovered at 21 °C. The composite was placed with the BC side on the tiles facing down. After 48 h, the composite was turned over, exposing the BC on the tiles. After further 48 h, the BC had turned translucent and fully dried. For experiment 3, the samples that showed signs of mold contamination were not dried on a bamboo board but dried in an industrial oven at 60 °C for 4–6 h.

3. Results

3.1. Results of Experiment 1 (Oven-Dried Mycelium with BC)

In experiment one, good BC growth occurred; however, the resultant growth was uneven across the surface. Growth was measured on day 14. The BC thickness ranged from 12 mm at the outer edges of the cellulose to 2 mm around the SCOBY (where a bubble had formed). The thicker areas of the pellicle showed good connection to the mycelium tile and did not detach while handling; however, the 2 mm BC ripped during the harvesting process. After drying, the tiles were firmly integrated into the pellicle, and the composite could be held up from the tiles without the cellulose detaching or ripping (Figure 5).

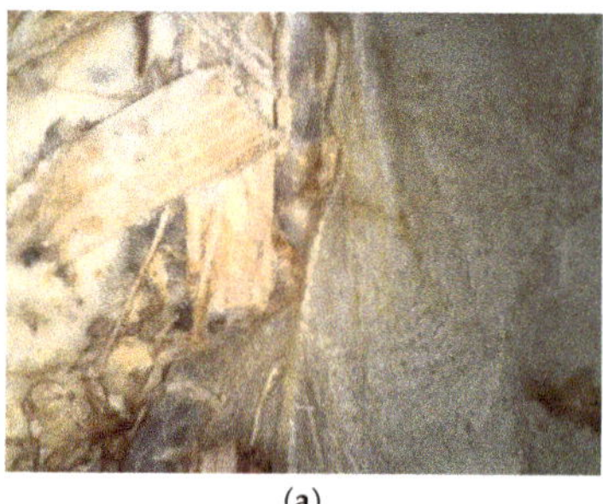

(**a**)   (**b**)

Figure 5. (**a**) Dried BC around oven-dried mycelium tile under the Dino-Lite digital microscope. (**b**) Large BC sheet dried around mycelium tile (image credit: authors).

The oven-dried tiles were very hydrophobic and initially moved freely within the liquid medium. On day three of the fermentation, a thin and clear layer of cellulose had grown on the surface, which stopped the mycelium tiles from moving around. In the center of the pellicle, a large bubble of gas built up where the SCOBY had been positioned during the fermentation. This bubble lifted the cellulose off the liquid and stopped the growth of the cellulose at this area after less than 7 days. One of the tiles was also lifted into an angled position.

The formation of CO_2 during the fermentation is a normal occurrence during the kombucha method; however, the visible amount of trapped CO_2 produced in this experiment was greater than in a control setup without mycelium present. The gas bubble was problematic because it limited BC growth in particular areas and disturbed the position of one of the mycelium tiles. This led to a poor connection between the BC and mycelium close to the gas bubble.

3.2. Results of Experiment 2 (Air-Dried Mycelium with BC)

In experiment two, good BC growth occurred producing a more even pellicle after 14 days of growth. Growth was measured on day six and day 14. On day six, there was a visible layer of cellulose on the surface that measured 2–3 mm. By day 14, growth measured between 10–13 mm across the whole surface. During the growing stage, there was a very weak attachment between the mycelium and the BC. This was clear during harvest because the BC fully detached from the mycelium. Despite this, during drying the BC adhered to the mycelium. The attachment was maintained even with heavy handling.

During BC growth, many small bubbles formed underneath the pellicle. To maintain growth, the bubbles were gently pushed to the side of the pellicle using sterile hands. This was continued each day until day 10, when the stiffness of the pellicle prevented intervention. Cellulose growth was thicker around dried fruiting bodies (the side of the mycelium tiles). This formed a patchy appearance overall (Figure 6a). In addition, the BC had formed a skin-like film under the submerged half of the tile, even without direct access to oxygen (Figure 6b). This indicates that the mycelium can provide a small amount of oxygen to the bacteria.

3.3. Results of Experiment 3 (Living Mycelium with BC)

In experiment three, BC growth was changed by the presence of a living mycelium. Growth was measured on day six and day 14. On day six, the pellicle thickness was 2–3 mm in line with previous results. However, by day 14 the pellicle had only grown to a thickness of between 3–5 mm. BC growth was observed across the mycelium tile in gaps within the material, and this led to good attachment during the growing stage. Due to contamination, this sample was oven-dried, and this drying process also created the strongest connection between the mycelium and the BC.

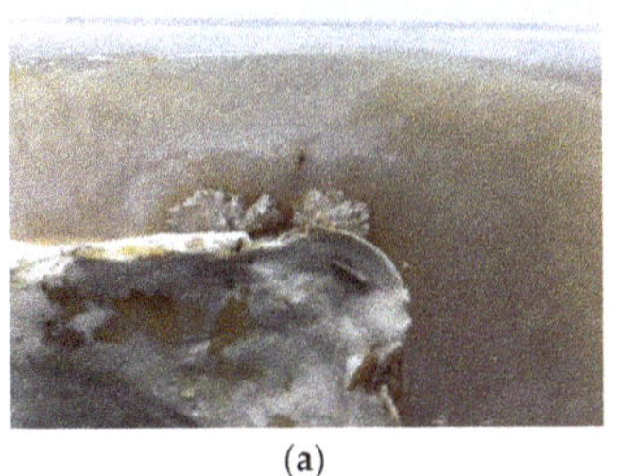

(a)

(b)

Figure 6. (a) Thicker BC growth around dried fruiting body. (b) BC skin around submerged half of mycelium tile after harvest (image credit: authors).

Aeration bubbles again appeared underneath the pellicle; however, these were much smaller and greater in number, leaving a foam-like appearance. Larger bubbles also formed, and these were pushed to the side. The living mycelium tiles had fruiting bodies that grew out on the side and were either hovering just above or within the air–liquid interface of the medium. The BC grew around the fruiting bodies and enclosed them. The mushrooms sitting half-submerged on the surface began to dry out and had shrunk in size by day seven, but the connection to the BC remained. Unlike the mycelium tiles in experiments one and two, these tiles were not hydrophobic, and they soaked up the liquid medium. During the experiment, the top of the tiles also became submerged. Where the top of the tile soaked up the medium, some BC growth occurred. Around the side of the tiles, the BC growth was uneven, with thinner parts forming the connection to the mycelium (2–3 mm) and the outside edges and corners being the thickest parts. As in the previous experiment, the cellulose growing around bunches of primordia formations was thicker. During the first 7 days, the area of the tile that was covered with the upside-down jar continued to grow white patchy mycelium. This growth slowed down and eventually stopped once the jars were removed.

Contamination in the form of green mold occurred on the top surface of one of the tiles where the medium had been soaked up. This meant that the composite could not be air-dried after harvest but needed to be oven-dried at 100 °C for 1 h. Compared to the air-dried versions, the BC was darker in color and significantly more brittle. During the harvest, the tiles soaked up the washing water and needed to be handled very carefully to not fall apart. The BC did not hold the tile substrate in shape.

Figure 7 shows a visual comparison of experiments one, two, and three in a hydrated and dried state.

3.4. Results of Experiment 4 (BC with Living Mycelium inside Soft Scaffold)

In experiment four, mycelium growth was contained in a knitted pocket submerged in BC. The results showed that the mycelium grew alongside the BC; however, mycelium growth was uneven (in comparison to controls with no BC present). In this experiment, the attachment between the mycelium and BC was created by the knitted pocket, and this was effective in bringing the two organisms together. After 12 days, the BC showed signs of contamination in the form of mold; therefore, the composite was oven-dried (see Figure 8). The connection between the mycelium and BC as a dried composite was strong, with the texture of the fabric clearly visible through the BC.

The first stage of the experiment resulted in an even growth of BC (5 mm) attached to the bottom half of the knitted pocket. Due to the high absorbency of the fabric used to create the pocket, the starter liquid for the kombucha culture partly saturated the top layer, which was suspended over the air–liquid interface. No visible layer of the BC formed on those parts of the pocket.

(a)   (b)

Figure 7. Comparison of experiments 1–3 in the order oven-dried, air-dried, and living from top to bottom in (**a**) wet state and (**b**) dried state (image credit: authors).

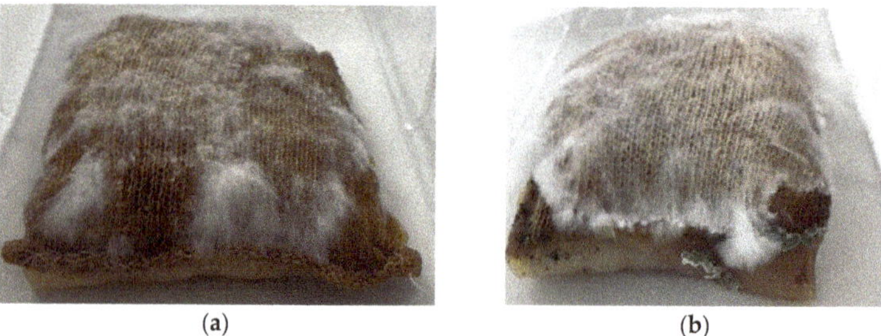

(a)   (b)

Figure 8. (**a**) Mycelium growth on the knitted pocket on day 7. (**b**) Mycelium growth on the knitted pocket on day 14 (image credit: authors).

During the second stage of the experiment, the growth behavior of the mycelium was observed. In the first 7 days, the organism started to visibly grow around the sides of the pocket, as seen in Figure 7. The larger patches of mycelial growth were located predominantly along the edges due to the level of moisture on those parts. However, the knitted pocked was not fully covered by a dense layer of mycelium. This was possibly due to the linen being difficult to digest for the mycelium. The BC had not begun drying at this time. The fabric on the top side had also not fully dried from the washing.

After a total of 12 days, the wet BC showed signs of contamination in the form of mold. The mycelium had continued to grow along the edges of the pocket and onto the plastic container. The very top of the fabric showed less growth than before. The corners of the pocket, where the liquid medium had been soaked up during the first stage, showed no mycelium growth but contamination reaching from the BC. Those areas of the fabric showed dark discoloration.

After drying the composite in an oven, the mycelium and BC both significantly decreased in thickness. Only one corner of the pocket showed signs of a beginning white coating (see Figure 9a), as have been observed in prior experiments. While the BC was very brittle, it showed strong attachment to the fabric and had molded onto it in a way that the texture of the fabric came through.

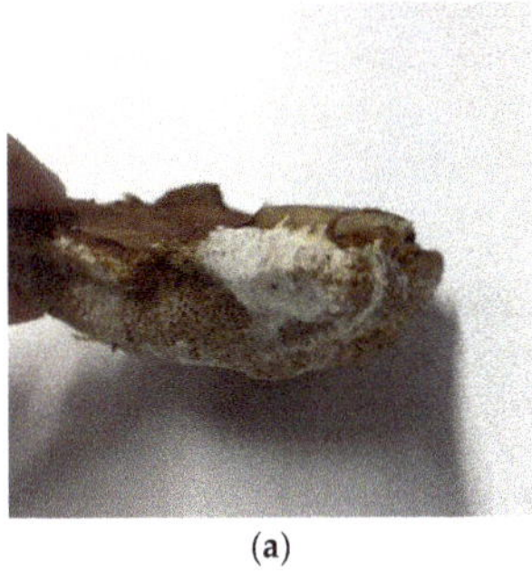 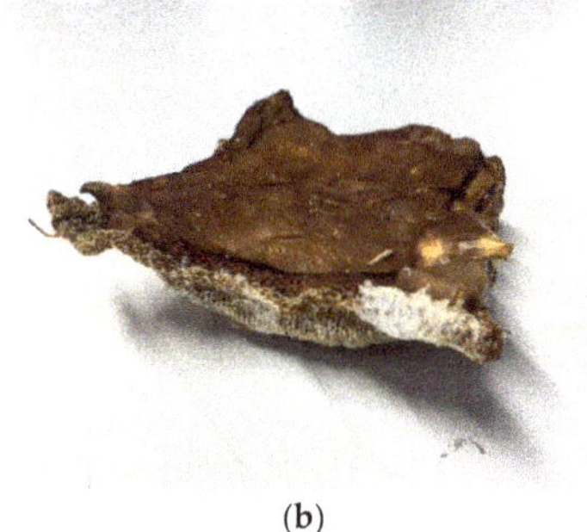

(**a**)   (**b**)

Figure 9. (**a**) Corner of the pocket with white mycelium skin. (**b**) BC dried onto the pocket, showing pattern of fabric underneath (image credit: authors).

4. Discussion

The primary purpose of this study was to test the compatibility of the two chosen biomaterials and identify methods of growing them into viable composites. The results of all four experiments demonstrated that reliable growth of a secondary organism was achievable in the presence of a primary organism (Table 1). However, we were not able to avoid contamination in the wet stage of the growth process in experiments three and four when both the mycelium and BC were alive. Although pre-grown mycelium composites are less receptive to contamination, the increased humidity and presence of sugar from the BC culture created a beneficial environment for mold, which was observed in experiment three. This problem could be solved by either adjusting the timeline of the setup, allowing the BC to dry partly before adding the mycelium in experiment four, or by adding active ventilation underneath the BC inside the mycelium growth box. For experiments 1–3, a more refined method to place the tiles inside the kombucha medium can be designed to maximize free air flow on the surface of the mycelium. Whilst contamination is evident in experiments three and four, it is notable that this has not prevented successful attachment between the mycelium and the bacterial cellulose; in fact, the best attachment was observed in experiment four, where mold was observed on both the BC and mycelium.

Table 1. The summary of the experiment.

	Oven-Dried (No. 1)	Air-Dried (No. 2)	Living (No. 3)	Pocket (No. 4)
BC growth	5–13 mm, <2 mm on gas bubble	5–8 mm	5–8 mm	4–6 mm and inclusion of knit
Mycelium growth	Dead, white film around substrate	Dormant pinheads, white film	Mushrooms and no white film	Little and on top side only
Attachment	Good	Weak to not at all	Stronger after oven drying	Strong
Contamination	No contamination	No contamination	Mold on exposed mycelium	Mold on BC and mycelium
Other observations	Uneven BC growth	BC around fruiting bodies thicker	BC grew around mushroom	-

The presence of mycelium in any state of aliveness caused greater production of fermentation gases, which led to visibly thinner BC in the areas where gas collected in bubbles. The BC also showed increased growth around fruiting bodies and primordia formations, indicating a greater availability of nutrients and/or oxygen in these areas through the mycelium composite.

The shape of knit in the form of a pocket was valuable as a scaffold for the BC to grow around and to hold the loose mycelium substrate in a compacted shape. The knit successfully offered a link between the two organisms in the setup. The growth behavior of mycelium with different types of yarn was tested in a separate set of experiments as part of the *BioKnit* project, which revealed linen to be less compatible [13]. This can partly explain the limited mycelium growth in experiment four.

5. Conclusions

This paper focused on the biological compatibility of mycelium and bacterial cellulose as two biomaterials that produce different and potentially complimentary qualities for a new generation of biohybrid materials for architecture. Through experimental study, the research investigated four different combined growth setups to test the potential for growing together, the use of a knit scaffold, and the challenge of contamination.

Growing Together: The research demonstrates that BC and mycelium can grow effectively in the presence of one another. The growth of the secondary organism is influenced by the growth requirements of the primary organism, particularly when both organisms are living. This influence was observed predominantly through uneven growth of the secondary organism. In two experiments, the presence of mushrooms growing from the mycelium increased the BC formation in the surrounding area, suggesting a localized increase in nutrition supply for the BC.

The use of a knit scaffold: Whilst BC and mycelium can grow together, producing a strong attachment during growth, forming a composite is more problematic. The knitted scaffold proved helpful in bringing the two organisms together in a way that supported both growth behaviors through the fabric's adjustable breathability and rough texture. As a result of this research, large-scale knit scaffolds have been implemented in the design of the *BioKnit* prototype. Further research investigating attachment strategies could guide the design of future textile scaffolds.

The challenge of contamination: The changing environmental conditions required by the different growth requirements of BC and mycelium increase the probability of contamination during the growth phase. The most challenging factors observed are humidity and ventilation leading to the growth of mold. Further research is required to optimize the environmental conditions, and active systems to control ventilation and humidity during growth will be developed to eliminate these problems in future projects.

Author Contributions: Conceptualization, A.H. and D.O.; methodology, A.H. and D.O.; validation, A.H. and D.O.; formal analysis, A.H. and D.O.; investigation, A.H. and D.O.; resources, A.H. and D.O.; data curation, A.H. and D.O.; writing—original draft preparation, A.H. and D.O.; writing—review and editing, J.S., A.H. and D.O.; visualization, A.H. and D.O.; supervision, J.S.; project administration, A.H. and D.O. All authors have read and agreed to the published version of the manuscript.

Funding: Research funded by Research England's Expanding Excellence in England (E3) Fund.

Institutional Review Board Statement: Not applicable.

Informed Consent Statement: Not applicable.

Data Availability Statement: Not applicable.

Acknowledgments: This research is part of a series of prototypes developed in the Hub for Biotechnology in the Built Environment (HBBE). HBBE is funded by Research England and is a joint initiative between Newcastle University and Northumbria University. The *BioKnit* prototype was further designed and built in collaboration with Armand Agraviador, Ahmet Topcu, Romy Kaiser, Elise Elsacker, Emily Birch and Ben Bridgens.

References

1. Karana, E. When the Mycelium Grows: A case study on Designing (with) Mycelium-based Materials. *Int. J. Des.* **2018**, *12*, 119–136.
2. Heisel, F. Design, Cultivation and Application of Load-Bearing Mycelium Components: The MycoTree at the 2017 Seoul Biennale of Architecture and Urbanism. *Int. J. Sustain. Energy Dev.* **2017**, *6*, 296–303. [CrossRef]
3. Biomaterial Building. Available online: https://www.biomaterialbuilding.com/mycocreate (accessed on 20 April 2022).
4. JDO Vaults. Available online: https://jdovaults.com/El-Monolito-Micelio (accessed on 20 April 2022).

5. Jones, M.; Mautner, A.; Luenco, S.; Bismarck, A.; John, S. Engineered mycelium composite construction materials from fungal biorefineries: A critical revies. *Mater. Des.* **2022**, *187*, 108397. [CrossRef]
6. Portela, R.; Leal, C.R.; Almeida, P.L.; Sobral, R.G. Bacterial Cellulose: A versatile biopolymer for wound dressing applications. *Microb. Biotechnol.* **2019**, *12*, 565–610. [CrossRef] [PubMed]
7. Sayfutdinova, A.; Samofalova, I.; Barkov, A.; Cherednichenko, K.; Rimashevskiy, D.; Vinokurov, V. Structure and Properties of Cellulose/Mycelium Biocomposites. *Polymers* **2022**, *14*, 1519. [CrossRef] [PubMed]
8. Ng, F.M.; Wang, P.W. Natural Self-grown Fashion from Bacterial Cellulose: A paradigm Shift Design Approach in Fashion Creation. *Des. J.* **2016**, *19*, 837–855. [CrossRef]
9. Scott, J. Knitted Cultivation: Textiling a Multi-Kingdom Bio Architecture. In Proceedings of the ICSA 2022, 5th International Conference on Structures and Architecture, Aalborg, Denmark, 6–8 July 2022.
10. Appels, F.V.W. The Use of Fungal Mycelium for the Production of Bio-Based Materials. Doctoral Dissertation, University Utrecht, Utrecht, The Netherlands, 2019.
11. Elsacker, E.; Vandelook, S.; Damsin, B.; Van Wylick, A.; Peeters, E.; De Laet, L. Mechanical characteristics of bacterial cellulose-reinforced mycelium composite materials. *Fungal Biol. Biotechnol.* **2021**, *8*, 18. [CrossRef] [PubMed]
12. Resnick, M.; Rosenbaum, E. Chapter 10-Designing for Tinkerability. In *Design, Make, Play*; Honey, M., Kanter, D., Eds.; Routledge: New York, NY, USA, 2013.
13. Scott, J.; Ozkan, D.; Hoenerloh, A.; Birch, E.; Kaiser, R.; Agraviador, A.; Elsacker, E. BioKnit Building: Strategies for Living Textile Architectures 2021. In Proceedings of the CEES International Conference Construction, Energy, Environment and Sustainability, Coimbra, Portugal, 12–15 October 2021.

 *biomimetics*

Article

A Modular Chain Bioreactor Design for Fungal Productions

Onur Kırdök [1],*[iD], Berker Çetintaş [2][iD], Asena Atay [2][iD], İrem Kale [3][iD], Tutku Didem Akyol Altun [3][iD] and Elif Esin Hameş [1,2,4][iD]

[1] Department of Biotechnology, Graduate School of Natural and Applied Sciences, Ege University, 35040 İzmir, Türkiye
[2] Department of Bioengineering, Graduate School of Natural and Applied Sciences, Ege University, 35040 İzmir, Türkiye
[3] Department of Architecture, Faculty of Architecture, Dokuz Eylül University, 35390 İzmir, Türkiye
[4] Department of Bioengineering, Faculty of Engineering, Ege University, 35040 İzmir, Türkiye
* Correspondence: onur@biopbiotech.com; Tel.: +90-5073721795

Abstract: Plastic bag bioreactors are single-use bioreactors, frequently used in solid culture fermentation. This study developed plastic bag bioreactors with more effective aeration conditions and particular connection elements that yield sensors, environmental control, and modular connectivity. This bioreactor system integrates the bags in a chain that circulates air and moisture through filtered connections. Within the present scope, this study also aimed to reveal that cultures in different plastic bags can be produced without affecting each other. In this direction, biomass production in the modular chain bioreactor (MCB) system developed in this study was compared to traditional bag systems. In addition, contamination experiments were carried out between the bags in the system, and it was observed that the filters in the developed system did not affect the microorganisms in different bags.

Keywords: modular chain bioreactor; solid-state fermentation; mycelium production; *Ganoderma lucidum*

Citation: Kırdök, O.; Çetintaş, B.; Atay, A.; Kale, İ.; Akyol Altun, T.D.; Hameş, E.E. A Modular Chain Bioreactor Design for Fungal Productions. *Biomimetics* **2022**, *7*, 179. https://doi.org/10.3390/biomimetics7040179

Academic Editor: Stanislav N. Gorb

Received: 16 July 2022
Accepted: 24 October 2022
Published: 27 October 2022

Publisher's Note: MDPI stays neutral with regard to jurisdictional claims in published maps and institutional affiliations.

1. Introduction

Today, mycelium growing plays an important role in mushroom production and various biotechnological studies. In recent years, bag containers have been investigated as promising tools for mycelium development, going beyond traditional tray-type bioreactors [1]. According to research, although a limited gas exchange and a limited heat transfer are observed in bag systems, this provides an ideal environment for mycelium production due to low moisture loss [2]. Since the beginning of the use of bag containers, studies have been carried out on edible mushrooms, such as *Ganoderma lucidum* [3,4], *Pleurotus ostreatus* [5,6], *Trametes versicolor* [7,8], and *Agaricus bisporus* [9,10]. As a result of these studies, bag systems are used for mushroom growing in many countries for academic and commercial purposes [11]. Such systems can be divided into two groups depending on the sterility. Commercial bag cultivator systems mostly use plastic bags with needle holes, or even straw sacks, that are often vulnerable to contamination risks. More recently, microholed plastic bags have become popular. However, the risk of contamination continues due to the humidity and perspiration on the plastic surface. Therefore, mushroom farmers mostly prefer pasteurizing the substrate before inoculation to ensure that the fungus is the dominant organism in production. On the other hand, in academic studies, sterile conditions are mainly created to provide controlled conditions, and cultivation is carried out with the selected pure microorganism. To achieve that, bags to be used in such systems should be able to be effectively sterilized. In most cases, the preferred sterilization method is high-pressure steam heated using an autoclave. These bags must be made of a high-temperature-resistant material, such as polypropylene, to be autoclaved. The sterilization of the substrate usually requires a temperature of 121 °C, and the bag used in the system must also be resistant to high temperatures.

Bag systems are also used for mushroom cultivation and microalgae systems. It is used worldwide in microalgae cultivation in liquid media due to its easy accessibility, low cost, and ability to be ordered in desired volumes [12]. In addition, it has application areas, such as biomass production, biofuel production, and wastewater treatment [13,14]. Accordingly, bag systems are frequently used for commercial and academic purposes due to their rapid growth and high productivity, easy homogenization, cheapness, easy addition to the nutrient medium, ease of work, easy cleaning, easy detection, and the intervention of possible contamination [15]. Using heat-resistant plastic bags as containers during production is common in industrial mushroom production. Specifically, micro-holed plastic bags have become popular among *P. ostreatus* producers. However, bag systems have some disadvantages compared to conventional tray bioreactors due to their limited gas exchange, insufficient air circulation, inability to remove heat caused by microbial growth, and limited heat transfer [16]. Therefore, bigger-scale mushroom production facilities mostly prefer tray-type bioreactors, and in some cases, a mixed use of bag systems and tray reactors is chosen. Tray systems are used in defined spaces (rooms, tents, greenhouses, etc.) with climate control, and they require vast space and expensive equipment. However, according to tray bioreactors, bag systems provide an advantage when limited air circulation and heat exchange issues are fixed (Table 1).

Table 1. Comparison of bag system and tray bioreactor.

Bag System	Tray Bioreactor
Easy to sterilize	Sterilization is more difficult than the bag system
Can be mixed during fermentation	Cannot be mixed effectively
Cheap	More expensive than bag system
Limited air circulation	Efficient air circulation
Limited heat exchange	Efficient heat exchange
Hard to remove microbial heat in fermentation	Microbial heat formed during fermentation can be removed
Low area requirement	Wide area requirement
Microbial growth is controllable	Microbial growth is uncontrollable

Conclusively, as stated, bag systems are relatively sufficient for biomass production on a laboratory scale or on a large scale. However, their most important disadvantage is their insufficient air circulation, which reduces production efficiency in biomass growth. Their advantages, such as ventilation, mixing, and moisture retention, make them an efficient tool for solid-state bioreactor designs for controlled mass production. However, such systems lack air circulation, controlled sterility, and humidity control.

In this study, a modular bioreactor was developed to eliminate the weaknesses of bag bioreactors, especially for the three main properties (air circulation, controlled sterility, and humidity control). This bioreactor, named in the study as the modular chain bioreactor (MCB), contains candy-bag modules connected with chain connectors (CCs). The candy bags are pouches with both sides open. A cotton filter was installed on both parts of the bag for high air circulation, and both cotton filters were fixed with a cardboard cup. Finally, this module was duplicated and connected by CCs to form the MCB.

To examine the effect of the developed modular bioreactor on the production of fungi in solid-state cultures, biomass production was investigated using *G. lucidum*. In addition, the prevention of the spread of contamination, which is the main risk factor in mushroom cultivation and causes great losses, was examined.

2. Materials and Methods

2.1. Substrates and Microorganism

G. lucidum, one of the most studied species of white-rot Basidiomycetes, was used for the pilot tests of this study. The fungus was maintained on malt extract agar (MEA) slants at 4 °C. In a different ongoing study by the authors, inoculum production was optimized for pH, time, and an initial amount of inoculum. Accordingly, inoculum production was prepared using malt extract broth (MEB) (pH 5) by incubation at 28 °C and 200 rpm for 11 days. After incubation, the inoculum (Figure 1a) was used as 1 mL/5 g of dry substrate.

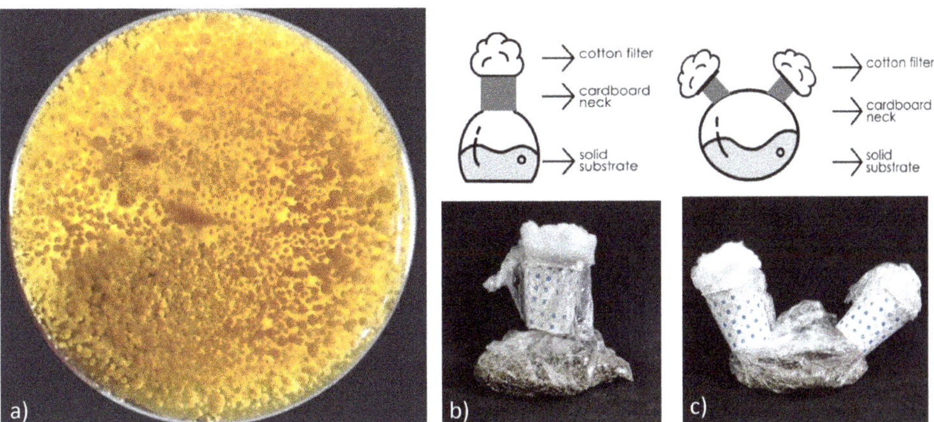

Figure 1. (**a**) Pellets (bottom the flask); (**b**) cotton-filtered bag container; (**c**) Candy-bag bioreactor (right).

In order to see the effect of the system designed for biomass production, three candy bags were prepared for the MCB. One candy bag and one traditional bag were prepared as a control group. Both bags were filled with 200 g of zeolite support material and were enriched with a liquid solution containing mineral salts and nutrients (Table 2). Zeolite is an inert support material, consisting of $SiO4$ and/or $AlO4$, and is ideal for mycelium production in solid-state fermentation, thanks to the large voids in its structure [17]. When zeolite reaches the ideal humidity with a liquid solution containing nutrients, it has a structure that triggers mycelium production and keeps the pH value of the microorganism constant [18]. Therefore, this solution is also used as a moistening liquid to trigger microorganism growth.

Table 2. Nutrient-mineral salt solution.

Composition	Formula	Amount (g/L)
Glucose	$C_6H_{12}O_6$	30
Peptone		2
Ammonium nitrate	NH_4NO_3	2
Trisodium citrate dihydrate	$C_6H_5Na_3O_7 \cdot 2H_2O$	2.5
Potassium dihydrogen phosphate	KH_2PO_4	5
Magnesium sulfate heptahydrate	$MgSO_4 \cdot 7H_2O$	0.2
Calcium chloride dihydrate	$CaCl_2 \cdot 2H_2O$	0.1
Di ammonium hydrogen phosphate	$(NH_4)_2HPO_4$	4

2.2. Modular Chain Bioreactor (MCB) Prototype & Arduino Setup for Environmental Control

Solid-state bioreactors are an engineering and design issue for cultivating these organisms for research and industrial purposes. Although there are many different types of bioreactors, the main design principles of solid-state bioreactors focus on several points. These designs should be able to:

- Keep environmental conditions at optimum levels that are favored by the organisms (temperature, water activity, oxygen concentration),
- Carry on solutions for ventilation and mixing required by the production process,
- Block any organism from entering inside to prevent contamination and keep colonized organisms inside to expel any harmful effect that the organism can cause,
- Be produced by a durable and corrosion-resistant material, which should also not cause any toxic effect on colonized organisms-,
- Allow sampling and observation,

- Be suitable for any process required for solid-state fermentation (substrate preparation, inoculation, loading, unloading, sterilization),
- Be economic [19,20].

Traditionally used bag containers are temperature-resistant bags with a substrate added. The required air is supplied to the mouth of the bag with a cotton filter supported by a cardboard neck collar. Thus, while the containers can reach the air needed by the microorganisms, the substrate bed is protected from contamination. In order to enhance the properties of air circulation, controlled sterility, and humidity, a secondary cotton-filtered opening on the opposite side were examined first. It was observed that this "candy"-like bag container increased the colonization speed of the fungus (Figure 1). Enzymes are one of the most important products produced in solid-state fermentation. Microorganism growth in solid-state fermentation also affects enzyme production [21]. Culture conditions (pH, temperature, humidity, etc.) should be at optimum levels for the microbial used for rapid microbial growth [22]. However, one of the most fundamental limitations of enzyme production in solid-state fermentation is the measurement accuracy of the enzyme obtained. This may be due to the extraction method, the difference in the substrate used in the enzyme determination, and the fermentation parameters [23]. The developed candy-bag system provides better air circulation compared to traditional bag systems. The two-mouth candy-bag system proposed in the author's ongoing solid-state fermentation and enzyme production study was compared with the traditional single-mouth bag system, and 1.8 times higher enzyme activity was observed in the candy-bag system compared to the bag system (data not shown).

This study used 45×50 cm polypropylene bags as candy-bag bioreactor systems for biomass production. Cotton was used as an air filter and cardboard cups (with the bottoms cut off) were used as scaffolding to keep the cotton stable.

However, fixed humidity control and problems in air circulation still occurred due to the crumpled structure of plastic bags and sterility limits. To solve these problems, the MCB was generated using separable, filtered, interlocking Chain Connectors (CCs) between the candy bags (Figure 2). It was designed in Fusion 360, and prototypes were developed by 3D printing. In order to test the design proposal, a system prototype was installed for 7 days and observed daily.

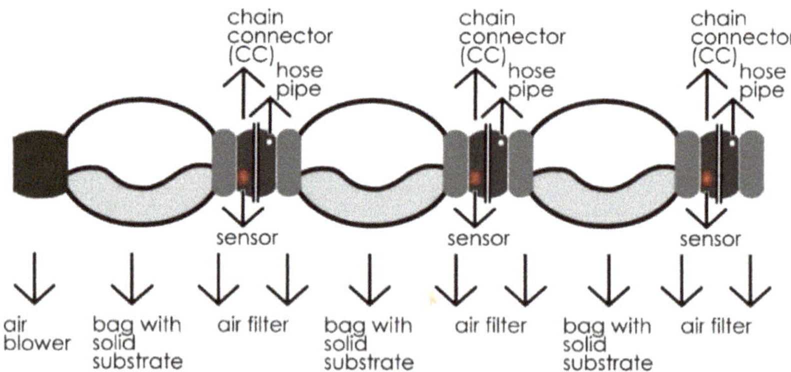

Figure 2. MCB—3 Unit chain scheme.

The proposed bioreactor aims to offer an improvement on the following qualities:

- Humidity-controlled environment
- Better air circulation
- Modularity (each colonization unit can be separated from the chain without risking others with contamination)
- Sterile conditions

- Low cost

Three heat-resistant colonization bags (candy bags) with identical substrates were linked for the testing process. A specially designed CC unit connected these bags in a chain. The main purpose of these connectors was to achieve better environmental control in the bags, while yielding the removal of desired bags without contaminating the system. An air blower with a 100 mm air outlet and a DC motor of about 100 W accommodated in the chain circulated the air through bags to ventilate the heating caused by the fermentation process. This motor blows 15 s of air per 3 min to circulate the air held inside the candy bags connected in chains through CCs. Meanwhile, a central humidifier mechanism feeds the system via silicon pipes connected to CCs with steam to keep the system in the range of 60–80% moisture, responsively to sensor readings. The relative humidity was adjusted simultaneously with the data collected from humidity sensors located within the CCs. In this study, Arduino Mega was used as a microcontroller to process the data gathered from sensors (DHT11) to maintain the relative humidity and air circulation at a desired level inside the bags by controlling air blowers and humidifier mechanisms, with the help of DC relays. The whole setup was constructed in a room with an air conditioner to cool the place to desired production temperatures between 25 and 28 °C. The temperature levels inside the bags were constantly monitored and recorded on an SD card loaded on the Arduino.

The humidifier mechanism consists of an air blower with a 75 mm air outlet and a DC motor of about 60 W, a sterilized humidifier with a 4 L water capacity, and pipes. The working principle of the humidifier is simply that an air blower is connected to the tank's top level, forcing moisture to circulate inside the colonization bags. This process is controlled with the microcontroller. Figures 3 and 4 illustrate a simplified block diagram of the electrical and mechanical system of the MCB and the humidifier.

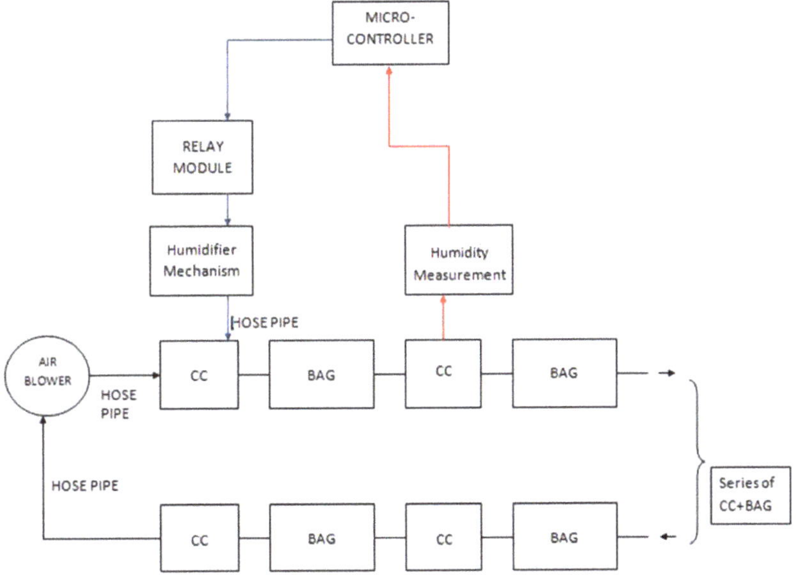

Figure 3. Simplified block diagram of MCB.

The MCB system can be considered an example of lean entrepreneurship to prevent contamination with its modularity. The most important part of this system is the CC apparatus and the filters between this apparatus. Unlike traditional bag systems, the additional cost of filters is $1.80 per bag.

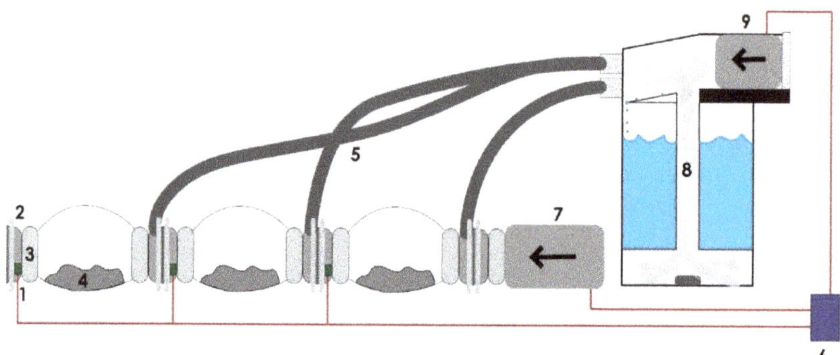

Figure 4. The simplified electrical and mechanical system of the MCB and the humidifier: (1) humidity/temperature sensor, (2) CC, (3) air filter, (4) solid substrate, (5) hose pipe, (6) Arduino, (7) 100 W air blower, (8) vapor chamber, (9) 50 W air blower.

2.3. Fungal Biomass Production Experiments

The efficiency of the MCB prototype was examined by colonizing a pure culture of *G. lucidum* from the author's previous research [24]. White-rot fungi, *G. lucidum*, with easy colonization and popular in academic research, was chosen as the control culture for the repetition of the tests. The zeolite was used as an inert support material to absorb the nutrients and minerals required for mycelium production and to be used at optimum pH levels [17,25]. Zeolite obtained from volcanic rocks is a mixed mineral-salt-medium solution as a nutrient broth, used as the substrate to achieve fast colonization and easy determination of the biomass amount. A pure culture of fungus was inoculated on the substrate for colonization. As can be expected from inorganic zeolite, its mass did not change during colonization, while fungi increased their mass by using nutrients. The dry mass obtained at the end of the incubation was proportional to the biomass of the fungus. The following equation was used in the calculations:

$$m_{(\text{final dry weight})} = m_{(\text{zeolite dry})} + m_{(\text{mycelium biomass dry})} \tag{1}$$

2.4. Contamination Spread Pilot Experiment

In the MCB, a three-bag experiment was designed to measure the effect of the filter system and the contamination spread between the bags. The first and third bags were prepared under sterile conditions, and *G. lucidum* was inoculated. The second bag, which had filtered connections with other bags, was prepared under non-sterile conditions and was kept in the solid-state fermentation laboratory for half an hour before being included in the MCB. The bags were then connected to the MCB and supplied with humidity and air.

3. Results and Discussion

3.1. Contamination Control Experiments

In the three-bag experiment, the second bag in the middle was kept in a non-sterile area and added to the system to control the spread of contamination between modular systems. Three observations were made on the 3rd, 5th, and 7th days (Figure 5). Visual inspection shows apparent white hyphae (*G. lucidum*) colonizing the first and the third bags without any contamination; however, an unknown microorganism (orange-pink color) grew predominantly in the non-sterile bag (middle unit). The contamination in the second bag did not spread to other bags, indicating that the modularity of the filtration system and the developed MCB system were successful. If the filtration system had been inadequate and the contamination had spread to bags one and three, it would have also inhibited the mycelial production of *G. lucidum*. On the other hand, when Figure 5 is examined, the development of *G. lucidum* in bags one and three on the 5th and 7th days, and the contamination in the 2nd bag, can be clearly distinguished in terms of mycelium production and color. This shows that the filtration between modules successfully prevented the spread of contamination as intended (Figure 6).

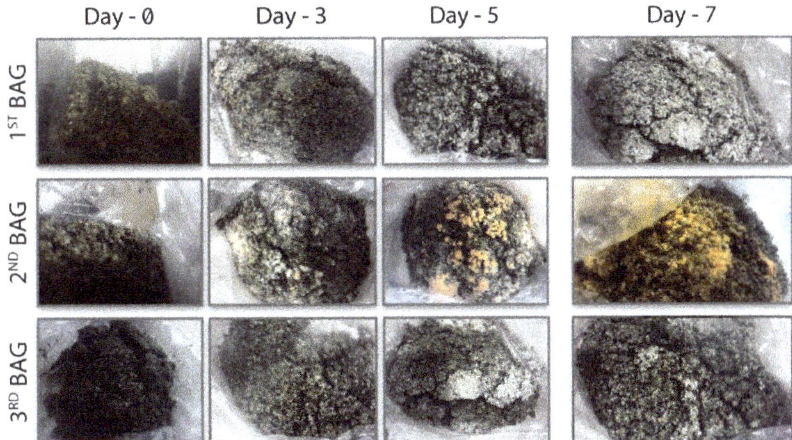

Figure 5. MCB contamination control experiments.

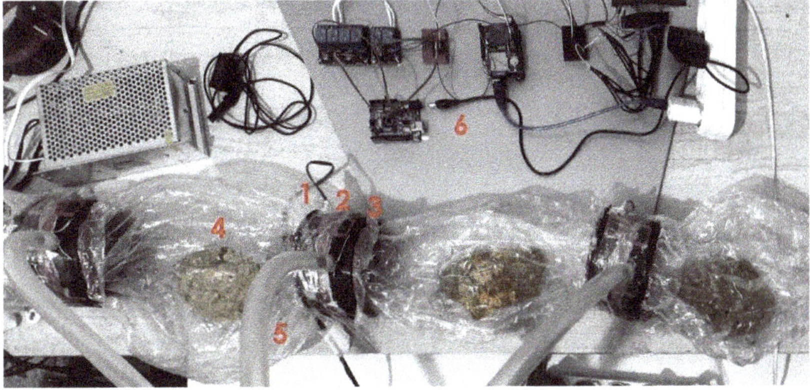

Figure 6. MCB contamination control experiment setup: 1, humidity/temperature sensor; 2, CC; 3, air filter; 4, solid substrate; 5, hose pipe; 6, Arduino).

As a result of the biomass production, 20.7 g and 11.3 g of mycelium production was observed in the bags (the 1st and 3rd bags, respectively). The 3rd bag did not obtain sufficient moisture due to water accumulation inside the humidifier's silicon tube. A biomass production of 55.7 g was observed in the contaminated bag (Table 3).

Table 3. Contaminant control experiment assay.

Bag	Final Weight (g)	Final Dry Weight (g)	Zeolite (g)	Biomass (g)
1	264.4	220.7	200	20.7
2	285.2	255.7	200	55.7
3	255.4	211.3	200	11.3

3.2. Modular Chain Bioreactor System Results

Mycelium development as a result of biomass production in the MCB is given below (Figure 7). As a result of the experiment, it was seen that mycelium production was relatively rapid with a high humidity and high air supply. In environments where the amount of free water is high, the mycelium formation cannot increase, and the microorganism loses its viability. Because solid culture fermentation occurs in static conditions, if there is excess water in the environment, oxygen cannot be included in the substrate as there is no mixing as in submerged fermentation. Therefore, the dissolved oxygen contained in the excess water is consumed in a short time, and anaerobic conditions are formed. As a result, the diffusion of O_2, which is needed for mycelium production, is prevented [21]. It should be taken into account in future studies that anaerobic conditions may have prevented aerobic contaminants. The main issue to be considered here is the O_2-CO_2 diffusion between the biomass and the substrate. In the presence of free water, this diffusion is restricted and inhibits aerobic microbial growth. In this case, the mycelium cannot reach the oxygen it needs, and microbial growth remains weak. However, because the moisture in the 3rd bag was too high, free water accumulation occurred in the environment. While the amount of free water was 0 mL in the inoculum fluid at the beginning of the experiment, approximately 25 mL of water accumulation was measured at the end. In environments where the amount of free water is high, the mycelium cannot grow, and the microorganism loses its viability. Therefore, the moisture supply to the environment during fermentation must be optimized so that it is neither too low nor too high to cause free water accumulation. Compared to conventional bags, a more successful result was obtained with the mycelial biomass developed in the MCB candy bag. Mycelial production in the moisture- and air-fed MCB was three times higher than that in the candy bag and mycelial biomass formation was 3.5 times higher than in conventional bag systems (Table 4).

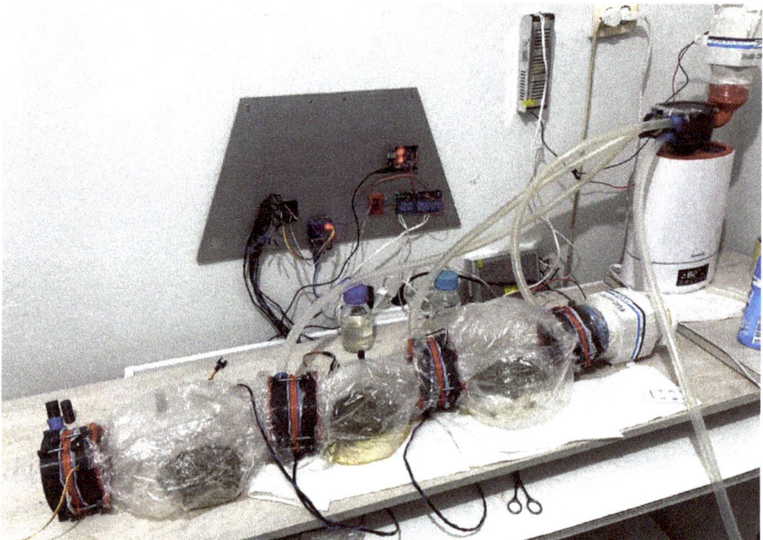

Figure 7. MCB setup for biomass experiments.

According to the results, the conditions provided for mycelium production during the 7 days showed a positive effect. The provided air supply and humidification were shown to provide early microbial activation. In addition, the fact that mixing can be achieved manually accelerates the growth of microorganisms that are activated early. However, during the experiment, it was observed that water accumulation occurred in the second bag. This negative situation can prevent mycelial production in solid-state fermentation. For this reason, as a result of the obtained biomass, the production in the 2nd bag was lower than in the other bags (Figure 8).

Table 4. MCB experiment biomass assay.

Bag	Final Weight	Final Dry Weight	Zeolite	Biomass
1	263.88	221.88	200	21.88
2	251.1	217.4	200	17.4
3	271.5	220.3	200	20.3
Control candy bag	261.3	209.7	200	9.7
Control traditional bag	248.2	206.1	200	6.1

Figure 8. MCB biomass experiment observations.

Another outcome of the experiments can be derived as follows: the design of the MCB, yields the potential to grow different kinds of organisms without affecting or contaminating each other directly. This way, different stages of mycelium development can be kept together, and the faster colonies can be harvested easily. Another outcome of the MCB would be the ability to multi-use. For example, while one bag houses spore-producing fruiting of medicinal *G. lucidum*, the neighbor bag can be a resident for *T. versicolor*, produced for enzyme subtraction. This chain can be continued for any use requiring similar environmental conditions, ranging from commercial to personal use.

4. Conclusions

In this study, a modular bioreactor chain system was produced to eliminate the disadvantages of bag systems frequently used in solid culture fermentation. In a modular system, parameters, such as environment control, sensor systems, and modular connectivity, were discussed, and a particular connection system named CC was developed to connect the modules. The MCB is constituted of connected modules and is a modular chain bioreactor prototyped to test the efficiency of the setup.

Within the scope of the study, the prototype produced with three bag modules with a two-filtered opening contained equivalent solid culture samples connected by CCs. There was a humidity/temperature sensor and a hose-pipe connection in the CCs. In addition, air blowers were added to provide air and humidity flow to the chamber, and these systems were connected with Arduino to provide environmental control. A double-sided filtration system was developed to minimize the risk of contamination between the CCs and bags to provide humidity control. Typical bag containers mostly have one hole for air circulation, and in the case of mass production scenarios, each bag is vulnerable during operation times. However, the double-sided filtration system fixes this vulnerability by connecting all openings in one filtered line for airflow. It allows the chosen unit to be removed without disturbing the rest of the chain.

Biomass production and contamination spread experiments were performed to show that the system works effectively in this study. In the pilot contamination experiments, the second module was left in a non-sterile environment, and the development of all three samples was observed. As a result of the 7-day mycelium-development observation in all modules, when contamination was observed in the second module as expected, the absence of contamination in other modules supported the system's efficiency.

According to the biomass production trial experiments, it has been predicted that the ventilation and humidification system designed in the MCB could be more effective than traditional bag systems. Furthermore, the system's modularity offers the potential to work with different microorganisms simultaneously, the option of harvesting at different times in the same production, and the loss of a single bag rather than the whole production in case of possible contamination. The results of the experiments and observations through the process revealed these advantages of the current MCB prototype in solid-state fermentation:

- High ventilation capacity
- Moisture level control
- Easy mixing
- Efficient contamination control
- Permitting cultivation and harvesting without risking the rest of the bags through modularity
- Cultivability of different microorganisms for different purposes
- Observability of mycelial growth through the process
- Low cost

However, there are still some issues with the system that need improvement. The accumulation of water in the feed hoses during humidification in the experiments and the flow of this accumulated water to the environment cause the risk of preventing mycelium production. The length of the silicon tubing should be adjusted to prevent water accumulation in further tests. In addition, the air blowers' response sequences and the optimization of the sensor reading times are required. The air blower's power capacity should be adjusted according to the filter's number and resistance to the direct blow of the wind. In addition, another factor to be considered in future studies is the determination of how long the prepared system can be used by calculating the efficiency in terms of gram biomass/gram substrate. This study used solid support materials to compare the MCB system and traditional bag systems. However, using a solid substrate instead of a solid support material will increase the yield for more extended use. Finally, the MCB needs more space than the standard bag containers, trained personnel, and the risk of damage to sensitive system elements due to moisture and disinfection chemicals.

Author Contributions: Conceptualization, O.K. and B.Ç.; methodology, O.K., B.Ç., T.D.A.A. and E.E.H.; software, O.K.; validation, O.K., B.Ç. and A.A.; formal analysis, O.K., B.Ç. and A.A.; resources, O.K., T.D.A.A. and E.E.H.; data curation, O.K. and B.Ç.; writing—original draft preparation, O.K., B.Ç., İ.K., T.D.A.A. and E.E.H.; writing—review and editing, O.K., B.Ç., İ.K., T.D.A.A. and E.E.H.; visualization, O.K. and İ.K.; supervision, T.D.A.A. and E.E.H.; project administration, O.K. All authors have read and agreed to the published version of the manuscript.

Funding: This research received no external funding.

Data Availability Statement: Data are available upon request.

Acknowledgments: O. Kırdök and B. Çetintaş would like to acknowledge The Scientific and Technological Research Council of Turkey (TUBITAK-BIGG) for the grant (no: 2211182), which is partially related to this study.

Conflicts of Interest: The authors declare no conflict of interest.

References

1. Battaglino, R.A.; Huergo, M.; Pilosof AM, R.; Bartholomai, G.B. Culture requirements for the production of protease by *Aspergillus oryzae* in solid state fermentation. *Appl. Microbiol. Biotechnol.* **1991**, *35*, 292–296. [CrossRef] [PubMed]
2. Lopez-Perez, M.; Rodriguez-Gomez, D.; Loera, O. Production of conidia of *Beauveria bassiana* in solid-state culture: Current status and future perspectives. *Crit. Rev. Biotechnol.* **2015**, *35*, 334–341. [CrossRef]
3. Khoo, S.C.; Ma, N.L.; Peng, W.X.; Ng, K.K.; Goh, M.S.; Chen, H.L.; Tan, S.H.; Lee, C.H.; Luang-In, V.; Sonne, C. Valorisation of biomass and diaper waste into a sustainable production of the medical mushroom Lingzhi *Ganoderma lucidum*. *Chemosphere* **2022**, *286*, 131477. [CrossRef]
4. Fatima, S.; Khan, F.; Asif, M.; Alotaibi, S.S.; Islam, K.; Shariq, M.; Khan, A.; Ikram, M.; Ahmad, F.; Khan, T.A.; et al. Root-Knot Disease Suppression in Eggplant Based on Three Growth Ages of *Ganoderma lucidum*. *Microorganisms* **2022**, *10*, 1068. [CrossRef] [PubMed]
5. Raman, J.; Jang, K.Y.; Oh, Y.L.; Oh, M.; Im, J.H.; Lakshmanan, H.; Sabaratnam, V. Cultivation and nutritional value of prominent *Pleurotus* spp.: An overview. *Mycobiology* **2021**, *49*, 1–14. [CrossRef] [PubMed]
6. Tesfay, T.; Godifey, T.; Mesfin, R.; Kalayu, G. Evaluation of waste paper for cultivation of oyster mushroom (*Pleurotus ostreatus*) with some added supplementary materials. *AMB Express* **2020**, *10*, 1–8. [CrossRef]
7. Jones, M.P.; Lawrie, A.C.; Huynh, T.T.; Morrison, P.D.; Mautner, A.; Bismarck, A.; John, S. Agricultural by-product suitability for the production of chitinous composites and nanofibers utilising *Trametes versicolor* and *Polyporus brumalis* mycelial growth. *Process Biochem.* **2019**, *80*, 95–102. [CrossRef]
8. Ramezan, D.; Alizade Jahan Abadi, B.; Samzade Kermani, A.; Pirnia, M.; Farrokhzad, Y. Cultivation of Turkey Tail Mushroom (*Trametes versicolor*) on Lignocellulosic Wastes and Evaluation of Substrate Bioconversion. *Proc. Natl. Acad. Sci. India Sect. B Biol. Sci.* **2021**, *91*, 777–787. [CrossRef]
9. Leiva, F.J.; Saenz-Díez, J.C.; Martínez, E.; Jiménez, E.; Blanco, J. Environmental impact of *Agaricus bisporus* mycelium production. *Agric. Syst.* **2015**, *138*, 38–45. [CrossRef]
10. Kumar, P.; Kumar, V.; Goala, M.; Singh, J.; Kumar, P. Integrated use of treated dairy wastewater and agro-residue for *Agaricus bisporus* mushroom cultivation: Experimental and kinetics studies. *Biocatal. Agric. Biotechnol.* **2021**, *32*, 101940. [CrossRef]
11. Samp, R. The bag or block system of *Agaricus* mushroom growing. *Edible Med. Mushrooms Technol. Appl.* **2017**, *8*, 175–195.
12. Chen, M.; Chen, Y.; Zhang, Q. A Review of energy consumption in the acquisition of bio-feedstock for microalgae biofuel production. *Sustainability* **2021**, *13*, 8873. [CrossRef]
13. Cui, J.; Purton, S.; Baganz, F. Characterisation of a simple 'hanging bag' photobioreactor for low-cost cultivation of microalgae. *J. Chem. Technol. Biotechnol.* **2022**, *97*, 608–619. [CrossRef]
14. You, X.; Yang, L.; Zhou, X.; Zhang, Y. Sustainability and carbon neutrality trends for microalgae-based wastewater treatment: A review. *Environ. Res.* **2022**, *209*, 112860. [CrossRef] [PubMed]
15. Hassan FR, H.; Medany, G.M.; Hussein, S.D. Cultivation of the king oyster mushroom (*Pleurotus eryngii*) in Egypt. *Aust. J. Basic Appl. Sci.* **2010**, *4*, 99–105.
16. Mitchell, D.A.; Krieger, N.; Stuart, D.M.; Pandey, A. New developments in solid-state fermentation: II. Rational approaches to the design, operation and scale-up of bioreactors. *Process Biochem.* **2020**, *35*, 1211–1225. [CrossRef]
17. Sargin, S.; Gezgin, Y.; Eltem, R.; Vardar, F. Micropropagule production from *Trichoderma harzianum* EGE-K38 using solid-state fermentation and a comparative study for drying methods. *Turk. J. Biol.* **2013**, *37*, 139–146. [CrossRef]
18. Ivarsson, M.; Bengtson, S.; Skogby, H.; Lazor, P.; Broman, C.; Belivanova, V.; Marone, F. A fungal-prokaryotic consortium at the basalt-zeolite interface in subsea floor igneous crust. *PLoS ONE* **2015**, *10*, e0140106. [CrossRef]
19. Doran, P.M. *Bioprocess Engineering Principles*; Elsevier: Amsterdam, The Netherlands, 1995.
20. Liu, S. *Bioprocess Engineering: Kinetics, Sustainability, and Reactor Design*; Elsevier: Amsterdam, The Netherlands, 2020.
21. Raghavarao KS, M.S.; Ranganathan, T.V.; Karanth, N.G. Some engineering aspects of solid-state fermentation. *Biochem. Eng. J.* **2003**, *13*, 127–135. [CrossRef]
22. El Sheikha, A.F.; Ray, R.C. Bioprocessing of horticultural wastes by solid-state fermentation into value-added/innovative bioproducts: A review. *Food Rev. Int.* **2022**, 1–57. [CrossRef]
23. Teigiserova, D.A.; Bourgine, J.; Thomsen, M. Closing the loop of cereal waste and residues with sustainable technologies: An overview of enzyme production via fungal solid-state fermentation. *Sustain. Prod. Consum.* **2021**, *27*, 845–857. [CrossRef]
24. Kırdök, O. A Biodesign Collaborator in Architecture: Mycelium. Master's Thesis, Dokuz Eylül University The Graduate School of Natural and Applied Sciences. Available online: https://tez.yok.gov.tr/UlusalTezMerkezi/TezGoster?key=wf-FPgY-5qjHEzEoOgvMs0KR1Q7W50JJXVMX7hhzba8-_3UHnWKVJJreSwsNbMmb (accessed on 25 May 2022).
25. Guerra, G.; Casado, G.; Arguelles, J.; Acebal, C.; Castillón, M.; Ramos-Leal, M.; Gómez, B.; León, T.; Sánchez, M.; Torrija, E.; et al. Zeolite as source of saline nutrients in solid state fermentation of sugarcane straw B*ytrichoderma citrinoviride*. *Sugar Tech.* **2003**, *5*, 243–248. [CrossRef]

 *biomimetics*

Article

Mycelial Beehives of HIVEOPOLIS: Designing and Building Therapeutic Inner Nest Environments for Honeybees

Asya Ilgun * and Thomas Schmickl

Artificial Life Laboratory, Institute of Biology, University of Graz, 8010 Graz, Austria; thomas.schmickl@uni-graz.at

* Correspondence: zeliha.ilgun@uni-graz.at

Abstract: The perceptions and definitions of healthy indoor environments have changed significantly throughout architectural history. Today, molecular biology teaches us that microbes play important roles in human health, and that isolation from them puts not only us but also other inhabitants of urban landscapes, at risk. In order to provide an environment that makes honeybees more resilient to environmental changes, we aim for combining the thermal insulation functionality of mycelium materials with bioactive therapeutic properties within beehive constructions. By identifying mycelial fungi's interactions with nest-related materials, using digital methods to design a hive structure, and engaging in additive manufacturing, we were able to develop a set of methods for designing and fabricating a fully grown hive. We propose two digital methods for modelling 3D scaffolds for micro-super organism co-occupation scenarios: "variable-offset" and "iterative-subtraction", followed by two inoculation methods for the biofabrication of scaffolded fungal composites. The HIVEOPOLIS project aims to diversify and complexify urban ecological niches to make them more resilient to future game changers such as climate change. The combined functions of mycelium materials have the potential to provide a therapeutic environment for honeybees and, potentially, humans in the future.

Keywords: biohybrid architecture; bio fabrication; living architecture; beehive; 3D printing; mycelium materials; symbiosis; multispecies architecture; healthy materials

Citation: Ilgun, A.; Schmickl, T. Mycelial Beehives of HIVEOPOLIS: Designing and Building Therapeutic Inner Nest Environments for Honeybees. *Biomimetics* **2022**, *7*, 75. https://doi.org/10.3390/biomimetics7020075

Academic Editors: Andrew Adamatzky, Han A.B. Wösten and Phil Ayres

Received: 15 March 2022
Accepted: 26 May 2022
Published: 7 June 2022

Publisher's Note: MDPI stays neutral with regard to jurisdictional claims in published maps and institutional affiliations.

1. Introduction

The characteristics that define a healthy environment have changed significantly over human history. The Miasma Theory (400 BC), an obsolete scientific theory, suggested that diseases are caused by bad air [1]. Later, the discovery of germs [2,3] as the origin of diseases led to a public perception of all microbes as pathogenic. As a result, humans' indoor lifestyle and their yearning for hygiene have set the goals for the design of buildings, and criteria for the selection of materials for isolation. In addition, human activities, e.g., monocultures and intensified agriculture, caused the isolation of other living species from their coevolved symbionts, e.g., through habitat fragmentation. A significant "dewilding" activity is the exploitative domestication of wild animals, especially keystone species like honeybees [4].

Among other pollinators, honeybees play a crucial role in conserving biodiversity in flora and fauna. Besides that, they ensure food diversity for human society. Commercial honeybees used in agriculture are affected by a rich spectrum of stressors, such as overcrowded positioning or long-range transportation of hives, and agriculturally applied chemicals. Honey and other bee products crucial to honeybees' wellbeing are also removed from honeybee colonies. In the meanwhile, cities are becoming megacities due to human population growth. This comes at the expense of diminishing forests, marshlands, and other ecologically vital habitats. Finally, several essential habitat types are decreasing, directly and indirectly harming all bee species. All of these interventions have a significant impact on honeybees' indoor lifestyle by limiting the diversity of microorganisms within hive interiors and, as a result, the likelihood of symbiotic connections occurring.

Symbiosis is a commonplace relationship of microbial and host elements working together to ensure good nutrition, health, and resistance [5]. When it comes to living in harmony with microorganisms, social insects are excellent models [6]. Among many microorganisms, fungi are one of the key actors within the insect microbiota. Fungi and insects have coevolved a wide array of functional interactions over the past 400 million years [7]. Fungal volatiles, which play an important role in terrestrial ecosystems, also affect the nesting and reproductive behaviour of insects [8,9]. Many termite species actively cultivate fungus to digest the foraged organic materials that they cannot digest themselves internally. The fungus then becomes digestible food for the insects. For this natural form of agriculture, they also build specific rooms in their nests, where the fungus can grow and thrive, forming a prominent form of fungus–insect symbiosis [10]. Humans do not interact with their built environment as effectively as other social animals do. Humans' architectural product is one that provides some degree of isolation and protection from the "outside" world. It artificially frames human activity [11]. While in other ecosystems, habitats which can support the social lifestyle of animals are inherently permeable to the outside world. Still, the contemporary bio-inclusive architectures propose mechanisms to incorporate other lifeforms into the outer layers of building boundaries. This, once again, separates the wildlife from the human population inside living spaces and does not support the potential health benefits the other life forms have on indoor city dwellers [12–14]. Modern beehives are an example of manmade design thinking being imposed on the habitats of other living animals. In general, artificial habitats, such as beehives and human houses, pose challenges in designing how forms, spatial configurations, and materials can affect the microbial diversity that establishes itself in those artificial environments.

1.1. Honeybee/Hive as a Model Organism/Habitat

Honeybees are a semi-domesticated animal species: there still are wild colonies thriving in forests or urban areas without human interference [15]. Their natural nests are mostly found in tree cavities that differ from the artificially made habitats concerning their microclimates and microbiomes. Cavities that are large enough for honeybee colony preference (25 L to 40 L), have potentially been formed, thus being occupied by a variety of species for hundreds of years. Such cavities are incrementally formed by various types of other organisms. Mostly initiated with wood rot fungi, other micro and macro-organisms such as invertebrates continue colonising depending on the microclimatic conditions established as a result of the location and sizes of the cavities [16]. On top of pre-existing multispecies communities in these cavities, honeybee colonies continually add diverse types of organic and inorganic materials by collecting particles. These particles range from the size of fungal spores, pollen grains, road and coal dust, and sawdust, up to dead wood [17]. Chlorella alga is one example of an organic material beneficial to honeybees, which they bring to their nest, and which provides nutritional benefits. Researchers discovered that when colonies had access to foraging lands with this alga, they produced more honey than when they moved to a location without it [18]. Forager bees have also been found digging in soil and cattle dung [19] but the reason—if there is one—has not been identified yet.

Honeybee colonies exhibit community-level immunity that also includes the nest material which is essentially a microhabitat for their beneficial symbionts. Resin use and propolis provide self-medication for honeybee colonies with their antiviral, antibacterial, and antifungal properties [20]. Propolis is collected plant resins, mixed with saliva, and wax that the bees use to coat the inner nest surfaces [21]. Beeswax in honeybee nests does not support microbial growth. However, it can be a bioindicator of environmental toxins and colony health: particulates such as larval faeces, shed exuviae, lipophilic chemicals and environmental toxins have been found in beeswax [22]. Pollen is generally the medium in which foragers bring nutrition but also environmental hazards (chemical and biological) to their nests. Additionally, studies have shown that mycotoxin-producing moulds and yeasts that ferment pollen into digestible bee bread, thrive in the conditions in which pollen is stored in the hive [23,24]. Findings on such microbial dependencies advise against the indiscriminate use of pesticides in agriculture. In the case of honeybee colonies, even if the substances do not affect bees directly, they still may be harmful to the microorganisms they essentially need to thrive. It is important to note that determining the microbial communities and their interactions with other organisms or other communities is a complex ecological and technological study.

Honeybee nests differ significantly from other social insect nests, as the only adaptability found in bees' nest building is the way they fill pre-given cavities with their combs. They cannot (re-)shape the cavity itself. In contrast to that, termites and ants can dynamically alter their nest enclosures. Still, honeybees can actively regulate the climate and maintain homeostasis within their "prefabricated" nest enclosures. They fill the cavity adaptively as an efficient movement and storage platform for specific behaviour such as fanning, heating, clustering, etc. [25]. Inhabiting a structured but rather static nest topology requires high connectivity between different functional areas, for example efficiently connecting transport paths between the nectar handover area—in the front the near the entrance—and the honey storage area in the back of the hive. The duration of nectar-storing trips between these areas is one of the key regulatory types of feedback of honeybee foraging, ultimately affecting the colony's pollination activity [26–28]. This is just one out of many prerequisites to be considered in a colony's "hive architecture". The ventilation and gas exchange should support a fluent transition from a colony's foraging season mode to the colony's winter mode and vice versa. However, again, honeybees are highly adaptive and resilient, they flourish in a variety of habitats given the warmth and darkness. Nature-inspired designs imitating or literally being tree trunks have been adopted by beekeepers and designers throughout history, employing a broad range of regionally and organically derived materials (Figure 1a–e). Belgian artist Annemarie Maes uses microorganisms, digital design, and fabrication to connect advanced technology with a living biosystem such as a beehive [29]. There are also projects that use mycelium-grown materials as part of beehive enclosures, but these ideas have never been tested against the risks of long-term colony habitation [30,31].

Figure 1. Diversity of man-made beehives and a natural honeybee nest. (**a**) Mediaeval bee haven found in Rosslyn Chapel, Scotland. The north-facing side of the pinnacle when bees are returning to their haven in 2015. (**b**) Fossilised bees' nest within the pinnacle of Rosslyn Chapel. Photo Credit: Rosslyn Chapel Trust. (**c**) An apiary of stacked mud hives in central Egypt. Photo Credit: Gene Kritsky. (**d**) Bee bole embedded in a historic cottage, UK. cc-by-sa/2.0 by Oast House Archive. Source: https://www.geograph.org.uk/photo/1296874 (accessed on 8 February 2022). (**e**) A handwoven basket hive, coated with ash and cobb, photo taken Cine Beekeeping Museum, Aydin, Turkey. (**f**) Urban beehives near an industrial area, 2021, Graz, Austria. (**g**) A feral honeybee nests. Source: https://forum.canberrabees.com/t/mount-taylor-act-wild-feral-bee-hive-in-a-tree/309 (accessed on 8 February 2022).

1.2. Fungal Biofabrication

In nature, fungi take the role of primary microbial decomposers, meaning they decompose material in the world's ecosystems. However, fungi can also be used not only to decompose but also to compose new structures. Mycelium is the vegetative growth of filamentous fungi that bonds organic matter through a network of hyphal microfilaments and it is currently a competitor of several synthetic materials [32]. There are two main types of mycelium-based materials: pure mycelium materials and mycelium-based composites. Pure mycelium materials are generally used and studied for smaller-scale applications such as paper or textile making [33,34] and biomedical applications such as wound healing [35] and tissue engineering [36]. Mycelium-based composites are made by growing the mycelium homogeneously in and around organic waste materials and are generally used for mesoscale modular applications such as bricks [37], thermal insulation, or acoustic panels [38,39], and low-value materials such as packaging [40]. Mycelium materials can be compared to one of the oldest types of composite materials, cob. Cob has organic fibrous material reinforcement, such as straw, which is bound by subsoil. In mycelium composites, instead of the extracted soil, mycelial hyphae are the fully organic and naturally grown binders. This means, if not treated additionally, the whole yielding material is organic. One can also grow these materials with zero waste and tune them to be mechanically intact, thermally insulative and 100% biodegradable. These properties highly depend on the growth factors including the type of lignocellulosic substrate base selected, fungal strains, and climatic control of the growth medium [41]. On the downside, an increased natural degradability means faster degradation or decay, a feature that is usually avoided in traditional building materials. Thus, the use of these materials demands solving additional challenges, e.g., repair or replacement regimes, or finding specific use cases, where such dynamics are not detrimental or even desired.

During the degradation processes, mycelial fungi's metabolic activities lead to self-healing [42], beneficial volatile production, and detoxification. However, especially inside buildings with no sufficient natural air ventilation and humidity control, competing bacteria and fungi can harm the mechanical structure of the mycelium. Additionally, mycelium can produce mycotoxins and sporulate which can harm its co-inhabitants. As a result, prior to use, mycelium biofabrication methods and design application scenarios include desiccation of the mycelium biomass.

To date, several biofabrication technologies have been developed to achieve desired shapes and functionalities in mycelium materials. To visually measure growth, the simplest, least complicated, and most generally used approach is to cast a mixture of organic substrate and mycelium inoculum into premade moulds, usually translucent plastic enclosures [43]. This method works effectively for specific design scales where the end product can be grown uniformly. Biofabrication techniques, on the other hand, that specify and generate mycelia's growth boundaries using computational design and digital fabrication tools, can allow for local variation in material qualities and result in more complex geometries. Textiles can be used to define stay-in scaffolds for mycelium-based composites [44]. Textile logics can be translated into the filament scale, such as the structural stay-in scaffolds produced using the Kagome weaving method in the FUNGAR Project's building elements [45]. In another recent work, computationally generated scaffold morphologies have been 3D printed and inoculated via a robot arm equipped with sensors [46]. Furthermore, researchers and designers have been successful in 3D printing pre-inoculated viscous materials directly [47,48].

1.3. A Therapeutic Design Problem

In this article, we propose a hybrid construction method for building more bioreceptive and bioactive beehives using living fungal mycelia formed via 3D printed stay-in scaffolds. Our main goal is to combine the thermally insulative properties of mycelia with its medicinal, potentially microbiome modulating properties. We use parametric design tools and fused deposition manufacturing to produce these mycelium scaffolds. Our hive morphologies are designed aiming at honeybee colonies that self-organise similar to how they reside in hollow trees (Section 3.3) while reducing the energy loss of the hive. We used the quantitative and qualitative aspects of tall and narrow tree cavities as a design reference since the community-level immunity of honeybees has evolved in such environments. The main function of the overall hive morphology is to be durable supporting a full bee colony, living mycelium body and against changing weather conditions. This "therapeutic design problem" is a challenging task in creating and testing artificial habitats. It requires setting up empirical experiments to study one-to-one scale hive designs, to compare morphologies that are successfully occupied long-term (minimum one year) and both by honeybees and mycelia.

The design of therapeutic inner nest environments starts with a bioreceptive strategy for the overall morphology. There are two layers of bioreceptivity to be considered in fungal architectures. First is the receptivity (to mycelia) strategy used in designing the overall morphologies. The overall morphology of the fungal construct is primarily the morphology of the reusable or sacrificial formworks in/on which the mycelia grow. This first layer affects the second layer, which is the receptivity (to any other microorganisms and insects) of the pre-established mycelium arising from variation in surface qualities, or density differentiation throughout mycelial volumes. For functional applications, the second layer of bioreceptivity is minimised when the mycelium is heated and desiccated after its dense network formation. However, our goal with therapeutic mycelial beehives includes both layers of bioreceptive design. First is to enable the release of beneficial fungal compounds towards the hive interior. This is only possible with a living—not necessarily growing—mycelium structurally supported by 3D printed stay-in scaffolds. This approach can be a counteraction to the modern beekeeping sector. The modern beehives fall behind in terms of design characteristics that affect the climate conditions that are most relevant for honeybees and their symbionts. For example, one study shows that tree cavities provide better humidity levels compared to traditional modular box hives [48].

A targeted approach to therapeutic design problems can be about designing with specific and known medicinal properties of some fungi. Recent research shows that fungus species with antiviral and antibacterial compounds modulating microbial communities that are beneficial to humans can also boost the immune system of honeybees against specific viruses and bacteria when fed to the bees [49–51]. For embedding such properties in the enclosure material by growing specific fungal strains into hive morphologies the mycelia need to be kept alive during the honeybee colony inhabitation. In a targeted case like this, the challenge is to find a way to match the environmental conditions—temperatures, humidity, pH and oxygen levels—of the fungal habitats with those of the harmful organisms. For instance, a directed evolution strategy is used to breed an entomopathogenic fungus *Metarhizium brunneum*. This fungus is known to inhibit the growth of a honeybee pathogen Varroa Destructor, but naturally lives in lower temperatures, so it is bred to thrive in mostly affected brood areas in the hive (avg. 35 °C) [52]. Genetic modification or breeding of mycelial fungal species are interesting for these applications. In addition to what mycelia can do for honeybees, one example of how honeybee activities might support mycelial life can be the propolis enrichment of habitats. It has been shown that propolis can be used as a growth supplement for decreasing contamination risk in the mushroom production industry [53].

Moreover, novel therapeutic properties of mycelial habitats for honeybees might lay hidden in plain sight, as microorganisms embedded in the construction material may be able to perform certain beneficial support functions for the honeybee collective. However, these properties and abilities may well depend on the environmental conditions they are growing in the microclimate of the hive, which is a special ecological niche for microorganisms, precisely controlled by the bees but still affected by the environmental conditions of the hive surrounding. This creates the main aspect of complexity that apply to the case of fungal biodesign for social animals. A circular feedback loop is in effect here, as the bees control the inner nest climate to a high extent, affecting the microbial communities, which in turn can affect the bees in return. The mycelium hive body with a large number of fungal cells also has an impact on the inner climate when they respire, degrade and regenerate. Even though this might not alter the inner hive climate as much as or as rapidly as the honeybee colony can, it would trigger the highly sensitive bees to more actively regulate the inhive climate in a homeostatic state. This would in turn affect the mycelium's morphology, molecular composition, and survival.

2. Methods

2.1. Bio (Material) Coupling

To investigate the interactions between honeybee nest-related materials, honeybee pathogens, and mycelial fungi, we employ methods from classical microbiology in-lab and honeybee behavioural biology on-field. For measuring the therapeutic properties of mycelia, we set up microbiological assays commonly used in insect immunity studies such as the lytic zone assay. These assays are used to measure the ability of any substance to break apart bacterial cell walls. Bioassays are controlled experiments that are commonly used to assess the potency of a bioactive agent—in our case, living or inactive mycelium—as inhibitors of pathogenic microorganism growth in comparison to standard measures. These experiments are typically designed in such a way that the environmental conditions are suitable to the pathogenic matter whose growth the test matter is hypothesised to inhibit. In addition to the inhibition assays, we make Petri dish experiments where a selection of organic nest matter—propolis, wax, pollen, honey, bee bread, etc. —are placed next to a mycelium patch and incubated at temperatures in which the nest materials would exist naturally. For coupling the living bees and our hive material designs, we make field experiments with full sized bee colonies. In these experiments, the prior criterion for the mycelium material is the selected fungus species being non-pathogenic to humans, bees, and plants, as well as to the local ecosystem.

2.2. Bio (Scaffolding) Design

We developed a design-to-fabrication methodology (digital design tailored for additive manufacturing) to make structural and nutritional scaffolds in/on which mycelia attach and grow. This method promotes a homogenous and fast growth of the selected fungal matrix (mycelium of fungus species and strain). In our case, we used fused deposition modelling, a sub-caste in the 3D printing family holds forth the promise of "digital craft" [54]: a set of topological and geometrical operations to produce patterns as continuous extrusion paths and overall morphologies which at the end are represented as a set of instructions for the 3D printer.

Topological Operations: Our current toolpath drawing pattern is based on radial or mesh hexagonal grid topologies. Following a continuous weaving sequence of hexagonal cell control points, one polyline is formed. We call it "continuous weaving" because the toolpath is not interrupted, extrusion is continuous. The deposited material is like a continuous thread (Figure 2).

Geometrical Operations: In the first digital method, which we call the "variable-offset (VO)" method, multiple values are used to offset the contour lines of the user-specified geometry (distances), defining the outer boundary as well as the inner material density with consideration of the spaces for mycelium inoculate. The spaces between the offset lines are then intertwined using hexagonal weaving (Figure 2A). The initial geometries can be created with the intuitive top-down control of parameters. From there on a cluster-oriented Genetic Algorithm, using the Biomorpher [55] plug-in for Grasshopper3D [56] rapidly explores a confined design space inside a parametric design model and provides interim 3D representations of design varieties. With an interactive evolution design tool, it is possible to rapidly generate morphological varieties while dealing with competing quantitative factors, such as the sizes of spaces required for the occupation of different scales of organisms and the time/material needed for 3D printing [57,58].

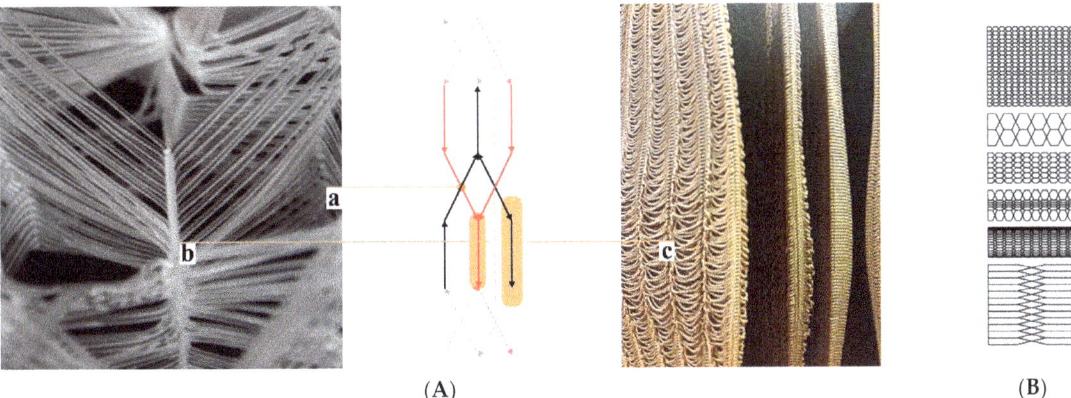

Figure 2. Topological operations for continuous toolpath drawing. (**A**) Hexagonal weaving. (a) Point bonding areas. (b) Linear bonding area. (c) Drape zone. (**B**) Density differentiation via the VO method.

The second digital method, "iterative-subtraction (IS)", allows for the distribution of voxels in 3D space according to specified load and support requirements. This is an iterative optimisation process using finite element functions. Density distribution throughout this voxelised space is defined in accordance with benchmarks, which in our fungal hive case are drawn in consideration of honeybee occupancy loads, spatial layouts for its landing zone, protected spaces or dark zones, and minimisation of material used for production [59]. The voxels are then replaced with hexcells and translated into a toolpath with the hexagonal weaving drawing method (Figure 3). Numerous toolkits can significantly assist in the topology design process in Grasshopper3D, we used Millipede [60]. This mode of operation yields a material microstructure with uniform porosity. So, if needed, geometrical attractors—lines or points—in the parametric hive model can be used to achieve density gradients. We previously presented this application of topology optimisation in a speculative design and construction scenario for multispecies architectural boundaries, the *Co-occupied Boundaries* Project [61].

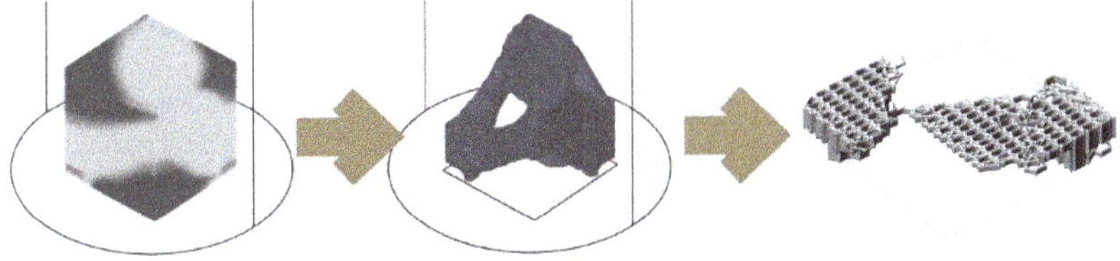

Figure 3. IS method. From left to right: density distribution in voxel space using Millipede, isomesh visualisation, and hexagonal weaving for voxel definition.

Using our design approach, we are able to provide more direct lines of communication between the digital geometries, machine parameters and physical model. Finally, we may also adjust both surface and inner porosities of the scaffold based on performance parameters for temperature and humidity control. Because it enables structural design with physics calculation, the IS technique may be more beneficial for designing overall structures capable of housing a beehive colony in an elevated position above the ground-similar to forest habitats in large trees.

2.3. Biofabrication

Depending on the material and microstructure resolution defining the overall morphology, we use different fungal inoculation methods. In what we call the "infill-feed" method, we manually fill the vertical tubular cells with mycelium inoculated solid substrates (Figure 4a). In the second method, the "self-feed" method, the liquid fungal culture is poured directly onto the printed scaffold. In this case, the printed scaffolds should have higher grid resolution and therefore must provide enough nutrition for the mycelia, nutrition providing surface area for mycelium growth (Figure 4b).

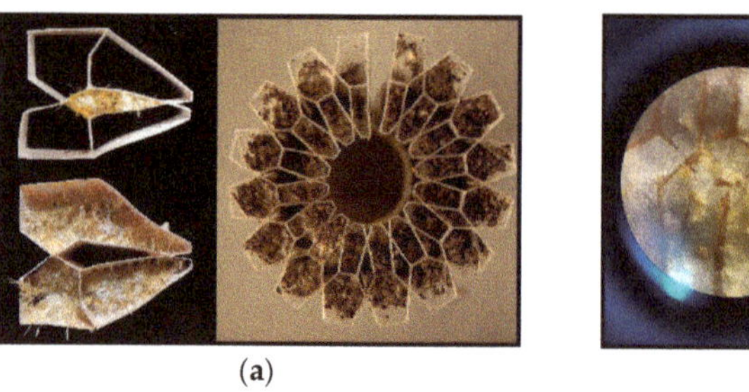

(a)   (b)

Figure 4. Artefacts produced with different methods. (**a**) "Infill-feed". (**b**) "Self-feed".

3. Experiments

3.1. Materials

The surface area of the nutritional matter that a fungus can attach to is an important factor that influences hyphae breakdown of specified material. It can be increased, or controlled, by using extrusion of lignocellulose rich substances. First, we took a preliminary investigation into a wide range of manufacturing parameters that affect the crucial geometrical attributes necessary for the growth of mycelium. In a continuous deposition, variables like print head speed and distance from the previously printed layer greatly impact the thickness of extruded lines, thus the surface area of the organic molecules on which fungal cells can attach. An experimental composite filament (GrowLay™) [62] proved to be the best candidate in our commercially available polymeric material assortment. GrowLay™ is made of cellulose particles, polyvinyl alcohol or PVA which is a water-soluble synthetic polymer and another backbone polymer –that the producer does not prefer sharing. Once the PVA is removed, the printed structures retain microcapillaries and increased surface area of the cellulose. Using the "self-feed" method, we prepared samples with TV hyphae growing on GrowLay™ and *T. Versicolor* (TV) hyphae growing on lignin infused polylactic acid (lignin PLA). In Figure 5, we demonstrated our observation of TV hyphae growth on these two 3D printing materials using microscopic scanning technology. As foreseen, hyphae could grow across the GrowLay™ material (Figure 5), which remains with microcapillaries after the removal of the water-soluble polymer. We also explored paper clay as a stay-in scaffold material [63] and clay extrusion to build. When compared to synthetic and engineered bioplastics, clay printing comes with its own set of challenges due to its organic nature. Therefore, we first experimented with the toolpaths and the 3D printing parameters in order to establish porous boundaries which are able to hold the mycelium substrates and avoid layer collapses (Figure 6). However, we did not make a microscopic scan of clay mycelium samples.

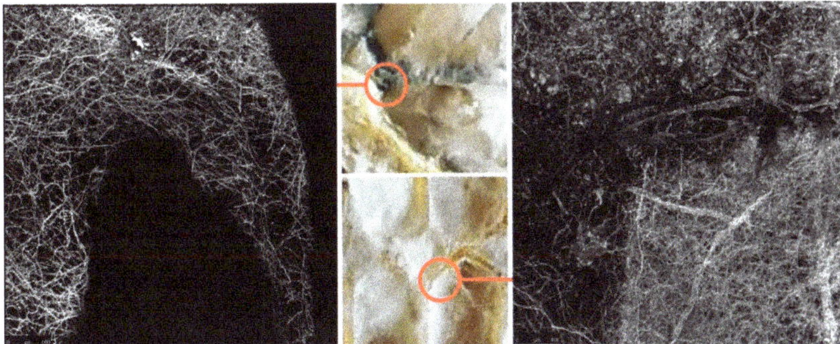

Figure 5. Autofluorescence of fungal hyphae (white) was captured with confocal laser-scanning microscopy and used for visualising the virtual slices of material. Left: TV mycelium and lignin PLA: Right: TV mycelium and GrowLay™.

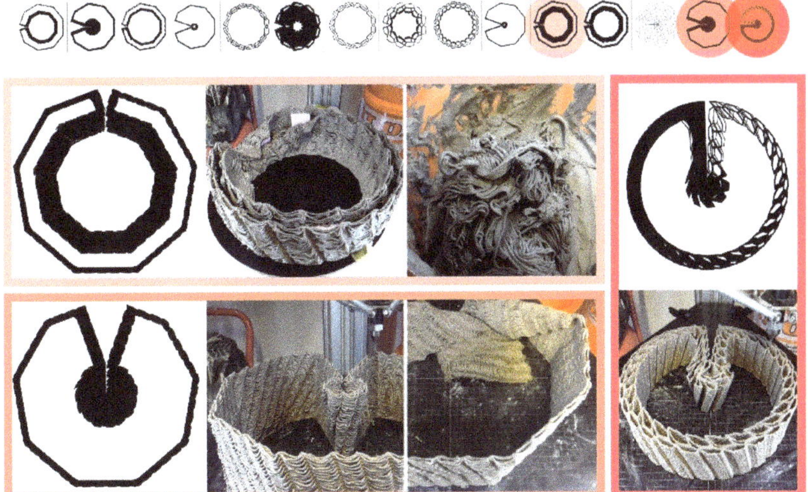

Figure 6. Clay extrusion toolpath tests to find the hex-weaving parameters and overall geometry. The darker red colour shows the final iteration which we used to 3D print the whole clay scaffold.

For the large-scale prototypes, we used a large thermoplastic 3D printer, Reprap BIG, using the GrowLay™ filament described above. We refer to them as GrowLay Hive-1 and GrowLay Hive-2. The third and the most recent one was printed with clay, with a liquid deposition modelling printer, Delta WASP 40 100 Clay, and we refer to it as the Mycelial Clay Hive. Both printers were placed in room conditions in springtime without indoor climate control. These hives are described in more detail in Section 3.4.

For measuring the fungal mycelia's potency as bioactive agents against potential honeybee pathogens, we used two bacterial players. *Micrococcus luteus* is a gram-positive bacterium commonly used in the initial stages of inhibition zone assays in insect immunology studies. For more targeted studies we used Paenibacillus larvae cells—the causative bacterium of deadly American Foulbrood (AFB) disease in honeybee colonies. Both agents were pre-initiated and grown to an active stage in agar Petri dishes. As fungal players in these assays, we used mycelia of five different species: *Trametes versicolor* (TV), *Pleurotus ostreatus* (PO), *Ganoderma lucidum* (GL), *Hericium erinaceus* (HE), and *Grifola frondosa* (GF). However, for the targeted assays we mainly focused on the mycelium of *Trametes versicolor* (TV) which is a common mushroom producing polypore fungus and is commercially known for its anticancer ingredient Krestin (PSK, a protein-bound polysaccharide). It has also been found that Krestin is also a strong antibiotic against microbes pathogenic to humans. This and other medicinal compounds are present in the mycelium of TV. For the large-scale design prototypes, we used PO, GL, and TV mycelia. PO, also known as oyster mushroom, is a widely grown edible mushroom with a rapidly spreading mycelium that efficiently utilises substrate resources, making it a good material maker [64]. GL belongs to the *Ganoderma* species which is broadly studied and showed antiviral properties effective against honeybee deformed wing virus [49]. In general, our biological organism selection criteria were mostly about the honeybee related properties.

3.2. In Vitro Coupling

We had a series of laboratory experiments to characterise the antibacterial activity of specific fungal species. First assays showed us the ability of TV mycelium grown on GrowLay™ to lyse bacterial cell walls, based on the lytic plate assay, using freeze-dried cells of *M. luteus*. Following this, we carried out the inhibition assay with living *P. larvae* cells. This is different from lytic activity, in that it shows the ability of the samples to inhibit the proliferation of bacterial cells, leading to the death of the bacterium. However, we had yeast contamination in all our plates potentially due to the incubation temperatures (35 °C), a much higher temperature than what living mycelium needs to grow. As the third testing method, we homogenised the patches of mycelia of different fungal strains. This process breaks down the cell walls of fungi and frees compounds which might be bioactive and antibacterial, and it potentially kills all living agents in the solution. We first cultured the mycelia of TV, PO, GL, HE, and GF. After we scraped 1 cm × 1 cm mycelium patches from the agar cultures, we diluted them in 1 mL PSB (phosphate solubilising bacteria) water. Then, we used a TissueLyser II to break down the fungal cells, for 15 min and shake with maximum frequency. Since the scraped mycelium culture was mixed with the substrate, the solutions were too thick for the micro filtering process which we needed for the last phase. Therefore, we used a centrifugal shaker to have clearer liquids at the end. After we filtered the solutions, we incubated the *M. luteus* cultured media with these solutions in conditions where the *M. luteus* thrive (35 °C). However, this assay did not show us any positive results as opposed to the first assay where we have used a grown and living patch of TV mycelium.

In unsterile conditions, we placed living TV mycelium and propolis in six malt extract agar dishes. The propolis was collected directly from our hives at the HIVEOPOLIS Honeybee Research Field Laboratory at the Botanical Gardens in Graz, Austria and we did not pre-process it to pasteurise or similar. We did not have any control samples, yet from our experience, agar cultures should be prepared in highly sterile environments to avoid contamination. Yet, the mycelium grew fast and healthy in our dishes with propolis. This can be in favour of the mycelium survival in a living fungal honeybee hive. Then, in a closed plastic container, we placed living TV mycelium grown on a solid substrate together with a wax comb piece, collected from one of our hives and not pre-processed to sterilise. We observed a superficial mycelium coverage on beeswax without degradation of the wax, potentially caused by the fattiness of the wax. We have not performed any coupling assays with pollen, honey, or other substances such as bee bread or royal jelly. Among these, we think that the pollen storage area is a good candidate for the fungal mycelia that we introduced to find nutrition and water. Thus, if our mycelial fungi can survive on pollen as a bio-fungicide, the mycotoxin release can be inhibited in favour of the honeybee colony. This also opens mutualism, in terms of an organic coupling towards therapeutic hives indoors. In terms of specific honeybee bacterial pathogens, we concentrated on AFB disease, which can be detrimental to any colony as fast as three weeks. AFB only attacks the honeybee larvae, located in the brood comb area within the nest with stable temperatures between 32–36 °C.

The conditions that AFB and larvae live in, especially the temperature, are different from the conditions of basidiomycetes fungi habitats. However, in the longer term, as directed evolution techniques become more common, we can improve our material-maker fungi to survive in higher temperatures and train them to inhibit the growth of pathogenic bacteria.

3.3. First field Experiment with Living Bees: "Beeocompatibility"

To test if honeybees are tolerant to mycelium materials, we set up a controlled experiment with living bees. We adapted the natural beekeeping friendly and open-source BCN Wárre Hive design and fabrication models provided by the OSBeehives Project [65]. We selected the mycelium of GL. As growth substrates, we used coconut fibre mats since coconut is hostile to bacterial and yeast contamination and has acidity levels suitable for mycelium (5.5–6.5). The cut-out coconut mats were soaked in water for 12 h, hand pressed, and autoclaved at 121 °C for 20 min. We laid the pieces from a pregrown GL mycelium on the fibre mats, placed them in shallow plastic boxes without lids, enclosed in large plastic bags and then incubated them for 14 days at 23 °C. When we were satisfied with the mycelial colonisation, we dried the pieces in a kitchen oven at 50 °C for 2 h each. We used the mycelial side oriented towards the inside of the hive. We stapled the leatherlike edges on the wooden panels and placed a metal mesh to protect the fibres from being eaten by other animals (Figure 7).

Figure 7. BCN Wárre Hive with mycelium retrofit. (**a**) Retrofit hive wall with GL mycelium grown on coconut fibre mat. (**b**) Fixing details. (**c**) One of the three mycelium retrofit hives.

On 20 July 2019, we introduced six small-sized bee colonies (3000–4000 bees each) into each hive, three of them with mycelial walls and three with regular 18 mm thick wooden walls. We monitored the climate within the hives for two months from 26 July 26 to 6 September, with combined temperature and relative humidity sensors, Beebots, provided by one of the HIVEOPOLIS research groups, Pollenity. The ambient temperature and relative humidity measurements were from near mycelium walls in three of the hives, and from the same locations in the control hives. These measurements were essential for tracking the microclimatic differences that occurred near two materials, as well as for our long-term hive design programme where the mycelium remains alive.

This experiment gave us several insights into the honeybee and mycelium material coupling and what to consider in experiment set-ups with hives accommodating full colonies.

- We had many contamination problems during the mycelium wall preparation and learned that the contamination can be avoided only with premeditated clean lab protocols, especially in such experimental scenarios in which the materials must be ready to be tested on-site with strict deadlines. The preparation labour and errors cost us a late start of the whole field experiment.
- Bees chewed out the mycelium from areas that are softer than others. We can only hypothesise about what they did with the mycelium: they might have moved it out from the hive or consumed it. We think that the second possibility could be beneficial for the colony given the nutritional and therapeutic benefits of mycelium.
- Regarding the hive set-up, we realised that the bees should have been blocked from the empty quilt box (a shallow empty volume below the feeder designed to be filled with humidity capturing material). As this area gets warmer and nearer to the feeder, some colonies initiated their nests there, instead of the targeted mycelium and sensor-attached area. The temperature and relative humidity measurements did not provide a clear distinction between the climate within the mycelium retrofit hives and fully wooden hives, potentially because the nesting spots were different in each hive.

Only one of the six hives survived until the following summer season. This survivor hive was one with mycelium attached. We think that the reason is that we populated the hives late in the season, not giving bees enough time to reproduce, collect pollen and nectar, and build wax combs to store enough honey. Additionally, the weather conditions were particularly challenging that year, with heavy rainfall leading to floodings in many adjacent buildings. The colonies and their wax combs were so small that there was too much empty space in the hives before the winter and not enough insulation on the walls.

3.4. Design Experiments

3.4.1. Digital Hive Morphologies

Based on the studies describing the living conditions of honeybees in feral nests [15], we identified our geometrical benchmarks for the overall structure such as the inner nest shape and volume and entrance opening size. Figure 8 demonstrates a design study for generating stand-alone hive morphologies. Load-wise, the inner nest should accommodate approximately 60,000 honeybee workers plus one queen, including their colony's honey, the wax, and other nest materials, weighing as much as 80 kg in total. This cavity should be well insulated to avoid temperature fluctuations within the nest space. For providing the fungal mycelia's scaffold structure, a high porosity for oxygen distribution and structural stability—especially during growth by degradation—are necessary. In consequence, the honeybees' nesting volume should be surrounded by a highly thick and voluminous enclosure of mycelium composite to provide sufficient insulation and mechanical stability.

One of the lessons we learned by exposing bees to mycelium-based composites in the first field experiment was that they removed the soft parts of the material. Therefore, our scaffold serves as a barrier between the bees' nest and the dense yet soft mycelial zones, while allowing hyphal reach. We use honeybees' natural tree nests as a reference for the inner nest geometry of our one-to-one size mycelial hives (further described in the next section). Figure 9 shows two morphologies created using the IS method, which only takes a basic load and support conditions into account in the topology optimisation part and is then edited to have open (Figure 9a) and enclosed (Figure 9b) bee habitation spaces. We show the 1:5 prototypes because this design method development is directly related to the following 3D printing and self-feed procedures.

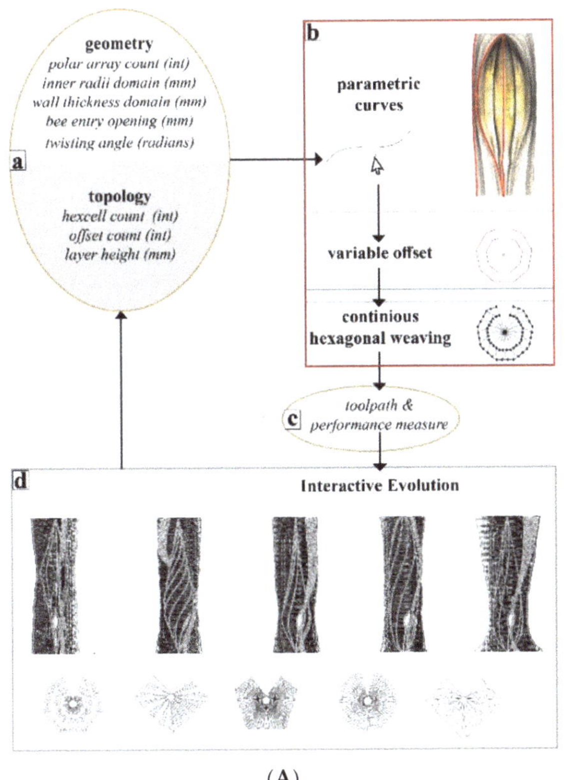

(A)

(B)

Figure 8. Design scheme for hive morphologies using VO. (**A**) Parameters (a) are fed to drawing algorithm (b), according to measurable performance criteria (c), we assess the 3D cluster representatives, (d) and reach a design iteration. (**B**) Output of one design iteration for a stand-alone hive morphology. (a) The tubular hex channel for "infill-feed". (b) Thick voluminous mycelium roof area for capturing moisture. (c) Warm and dark nest space. (d) Landing platform for the honeybees.

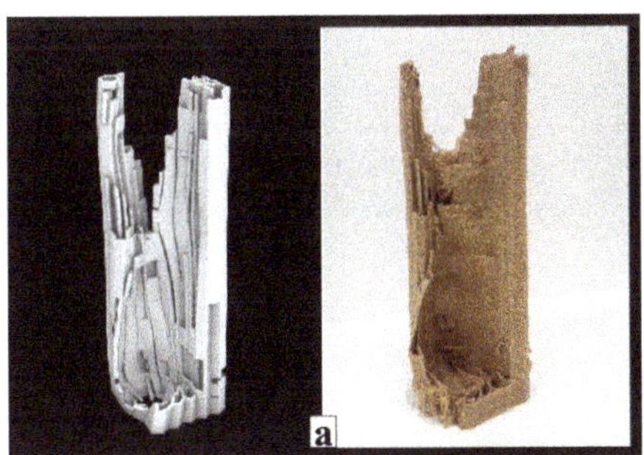

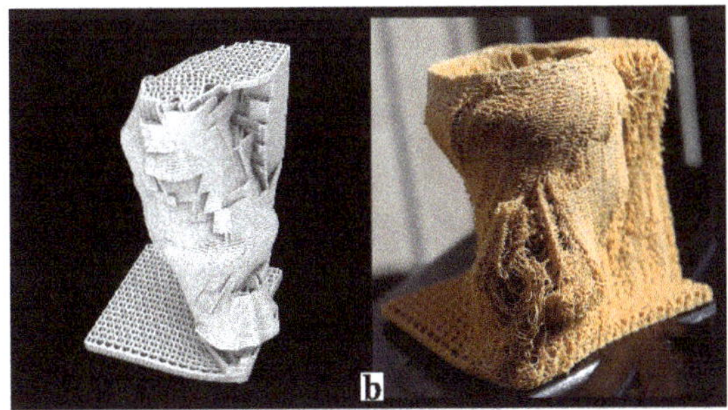

Figure 9. Morphologies created using IS method. (**a**) Open structure for a swarming honeybee colony to nest and 1:5 prototype 3D printed with a wood-infill filament. (**b**) Enclosed structure for a feral honeybee colony to nest and 1:5 prototype 3D printed with a wood-infill filament.

3.4.2. Fully Grown Mycelial Hives

Here we describe the full-scale mycelium-based hives produced by using the VO method. From summer 2019 to summer 2021, we produced three types of mycelium-based hives.

In the summer of 2019, our GrowLay Hive-1 was printed in three parts in 32.5 h using approx. 5 kg of filament. We soaked the printed parts in the purified water for 10 h each, refreshing the water every 2–3 h. This was a delicate and laborious process which also caused the thin extrusions to lose their stability, weakening the layer adhesion, and yielding rickety parts with decreased structural scaffolding capacity. After the removal of the PVA, we exposed the parts to ultraviolet light on a clean bench for 12 h (to kill the contaminating microorganisms). After 6 h of drying time on a clean bench, in separate clean plastic boxes, we filled the vertical channels of the three hive parts with grain seeds and beechwood dust inoculated with the PO mycelium. After three weeks of growth at room temperature (23 °C), these modules were taken out of their plastic boxes and left to dry in a kitchen oven for six hours. We moved the largest fresh and growing part to the open exhibition space of the Festival Headquarters, at the Vienna Design Week 2019. We let the hive fruit by exposing it to more oxygen, light and water as part of our exhibition "Biohybrid Superorganisms Diversify Urban Ecological Niches" (Figure 10).

Figure 10. GrowLay Hive-1 (**a**) 3D printed scaffold before the PVA removal. (**b**) A close up picture taken two years after: the powder like mycelium remains on the surface. (**c**) Vienna Design Week, 2019: mycelium kept degrading the hive scaffold and as we misted it, the fruit bodies emerge. (**d**) A picture taken right after the mycelium hive part is removed from its box and placed in the kitchen oven.

Within the GrowLay Hive-2—which had a slightly larger, 45 L inner nest volume compared to the GrowLay Hive-1—we aimed to introduce a honeybee colony in the spring of 2020 (Figure 11a). Therefore, the hive's stability and durability became priorities. Instead of increasing the structural stability via geometry and density of the printed structure, we kept the PVA, the water-soluble polymer, to support the scaffold with better particle adhesion. For the ease of mycelium infill and handling, we printed this hive in six smaller parts compared to the GrowLay Hive-1, in 45 h using approx. 6 kg of filament. The first prototype of GrowLay Hive-2 was disposed of due to *Trichoderma* fungus contamination. The next one could be produced only in mid-July 2020 which is almost the mid of the bee season. This time, the TV mycelium was grown in the vertical channel with birchwood chips. The inoculation process was in a laboratory but not on a clean bench.

After 4 weeks of colonisation, beginning of September 2020, we moved the hive under a small wooden protective shelter into an outdoor setting at the HIVEOPOLIS Honeybee Research Field Laboratory. However, it was too late in the season for a honeybee colony to start building their natural comb structures which require a lot of their energy. To avoid risking a foreseeable winter death of such a late-established colony, we left the hive empty (Figure 11b).

Figure 11. GrowLay Hive-2. (**a**) An exploded axonometric drawing of the whole construction, showing details of the fungal parts. (**b**) Outdoor assembly during late summer 2020.

The most recent hive scaffold was printed with clay. The overall form of the digital model was built using the VO method and after determining the toolpath drawing parameters. To achieve the required honeybee inhabitation volume within fabrication constraints (max. printing diameter = 40 cm and heights of each module kept to a maximum of 15 cm to avoid layer collapse), the walls of the mid-body parts had to be thinner, making the inoculation—filling the vertical gaps with mycelium inoculated flax fibres—more difficult, if not impossible in those areas. Furthermore, because we initially added water to adjust the viscosity, the printed clay shrank by nearly 20% during the drying and firing processes, as a result, the honeybee inhabitation volume decreased from 39 L to an average of 30 L. The more we expand the inner nest volume, the thicker the mycelial wall should be to maintain the thermal stability within the hive. This would require either dividing the ring-like modules into printable sizes across their cross sections or using a larger 3D printer. Additionally, this would result in longer toolpaths, therefore a larger surface area through which the clay would lose water and higher shrinkage rates.

Our Mycelial Clay Hive is made up of 13 ring modules and took 15 h to print. Before printing the clay scaffold, a commercially available stoneware paper clay was hand mixed with 15% of its weight with tap water to reach a suitable printing viscosity. When compared to the previous GrowLay Hive modules, the clay ones had a significantly higher weight, which was beneficial for overall stability, but it had to be kept to a minimum for subsequent single-person handling. To avoid cracks, the units were air dried for two days after 3D printing, loosely covered with plastic sheets. They were then fired at 1200 °C. We filled the voids in the modules with pregrown mycelium spawn and ground particles of entire flax plants that had been inoculated with TV's mycelium. The inoculation procedure was carried out as quickly and cleanly as possible without sterile conditions. They were incubated in a 23 °C ambient room temperature for the first ten days. During this time, we used a heat mat connected to a temperature sensor and controller, alternating the plastic boxes. During this time, one module that had cracked during transportation was stabilised along with its constituent mycelium.

The main challenge we had in making the first two hives was that the aeration of the mycelium inoculate within the scaffold channels was not sufficient. This created a fast formation of the thick mycelial skin on the surfaces, which were exposed to air while leaving the inner areas of the walls only marginally colonised. When hydrophobia and protection from intruding animals are required, this thick leatherlike differentiation of mycelium can be beneficial. However, when it occurs at the intersecting surfaces of modules that are stacked on top of each other, it inhibits the further hyphal growth for biowelding separate modules together. Another problem was the deformation of the module geometries during the incubation period with high moisture levels. This resulted in an ill-defined continuity in the overall hive geometry. We observed that PVA drips out, creating a strong chemical border that blocks the mycelium from penetrating into the extruded scaffold material. So, if GrowLay™ was a decided material, removing the PVA and improving the structural integrity of the hive via the scaffold design variables would have been a better solution. To compensate for these losses of deformation and stability, we used bamboo sticks as an inner reinforcement when we installed the hive in the garden. The hive was disassembled in Autumn 2021, and we observed that several other animals—such as snails, spiders, and soil insects—already occupied the hive. In the clay scaffolds, the mycelium infill grew faster and more uniformly than in previous GrowLay Hives due to the increased porosity created by the toolpath and the micropores formed after firing the fibres out of the modules. Mycelium's robust and uniform growth allowed for the biowelding of modules with a large enough surface area in contact (Figure 12).

Figure 12. Mycelial Clay Hive.

4. Discussion

To investigate the animals' (including humans') dependency on diverse microbial communities in their habitats (our built environment), we use the honeybee as a model organism and fungal architectures as a biodiversity maximising strategy. In this paper, we showed our hybrid construction method and a design framework for merging two complex and dynamic material systems—the honeybee superorganism and mycelial networks—in order to reassemble a potentially lost link between the social insects' wellbeing and bioactive inner nest spaces. We report our findings and insights on the following topics in this paper.

The architecture of honeybee nest enclosures and the architecture of their comb construction differ. Nevertheless, mycelial architecture is comparable to both. The honeybee colony builds a custom comb structure within established nest enclosures. These enclosures function as heat and light barriers, and bees themselves engage as material constituents of this dynamic multi-material system. Thus, the nest's material components are the bees' products and also their bodies. Mycelial material systems are similar to honeybee nests in that they are adaptive material systems that are entangled with their surroundings and actively manage their local environment, such as chemical and microbiological conditions, in order to survive. When placed in predetermined nest enclosure formworks or scaffolds, the microscale hyphae span the entire geometry. Like honeybees, mycelium largely remains within the physical domains of its prepared form while being able to adjust its behaviour dynamically. In addition, as mentioned earlier, wood-rotting fungi are already present in the natural tree hollows where robust wild bee colonies live, and mycelium architecture is already present in detectable levels in honeybee habitats and their nearby ecosystems.

We argued that the mycelia have been part of the social immunity of the honeybees while co-occupying the tree cavity nests throughout evolutionary history. However, the concentration of mycelia in the nest enclosure materials is significantly higher in the fungal hives we presented. This could imply that the natural balance of fungal metabolites, honeybee symbionts, and honeybees may not be established in fully grown mycelial hives. Our goal is not to reconstruct a tree cavity, yet we aim to seed sentient material systems in which mycelium takes part and renders it specifically receptive to microbial communities that have coevolved with the honeybee species over millions of years. Nothing exists in isolation in nature, especially in a honeybee hive. The adaptive mechanisms found in living materials arise due to their form, multiresolution microstructures, causing them to behave in nonlinear ways, responding to external stimuli in unpredictable ways. In the targeted microbiology coupling experiments, we isolate organisms and organic materials from the environments in which they occur in harmony with their complex surroundings. This entails increasing the size of the specimens from a Petri-dish scale to a fully functional on-site beehive structure, which alters the length and time scales in which mycelial fungi might act as a symbiotic agent for honeybee and vice versa. For example, we do not know if the antibacterial capacity of mycelia can inhibit the dispersal of beneficial bacteria in the honeybee nest microbiome. To effectively map the beneficial bioactivities of mycelia via designing its scaffolds, various compositions with various fungal species, growth substrates, and morphologies should be rigorously tested in both laboratory and field conditions. To grow many identical repetitions of mycelial parts and full-scale beehive prototypes for such experiments, fungal biofabrication processes should be improved to be more efficient in terms of human labour. However, even though some properties, such as surface qualities, bioactive agents, and densities may vary on different sizes and timescales, the primary goal of providing a good thermal environment for bees must be met. Therefore, the thermal dynamics of this coupled system in different hive morphologies, including heat distribution and water balance, are of particular interest to us.

We proposed two digital methods for generating nest morphologies. Devised specifically for fused deposition manufacturing, these methods aim to establish bespoke stay-in scaffolds for living mycelia and also for other biodesign narratives. As opposed to the commonly used casting technique where the nutrition substrate and mycelium culture are filled in a mould manually, by using fused deposition techniques to lay the mould as a stay-in scaffold with internal structure, we can radically increase the surface area within a given volume and encourage a more uniform distribution of hyphae within the composite material system. We continue to improve our digital design techniques in order to produce more versatile and efficient design models that facilitate the iterative and exploratory nature of designing with and for other living organisms. We further improve our digital design skills in order to create more flexible and efficient design models that support the iterative and experimental nature of designing with and for other living organisms.

It is an intriguing challenge to develop a fungal bioproduction process and ensure its healthy maintenance as an integral part of another complex entity: the technologically enhanced beehive of HIVEOPOLIS [66]. According to the HIVEOPOLIS design brief, the fungal materiality and scaffold morphologies should synergistically support a variety of physical, mechanical, and chemical properties. First and foremost, a self-sustaining mycelial hive should have at least three mycelium growth qualities to ensure the bioavailability of fungal metabolites and enzymes, as well as their safety in a durable and warm hive structure: (1) maintained healthy mycelium colonisation for its biological activities like enzymes and beneficial volatiles production, (2) thick mycelial skin on spots where water protection is needed, and (3) aerial growth of hyphae in order to mechanically connect separate modules and towards the beehive interior for bees exposure to mycelium in its purer state. In conclusion, the idea of utilising mycelium materials as a living material, coevolving hosts in the nests of other creatures, including humans, is an intriguing novel concept that guides our research. Yet, the meaningful re-integration of other living entities into the bee habitats, and human indoor spaces, is a challenging but promising task. It requires economic, cultural, and technological positioning of biohybrid architectures in human society, and eventually all ecosystems of Earth, this also demands the cultivation of multispecies living narratives and practices in our everyday lives. For us, the challenge was to bring the performance of smaller scale prototypes and qualities of small-scale samples to full-scale prototypes, ready in time for honeybee field experiments which are highly season dependent. When the technical challenges are overcome, more research is needed, however, to discover the long-term events that mycelium may initiate as part of habitat architectures. The increased microbial diversity in terms of quantity—the number of distinct species— does not satisfy the goal to reach well-balanced microbial diversity. It is still speculative whether we will be able to effectively modulate the indoor microbiomes of our habitats or other organisms by incorporating dense mycelial networks into our architecture. We need to learn more about "who is there, where they live, and what they are doing" in our fungal designs. State of the art biotechnologies can aid in collecting this information in relation to our design probes. We can then use this knowledge in combination with the tools and methods developed in our fungal architecture community, to re-establish these causalities between geometrical, topological, and tactile aspects of our designs and microbial activities. In general, all these goals require focused groups of biodesign researchers collaborating with people in other scientific and engineering fields.

Author Contributions: Conceptualization, A.I. and T.S.; methodology, A.I.; software, A.I.; validation, A.I. and T.S.; formal analysis, A.I.; investigation, A.I. and T.S.; resources, T.S.; data curation, A.I.; writing—original draft preparation, A.I. and T.S.; writing—review and editing, A.I. and T.S.; visualization, A.I.; supervision, T.S.; project administration, T.S.; funding acquisition, T.S. All authors have read and agreed to the published version of the manuscript.

Funding: This work was supported by the Field of Excellence "Complexity of Life in Basic Research and Innovation" (COLIBRI) at the University of Graz and the EU H2020 FET-Proactive project "HIVEOPOLIS" (no. 824069).

Institutional Review Board Statement: Not applicable.

Informed Consent Statement: Not applicable.

Data Availability Statement: Not applicable.

Acknowledgments: The authors thank the Mycology Lab and Dalial Freitak at the Institute of Biology at the University of Graz, and the Institute of Environmental Biotechnology at the Technical University of Graz. Open Access Funding by the University of Graz.

Conflicts of Interest: The authors declare no conflict of interest.

References

1. Last, J.M. *A Dictionary of Public Health*; Oxford University Press: New York, NY, USA, 2007.
2. Semmelweis, I.F. *Semmelweis' Gesammelte Werke*; G. Fischer: Jena, Germany, 1905.
3. Pasteur, L. On the Extension of the Germ Theory to the Etiology of Certain Common Diseases. University of Adelaide Library. Available online: http://ebooks.adelaide.edu.au/p/pasteur/louis/exgerm/ (accessed on 5 February 2022).
4. Ropars, L.; Dajoz, I.; Fontaine, C.; Geslin, B. Which Impacts of Domesticated Honeybee Introductions and Management Practices on the Pollination Ecosystem Service in Urban Habitats? Abstract presented to *SFécologie2016*, Marseille, France. 2016. Available online: https://www.researchgate.net/publication/309674183_Which_impacts_of_domesticated_honeybee_introductions_and_management_practices_on_the_pollination_ecosystem_service_in_urban_habitats (accessed on 5 February 2022).
5. Rohlfs, M.; Churchill, A.C.L. Fungal Secondary Metabolites as Modulators of Interactions with Insects and Other Arthropods. *Fungal Genet. Biol.* **2011**, *48*, 23–34. [CrossRef]
6. Janson, E.M.; Stireman, J.O., III; Singer, M.S.; Abbot, P. Phytophagous Insect–Microbe Mutualisms and Adaptive Evolutionary Diversification. *Evolution* **2008**, *62*, 997–1012. [CrossRef]
7. Biedermann, P.H.W.; Vega, F.E. Ecology and Evolution of Insect–Fungus Mutualisms. *Annu. Rev. Entomol.* **2020**, *65*, 431–455. [CrossRef]
8. Davis, T.S.; Landolt, P.J. A Survey of Insect Assemblages Responding to Volatiles from a Ubiquitous Fungus in an Agricultural Landscape. *J. Chem. Ecol.* **2013**, *39*, 860–868. [CrossRef]
9. Birkemoe, T.; Jacobsen, R.M.; Sverdrup-Thygeson, A.; Biedermann, P.H.W. Insect-Fungus Interactions in Dead Wood Systems. In *Saproxylic Insects*; Ulyshen, M.D., Ed.; Zoological Monographs; Springer International Publishing: Cham, Switzerland, 2018; Volume 1, pp. 377–427. [CrossRef]
10. Barcoto, M.O.; Carlos-Shanley, C.; Fan, H.; Ferro, M.; Nagamoto, N.S.; Bacci, M.; Currie, C.R.; Rodrigues, A. Fungus-Growing Insects Host a Distinctive Microbiota Apparently Adapted to the Fungiculture Environment. *Sci. Rep.* **2020**, *10*, 12384. [CrossRef]
11. Banham, R. *The Architecture of the Well-Tempered Environment*; University of Chicago Press: Chicago, IL, USA, 1984.
12. I'm Lost in Paris/R&Sie(n). Available online: https://www.archdaily.com/12212/im-lost-in-paris-rsien (accessed on 13 March 2022).
13. Bat Access Brick. Available online: https://www.nhbs.com/bat-brick (accessed on 12 March 2022).
14. Poikilohydric Living Walls Project. Available online: https://www.ucl.ac.uk/bartlett/architecture/about-us/innovation-enterprise/building-greener-cities-poikilohydric-living-walls (accessed on 23 March 2022).
15. Seeley, T.D. *The Lives of Bees, Untold Story of the Honeybee in Wild*, 1st ed.; Kalett, A., Zodrow, K., Eds.; Princeton University Press: Princeton, NJ, USA; Oxfordshire, UK, 2019; pp. 93–98.
16. Stokland, J.N.; Siitonen, J.; Jonsson, B.G. *Biodiversity in Dead Wood*; Cambridge University Press: Cambridge, UK, 2012; pp. 150–182. [CrossRef]
17. Spencer-Booth, Y. Feeding Pollen, Pollen Substitutes and Pollen Supplements to Honeybees. *Bee World* **1960**, *41*, 253–263. [CrossRef]
18. Benevenute Parish, J. Aspects of the Interactions between Honey Bees (*Apis mellifera*) and Propagules of Plant Pathogens. Ph.D. Thesis, School of Agriculture, University of Adelaide, Adelaide, Australia, 2019.
19. Tihelka, E. The Immunological Dependence of Plant-Feeding Animals on Their Host's Medical Properties May Explain Part of Honey Bee Colony Losses. *Arthropod-Plant Interact.* **2018**, *12*, 57–64. [CrossRef]
20. Simone-Finstrom, M.; Evans, J.; Spivak, M. Resin Collection and Social Immunity in Honey Bees. *Evol. Int. J. Org. Evol.* **2009**, *63*, 3016–3022. [CrossRef]
21. Anderson, K.; Sheehan, T.H.; Eckholm, B.; Mott, B.M.; Degrandi-Hoffman, G. An Emerging Paradigm of Colony Health: Microbial Balance of the Honey Bee and Hive (*Apis mellifera*). *Insectes Sociaux* **2011**, *58*, 431–444. [CrossRef]
22. Kostić, A.Ž.; Milinčić, D.D.; Petrović, T.S.; Krnjaja, V.S.; Stanojević, S.P.; Barać, M.B.; Tešić, Ž.L.; Pešić, M.B. Mycotoxins and Mycotoxin Producing Fungi in Pollen: Review. *Toxins* **2019**, *11*, 64. [CrossRef]
23. Free Living Bees. Bees Don't Make It Honey Team Do. Available online: https://www.freelivingbees.com/post/bees-don-t-make-honey-teams-do (accessed on 12 April 2022).
24. Turner, J.S. The Soul of the Superorganism. In *The Extended Organism: The Physiology of Animal-Built Structures*; Harvard University Press: Cambridge, MA, USA, 2000; p. 179.

25. Thenius, R.; Schmickl, T.; Crailsheim, K. Optimisation of a Honeybee-Colony's Energetics via Social Learning Based on Queuing Delays. *Connect. Sci.* **2008**, *20*, 193–210. [CrossRef]
26. Schmickl, T.; Thenius, R.; Crailsheim, K. Swarm-Intelligent Foraging in Honeybees: Benefits and Costs of Task-Partitioning and Environmental Fluctuations. *Neural Comput. Appl.* **2012**, *21*, 251–268. [CrossRef]
27. Zahadat, P.; Hahshold, S.; Thenius, R.; Crailsheim, K.; Schmickl, T. From Honeybees to Robots and Back: Division of Labour Based on Partitioning Social Inhibition. *Bioinspir. Biomim.* **2015**, *10*, 066005. [CrossRef]
28. Maes, A.M. ElbBienen in Hamburg, Art in Public Space. Available online: https://annemariemaes.net/presentations/bee-laboratory-presentations-2/2019-elbbienen-in-hamburg-machine-art-in-public-space/ (accessed on 24 February 2022).
29. Thompson, G. A Mycelium Hive—Growing Your Own. In *Bee Craft: The Independent Voice of British Beekeeping*; Bee Craft Ltd.: Los Angeles, CA, USA, 2022; Volume 6.
30. Turner, T. A Buzzworthy Beehive | Yanko Design. Available online: https://www.yankodesign.com/2017/04/07/a-buzzworthy-beehive/ (accessed on 17 May 2020).
31. Jones, M.; Mautner, A.; Luenco, S.; Bismarck, A.; John, S. Engineered Mycelium Composite Construction Materials from Fungal Biorefineries: A Critical Review. *Mater. Des.* **2020**, *187*, 108397. [CrossRef]
32. Jones, M.; Gandia, A.; John, S.; Bismarck, A. Leather-like Material Biofabrication Using Fungi. *Nat. Sustain.* **2021**, *4*, 9–16. [CrossRef]
33. Mazur, R. Mechanical Properties of Sheets Comprised of Mycelium: A Paper Engineering Perspective. Honors Thesis, Department of Environmental Resources Engineering, State University of New York, College of Environmental Science and Forestry, New York, NY, USA, 2015.
34. Su, C.-H.; Liu, S.-H.; Yu, S.-Y.; Hsieh, Y.-L.; Ho, H.-O.; Hu, C.-H.; Sheu, M.-T. Development of Fungal Mycelia as a Skin Substitute: Characterization of Keratinocyte Proliferation and Matrix Metalloproteinase Expression during Improvement in the Wound-Healing Process. *J. Biomed. Mater. Res. A* **2005**, *72*, 220–227. [CrossRef]
35. Antinori, M.E.; Contardi, M.; Suarato, G.; Armirotti, A.; Bertorelli, R.; Mancini, G.; Debellis, D.; Athanassiou, A. Advanced Mycelium Materials as Potential Self-Growing Biomedical Scaffolds. *Sci. Rep.* **2021**, *11*, 12630. [CrossRef]
36. Xing, Y.; Brewer, M.; El-Gharabawy, H.; Griffith, G.; Jones, P. Growing and Testing Mycelium Bricks as Building Insulation Materials. *IOP Conf. Ser. Earth Environ. Sci.* **2018**, *121*, 022032. [CrossRef]
37. Girometta, C.; Picco, A.M.; Baiguera, R.M.; Dondi, D.; Babbini, S.; Cartabia, M.; Pellegrini, M.; Savino, E. Physico-Mechanical and Thermodynamic Properties of Mycelium-Based Biocomposites: A Review. *Sustainability* **2019**, *11*, 281. [CrossRef]
38. Pelletier, M.G.; Holt, G.A.; Wanjura, J.D.; Bayer, E.; McIntyre, G. An Evaluation Study of Mycelium Based Acoustic Absorbers Grown on Agricultural By-Product Substrates. *Ind. Crops Prod.* **2013**, *51*, 480–485. [CrossRef]
39. Holt, G.A.; Mcintyre, G.; Flagg, D.; Bayer, E.; Wanjura, J.D.; Pelletier, M.G. Fungal Mycelium and Cotton Plant Materials in the Manufacture of Biodegradable Molded Packaging Material: Evaluation Study of Select Blends of Cotton Byproducts. *J. Biobased Mater. Bioenergy* **2012**, *6*, 431–439. [CrossRef]
40. Appels, F.V.W.; Camere, S.; Montalti, M.; Karana, E.; Jansen, K.M.B.; Dijksterhuis, J.; Krijgsheld, P.; Wösten, H.A.B. Fabrication Factors Influencing Mechanical, Moisture- and Water-Related Properties of Mycelium-Based Composites. *Mater. Des.* **2019**, *161*, 64–71. [CrossRef]
41. Van Wylick, A.; Monclaro, A.V.; Elsacker, E.; Vandelook, S.; Rahier, H.; De Laet, L.; Cannella, D.; Peeters, E. A Review on the Potential of Filamentous Fungi for Microbial Self-Healing of Concrete. *Fungal Biol. Biotechnol.* **2021**, *8*, 16. [CrossRef]
42. Grown.bio. Grow It Yourself Mycelium Packaging Kit. Available online: https://www.grown.bio/wp-content/uploads/2020/07/GIY_Manual_GrownBio.pdf (accessed on 1 December 2021).
43. Mycelium Textile. Available online: https://neffa.nl/portfolio/mycelium-textile/ (accessed on 12 December 2021).
44. Adamatzky, A.; Gandia, A.; Ayres, P.; Wösten, H.; Tegelaar, M. Adaptive Fungal Architectures. Available online: http://links-series.com/wp-content/revues/5-6-Proust/10-Adamatsky-Gandia-Ayresetal-Adaptive_fungal_architectures-LINKs_serie_5_6.pdf (accessed on 18 December 2021).
45. Alima, N.; Snooks, R.; McCormack, J. Bio Scaffolds: The Orchestration of Biological Growth through Robotic Intervention. *Int. J. Intell. Robot. Appl.* **2022**. [CrossRef]
46. Blast Studio 3D Prints Column from Mycelium to Make "Architecture that Could Feed People". Available online: https://www.dezeen.com/2022/01/18/blast-studio-tree-column-mycelium-design/ (accessed on 5 January 2022).
47. Goidea, A.; Andreen, D.; Floudas, D. Pulp Faction: 3d Printed Material Assemblies through Microbial Biotransformation. In *Proceedings of Fabricate 2020: Making Resilient Architecture*; UCL Press: London, UK, 2020; pp. 42–49. [CrossRef]
48. Mitchell, D. Ratios of Colony Mass to Thermal Conductance of Tree and Man-Made Nest Enclosures of *Apis mellifera*: Implications for Survival, Clustering, Humidity Regulation and Varroa Destructor. *Int. J. Biometeorol.* **2016**, *60*, 629–638. [CrossRef]
49. Stamets, P.E.; Naeger, N.L.; Evans, J.D.; Han, J.O.; Hopkins, B.K.; Lopez, D.; Moershel, H.M.; Nally, R.; Sumerlin, D.; Taylor, A.W.; et al. Extracts of Polypore Mushroom Mycelia Reduce Viruses in Honey Bees. *Sci. Rep.* **2018**, *8*, 13936. [CrossRef]
50. Evans, J.D.; Schwarz, R.S. Bees Brought to Their Knees: Microbes Affecting Honey Bee Health. *Trends Microbiol.* **2011**, *19*, 614–620. [CrossRef] [PubMed]
51. Reinbacher, L.; Fernandez Ferrari, C.; Angeli, S.; Schausberger, P. Effects of Metarhizium Anisopliae on Host Choice of the Bee-Parasitic Mite Varroa Destructor. *Acarologia* **2018**, *58*, 287–295. [CrossRef]

52. Han, J.O.; Naeger, N.L.; Hopkins, B.K.; Sumerlin, D.; Stamets, P.E.; Carris, L.M.; Sheppard, W.S. Directed Evolution of Metarhizium Fungus Improves Its Biocontrol Efficacy against Varroa Mites in Honey Bee Colonies. *Sci. Rep.* **2021**, *11*, 10582. [CrossRef]
53. Türkekul, İ.; Gülmez, Y. Propolis: An Enrichment Material for Mycelium Development of Oyster Mushroom (*Pleurotus Ostreatus*). *Nat. Resour.* **2016**, *7*, 103–107. [CrossRef]
54. Oxman, N. Digital Craft: Fabrication Based Design in the Age of Digital Production. In Proceedings of the 2007 International Conference on Ubiquitous Computing, Innsbruck, Austria, 16–19 September 2007; pp. 534–538.
55. Available online: https://github.com/johnharding/Biomorpher (accessed on 15 November 2021).
56. Grasshopper3D Is a Visual Programming Language (VPL) Platform for Use in the CAD Package Rhinoceros3D. Available online: https://www.rhino3d.com/download/grasshopper/1.0/wip/rc (accessed on 10 January 2022).
57. Takagi, H. Interactive Evolutionary Computation: Fusion of the Capabilities of EC Optimization and Human Evaluation. *Proc. IEEE* **2001**, *89*, 1275–1296. [CrossRef]
58. Harding, J.; Brandt-Olsen, C. Biomorpher: Interactive Evolution for Parametric Design. *Int. J. Archit. Comput.* **2018**, *16*, 144–163. [CrossRef]
59. Seeley, T.; Morse, R. The Nest of the Honey Bee (*Apis mellifera* L.). *Insectes Sociaux* **1976**, *23*, 495–512. [CrossRef]
60. Available online: https://www.grasshopper3d.com/group/millipede (accessed on 10 January 2022).
61. Ilgun, A.; Ayres, P. *Co-Occupied Architectural Boundaries as 3D Printed Scaffolds Embellished Through Self-Organised Construction*; Living Architecture Systems Symposium White Papers; Riverside Architectural Press: Waterloo, ON, Canada, 2016; p. 168. Available online: https://www.dropbox.com/s/zjx2uhlan4tuz5t/171206_White-Papers_with%20Cover.pdf?dl=0 (accessed on 5 February 2022).
62. 3d Printing. Kai Parthy Introduces GROWLAY Indoor Farming Material. Available online: https://3dprinting.com/news/german-filament-creator-introduces-indoor-farming-material/ (accessed on 2 May 2022).
63. Available online: https://shop.keramik.at/c/59/a/20116/ueber-1250%C2%B0C/Sio2-Paperclay-PCLI-weiss-125-kg-Pkg-.html (accessed on 5 December 2021).
64. Stamets, P. *Growing Gourmet and Medicinal Mushrooms*; Ten Speed Press: Berkeley, CA, USA, 1993; p. 384.
65. OSBeehives. Source Files. Available online: https://www.osbeehives.com (accessed on 16 May 2020).
66. Ilgün, A.; Angelov, K.; Stefanec, M.; Schönwetter-Fuchs, S.; Stokanic, V.; Vollmann, J.; Hofstadler, D.N.; Kärcher, M.H.; Mellmann, H.; Taliaronak, V.; et al. *Bio-Hybrid Systems for Ecosystem Level Effects*; MIT Press: Cambridge, MA, USA, 2021. [CrossRef]

Article

Are Mushrooms Parametric?

Dilan Ozkan [1,*], **Ruth Morrow** [1], **Meng Zhang** [2] **and Martyn Dade-Robertson** [1]

[1]   School of Architecture, Planning and Landscape, Newcastle University, Newcastle upon Tyne NE1 7RU, UK; ruth.morrow@newcastle.ac.uk (R.M.); martyn.dade-robertson@newcastle.ac.uk (M.D.-R.)

[2]   Applied Sciences, Northumbria University, Newcastle upon Tyne NE1 8ST, UK; meng.zhang@northumbria.ac.uk

*   Correspondence: d.ozkan2@newcastle.ac.uk; Tel.: +44-7798691165

Abstract: Designing with biological materials as a burgeoning approach in the architecture field requires the development of new design strategies and fabrication methods. In this paper, we question if designers can use a parametric design approach while working with living materials. The research uses fungi as a biomaterial probe to experiment with the parametric behavior of living systems. Running design experiments using fungi helps to understand the extent to which biological systems can be considered parametric and, if so, what kind of parametric systems they are. Answering these questions provides a method to work with complex biological systems and may lead to new approaches of fabricating materials by tuning the environmental parameters of biological growth.

Keywords: fungal fruiting bodies; parametric design thinking; plasticity; linearity; non-linearity

Citation: Ozkan, D.; Morrow, R.; Zhang, M.; Dade-Robertson, M. Are Mushrooms Parametric? *Biomimetics* **2022**, 7, 60. https://doi.org/10.3390/biomimetics7020060

Academic Editors: Andrew Adamatzky, Han A. B. Wösten and Phil Ayres

Received: 15 March 2022
Accepted: 29 April 2022
Published: 10 May 2022

Publisher's Note: MDPI stays neutral with regard to jurisdictional claims in published maps and institutional affiliations.

1. Introduction

There is a growing interest in harnessing living systems in the fabrication of materials and structures. Biological systems are capable of self- assembling complex materials and composites in highly energy efficient ways. While we make use of the materials provided by nature after the organism that created them is dead, utilizing living process may offer new methods of material assembly. These methods, however, will also require novel design tools and a new understanding of the relationship between the designer and their materials.

A promising group of organisms for biological material fabrication is fungi. For example, fungi can be manipulated at various scales for different purposes, such as in leather form with similar texture to animal leather and as a binder for bulk material (as mycelium composites). It also can act as a functional material when it is still alive, to form networks for microorganisms (see Fiber Highways Project [1]), or as a sensor (see Fungal Architecture project [2]). Most of these projects utilize fungal mycelium; however, few design projects address the fruiting body. Unlike mycelium, the fruiting bodies of many fungus species exhibit complex morphologies and self-assemble without the 'scaffold' of a substrate or aggregate. While we tend to harvest the fruiting bodies as food, the morphological complexity and their sensitivity to environmental conditions, as well as their speed of growth make them especially suitable for studies on how biological systems fabricate complex forms and materials. To this end we provide an early study in which the fruiting bodies of a well-studied fungal species (*Pleurotus ostreatus*) are shown to be somewhat controllable given their sensitivity to key environmental parameters. They were used as a biomaterial probe to test the concept of biological parametrics [3].

Biomaterial probes are defined as experiments that are carried out on biological materials or fabrication strategies without designed goals, but which are used to understand the factors influencing a biological system [4]. As Ramirez-Figueroa explains, it focuses on design explorations which show how the practice of design is transformed and redefined by using living systems. Although mycelium was used as a material probe in the preliminary experiments, the main design experiments were conducted here using fruiting bodies [5].

The goal of the design experiments was to intervene in the fruiting body formation of oyster mushrooms by altering the environmental factors for growth.

1.1. Biological Parametrics

Parametric design is a broad concept that connects data to the design of form and structure. Often synonymous with generative design, the role of the designer in a parametric design process is not to design the form of the object or system directly but rather to define the key controlling parameters and their relationships [6]. In Architecture, parametric design is often associated with the development of complex organic forms derived from initial conditions created by, for example, site mappings or simulations of use and function [7,8].

There is an analogy between parametric and biological processes in that in many examples of biological growth, especially in plants and fungi, the form of the organism is often, in part, derived from an interaction with environmental factors, including, access to sunlight and nutrients, physical constraints and barriers and interaction with other organisms. To some extent we already intervene in these biological processes in agriculture. A tomato, cultivated in the highly controlled, nutrient rich environment of a greenhouse, for example, could be described as 'parametrically designed'. Refined crafts such as bonsai tree growing are also examples of intervening in biological growth with specific forms in mind. The Bonsai tree is produced through direct and 'coercive' control through the 'directing' of branches and the severe limitations of nutrients to keep the trees in dwarf form. In design terms this cultivation approach is more akin to direct control than parametric design, which implies a separation between the intervention (through data) and the generated design outcome.

Our research into fruiting bodies (of *P. ostreatus* known as *oyster mushrooms*) has, however, suggested that, for certain biological systems, a parametric approach to their 'design' and cultivation may be possible. To this end the paper will introduce the concept of biological parametric design as a fabrication strategy through design experiments which investigate the relationship between environmental parameters and fungal fruiting body morphology.

1.2. Plasticity

While it is often stated that DNA is the 'blueprint of life', biological systems are only partially shaped by the information contained in genes. Biological systems are subject to epigenetic influences i.e., environmental conditions which will cause genes to activate or not [9]. This relationship between phenotype and environment is sometimes referred to as plasticity and can be measured in terms of the degree of variation between organisms given the same genome [10]. Plasticity is exhibited at different stages of an organism's life. Here, however, we will focus on developmental plasticity of mushroom fruiting bodies which lead to a variation in morphogenesis and final form. The concept of plasticity implies a pliability of developmental processes which may, we suggest, enable human intervention in direct parametric control.

As Dade-Robertson discusses, these indirect methods of affecting a living material through environmental parameters use "nature's own agencies" without human imposition through "forcible constraints" such as cutting and molding the organism, or genetic manipulation [10]. A question remains, however, as to what degree this plasticity is amenable to a parametric approach. Biological systems and processes often exhibit non-linear behavior with, for example small changes in environmental conditions creating tipping points and leading to developmental outcomes that are not easy to attribute to single or limited sets of parameters and/or where the same effect does not always cause the same results [11]. An organism growing under exactly the same environmental conditions can form different morphologies. It is the non-linear behavioral pattern of the living materials that leads to an abundance of variations in the final product. Biological systems are also subject to noise and exhibit cell-to-cell variation and emergence where outcomes are not easily reducible to the behavior of parts. This biological complexity, therefore, challenges a parametric approach, and at the same time requires designers to have deep knowledge about the biological materials and bioprocess for the fabrication of the materials. Designers need to explore the value ranges and tipping points where the organism presents a linear change, so (if applicable) they can apply parametric design principles.

1.3. Prior Work

There are a number of notable precedents for a parametric approach to fabricating with biological systems outside the context of fungus. The works of Jiwei Zhou et al., Thora Arnardottir et al., and Neri Oxman et al., using plant roots, bacteria, and silkworms, respectively, show approaches to influence the environmental conditions of living organisms to achieve a desired material [9–11]. For instance, Arnodottir uses urease producing bacteria to calcify sand, creating cemented columns of material, without including digital tools to control the parameters [12]. By altering the cast sizes, inlet positions for nutrients and reactants, she shows that parameters which affect biological growth can be influenced. The influence of the parameters can be predicted while creating cast materials and the final form of the cemented columns does not have to be dictated by the shape of the cast. More complex forms emerge because of the interaction of these biological and environmental factors. In the case of the Silkworm Pavilion-II project by Oxman et al., they guide silkworms to cover the woven surface of the pavilion [13]. The distribution of the silkworms was controlled by heat, gravity, and light as variables. Since environmental conditions were directly linked to silk production, they could spread the fibers homogenously as they intended. Zhou et al., uses plant roots to test digital biofabrication strategies for product design purposes [14]. They fabricate self-supported 3D structures by altering the growth media, direction of gravity and porosity of their digitally fabricated mold. These variables allowed them to manipulate plant roots, since the nutritional richness and the force of gravity have an impact on the root growth [14].

In each example above, designers initially define the environmental factors (in a parametric manner) as variables they can work with to manipulate the final outcome. In each case they have shown that, to some extent (within a value range) there is a somewhat predictable relationship between environmental parameters and specific material outcomes. The outcomes of these processes also exhibit variations, however, and this challenges notions of fabrication tolerances.

1.4. Focus

This paper extends these works on biological parametric design by reporting four design experiments using fungal fruiting bodies. In each case the objectives are to find the environmental parameters responsible for different fruiting body morphologies and to see whether such morphologies can be predicted. The fruiting bodies of the selected fungus have the benefit of being complex, in terms of morphology but also plastic, in that they exhibit significant phenotypic variation given the same genetic information. They can also be grown quickly. These experiments seek to answer the question: To what extent is mushroom growth parametric?

2. Materials

Oyster mushrooms (*P. ostreatus*) were used in this study because of their fast growth (compare to other species used in the design field such as *ganoderma resinaceum* and *trametes versicolor*) and the wide variety of known fruiting body morphologies due to their gastronomic use, indicating a high level of developmental plasticity [15]. In addition to the rapid growth rate and plasticity, fungal fruiting bodies possess totipotency. Totipotency describes the ability of a cell to divide and produce all the differentiated cells of an organism autonomously [16]. This means if even a tiny amount of mushroom tissue is transplanted onto a nutrient medium, it can initiate new growth [15]. Totipotency enables the harvested cells to be used as the basis for a new experiment.

In each experiment the mushroom growth followed a common and well described developmental pathway starting with the vegetative phase (hypha growth), which continued to the reproductive phase (fruiting body formation) (Figure 1) [17]. Hypha filaments transform to the fertile tissue of a fruiting body under suitable conditions. The organization of hyphae significantly changes while creating fruiting bodies. Normally, the filaments show positive autotropism by growing in an upwards direction; however while forming a fruiting body structure, they start to grow inwards and show negative autotropism [18]. This is due to hyphae forming a three-dimensional compound complex by interlocking with other hyphae structures, instead of simply forming an unconstrained mesh. The initial development of the fruiting body begins with a hyphal knot, which can be triggered by a disturbance such as an injury, edge encounter or changes in nutrient levels, temperature, or light exposure [18]. In the formation of *Basidiomycota* fungi, hyphae form knots by reducing their level of chitin and the knots become mushrooms by expansion and inflation of pre-existing hyphae. Depending on the species' phototropic requirements, the progress can proceed with the introduction of light that leads to cellular differentiation [17]. The formation of stipe (stalk), cap (pileus), and gill cells occur during this process, during which the mushroom takes on its characteristic appearance [15]. The spores are discharged from the surface of gills. Therefore, gills increase their surface area by folding, to allow the production of more spores.

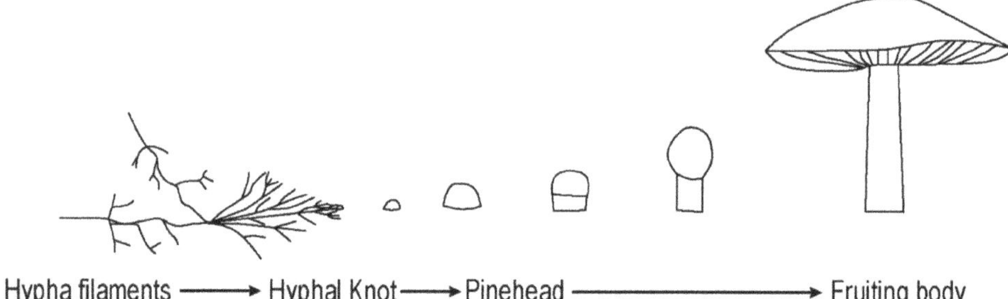

Figure 1. The developmental path of the fungal reproductive phase.

3. Methods

3.1. Factors in the Morphogenesis of Mushrooms

In previous literature it has been shown that different mushroom species adapt to the environment they occupy to maintain their life and chances of reproducing [19]. Mushroom morphology is connected to the transportation of spores, where the fungi adopt forms that optimize the diffusing of spores [15]. For instance, the umbrella shape of mushrooms comes from the upward development of stipes under the influence of light, whereas the gills that diffuse the spores develop downward and are affected by gravity [19]. However, the umbrella shape can be changed by altering the direction of light and gravity [19].

The major factors that affect the form of mushrooms depend on the species. Bellettini et al. has conducted experiments which show the key parameters affecting the mushroom morphology of oyster mushrooms: air temperature, light, humidity, CO_2 levels, gravity, substrate amount and size [20]. These factors influence the cap and stalk's shape, size, and surface finish of oyster mushrooms [21]. Therefore, in this study humidity, CO_2 level, gravitational force and substrate amount are used as variables to test the parametric qualities of mushrooms. Light duration and temperature are kept as constant values since we found across our interaction with fungi that they are more effective in initiating the mushroom formation rather than affecting mushroom morphology. High humidity environments provide favorable conditions for mushrooms to thrive in and bear fruit [18]. Different sources state that using 90–95% humidity or using 80–85% humidified culture room as well as spraying their fungi three times a day helps to achieve the optimal mushroom yield [20,22]. Stalk thickness tends to decrease with the decrease in the level of humidity, since there is not enough water for mushroom development [18,23].

A change in CO_2 concentration also triggers different stages of the fungal life cycle and affects the morphology of mushrooms. During the development of mushrooms, respiration activity increases, so the preferred CO_2 level decreases. While the preferred CO_2 concentration is 2000–2500 mg/L for mycelium growth, it decreases to 1500–2000 mg/L for fruiting body development. If the CO_2 level remains high, the cap formation may not occur [20]. High CO_2 concentration blocks pileus formation while boosting stalk elongation because the cell wall is affected by elevated CO_2 levels [19].

Many mushroom stalks possess negative gravitropism [24] as the fruiting bodies grow in the opposite direction of gravity and bending of the stalk occurs at the upper region closest to the cap [19]. In the literature, the substrate mass has often been studied as it affects the size and number of mushroom blooming because of the impact on nutrient availability.

3.2. The Experimental Design

From the literature above the effects of humidity, CO_2 levels, gravitational force and substrate amount were chosen as variables as these had the potential to have the most significant impact on mushroom morphology. To validate this decision, a series of experiments were conducted testing the effect of different conditions in isolation. The experiments were carried out during the COVID-19 period and hence some of the experimental setups were improvised around the available equipment and facilities.

Humidity and CO_2 levels in the experiments were controlled by a growth chamber that consists of an Arduino UNO (connected to a laptop), Arduino sensors (DHT11 air humidity and temperature sensor, SEN0219 infrared CO_2 sensor, V1.0 soil moisture sensor and HC-SP04 ultrasonic distance sensor) and devices (12V DC fan, humidifier, 450 nm LED blue light source and 75 watt heat bulb) [3]. The chamber also helped to keep temperature and light exposure stable. Only one variable was changed at a time and the others were kept constant for each experiment (Table 1).

Table 1. The variables used for four experiments.

		Experiments			
		Humidity	CO_2	Gravity	Sub. Amount
	Humidity (%)	95 85 80 75	80	75	80
	CO_2 (ppm)	2000	1000 3000 5000	5000	300
Variables	Sub. Amount (g)	55	55	55	40 80 120 160
	Gravity (degree)	90°	90°	90° 135° 180°	90°
	Light (nm)	4 h, 450 nm	4 h, 450 nm	4 h, 450 nm	4 h, 450 nm
	Temp. (°C)	20–22 °C	20–22 °C	20–22 °C	20–22 °C

Each set started with the same substrate ratios with 25% of strawbale, 25% of wood shavings, and 25% of coffee grounds. Straw was blended in a Nutri Ninja Blender & Smoothie Maker 900 W for 5 s to a homogeneous mixture. The wood shavings and coffee grounds were not blended since they already had uniform size. The substrates were prepared and sterilized in an autoclave at 121 °C for 15 min. This mixture was then seeded with 25% of oyster mushroom spawn (*P. ostreatus*) from GroCycle-UK, and sealed in (10 × 10 × 3 cm) plastic boxes, in the dark, at ambient temperature. The experiments ran for 29 days. After an initial three weeks of growth the samples were exposed to different environmental conditions for eight days in the growth chambers. All experiments were conducted in triplicate.

By altering the parameters incrementally across different experiments, as seen in Table 1, we were able to measure the scale effect of different environmental conditions and relate specific parameters with mushroom dimensions.

3.2.1. The Humidity Experiment

The variable of humidity level was set to four different levels as discussed in our previous paper and seen in Table 1 [3].

3.2.2. The CO$_2$ Experiment

The variable of CO$_2$ level was set to three different levels as explained in our previous paper and seen in Table 1 [3].

3.2.3. The Gravity Experiment

In this experiment the angle of growth was tested. The effect of gravity upon the growing mushroom was adjusted as a means of support by using the aforementioned plastic containers. After being removed from the containers, the mycelium tiles were kept in 90°, 135° and 180° angles, as seen in Figure 2. The samples with 180° angles were positioned on a box. Lifting them prevented moistening and mushroom growth on the contact surface.

Figure 2. The positioning of the mushrooms in the gravity experiment.

The experiment was repeated under 2000 ppm CO$_2$ level. In this way, it was possible to see the effect of gravity on caps in different sizes.

3.2.4. The Substrate Amount Experiment

In this set of experiments, the effect of substrate amount on mushroom size was tested. 40 g, 80 g, 120 g and 160 g mycelium and various substrates were mixed in the ratio of 25% of strawbale, 25% of wood shavings, 25% of coffee grounds, and 25% of mushroom spawn, as mentioned before. All mixtures were kept in (10 × 10 × 3 cm) plastic box and covered with aluminum foil with a 4 × 4 cm hole in the middle of one of the widest surfaces, as seen in Figure 3. The aim of guiding the mushroom growth from a single opening was to limit the number of fruiting bodies, thus, to prevent overcrowding, to focus on the size of the mushrooms.

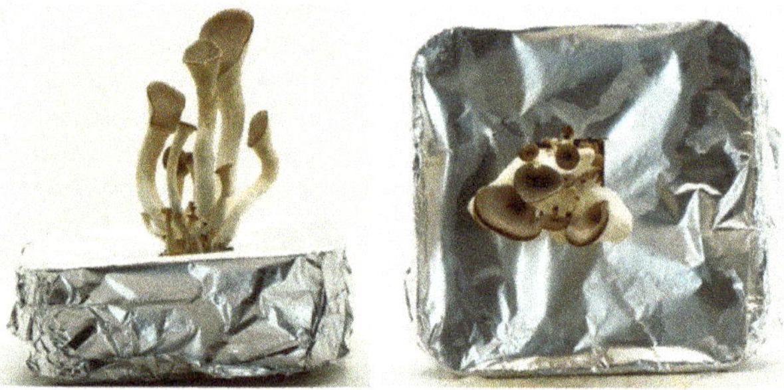

Figure 3. The preparation of the samples for the substrate amount experiment.

3.3. Measuring the Results

The mushroom morphology was documented at the end of day 27, through photography (Fujifilm X-T2 with 80 mm lens), microscopy (Dino-Lite digital microscope at 70× magnification) and 3D scanning (EinScan-SE desktop scanner). These tools helped to analyze the overall mushroom forms by allowing for the digital measurement of dimensions of the caps and stalks [3].

The biggest mushroom from each replicate was selected as the most mature specimen (Figure 4). Measurements were made digitally using Rhinoceros 3D due to the difficulty in measuring delicate mushrooms of a small size.

Figure 4. The locations of the measurement points.

The location of the measurement points for each specimen were standardized as follows:

- Capsize and stalk length are measured using curved lines. To measure the capsize ∣AB∣, point-A is selected arbitrary on the cap edge, and point-B is located on the opposite side of the edge/point-A. To measure the stalk length ∣EF∣, point-F is selected as the bottom of the stalk and point-E is selected as the lowest mid-point of the cap.
- The angle of the cap curvature ($D°$) is measured by:
  1. Drawing a line between the lowest and highest point on the cap edge.
  2. Measuring the angle between this line and the x-axis (parallel to the ground).
- The stalk curvature angle ($G°$) is measured by drawing two lines parallel to the stalk (one from underneath the cap, the other from the base of the stalk) and measuring the angle between these two lines.

4. Results

4.1. The Results of the Humidity Experiment

In line with the previous study, humidity influences the curvature of cap edges and the stalks [3]. The replicates grown in the in-between conditions exhibit in-between morphologies. As we know from the humidity experiment, cap edge and stalk curvature increase with the increase in humidity level, as seen in Figure 5 and Table 2.

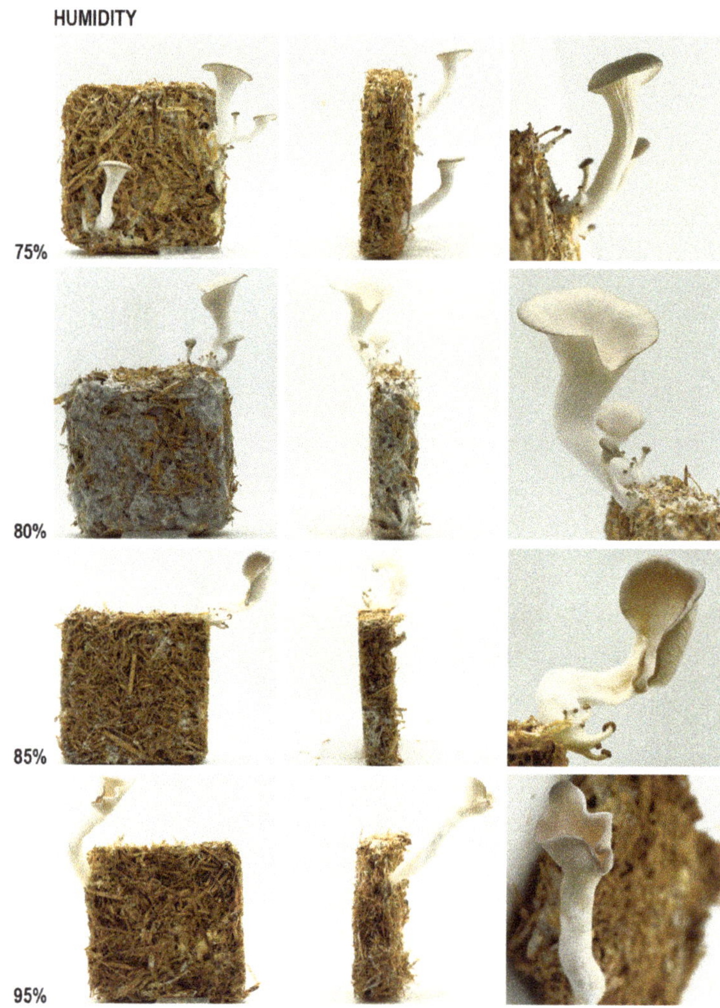

Figure 5. The results of the humidity experiment, front (**column 1**), side (**column 2**) and detailed (**column 3**).

Table 2. The measurements of the humidity experiment.

		95%			85%			80%			75%		
		Rep-1A	Rep-2A	Rep-3A	Rep- 1B	Rep-2B	Rep-3B	Rep-1C	Rep-1C	Rep-3C	Rep-1D	Rep-1D	Rep-3D
Cap	Size (cm)	2.8	3.1	3.1	1.8	3.2	3.4	2.1	2.3	2.4	3.9	1.6	3.6
	Average size		3 ± 0.1			2.8 ± 0.6			2.3 ± 0.1			3 ± 0.9	
	Curvature (degree)	120	127	145	53	64	60	30	38	31	31	32	36
	Average curv.		130.7 ± 7.4			59 ± 3.2			33 ± 3.3			33 ± 1.5	
Stalk	Length (cm)	5.6	5	4	4	4.9	4.4	3.5	4.1	4.7	4.7	4	3.2
	Average length		4.9 ± 0.5			4.4 ± 0.3			4.1 ± 0.3			4 ± 0.4	
	Curvature (degree)	104	133	140	85	83	90	51	78	81	54	61	76
	Average curv.		125.7 ± 2.9			86 ± 2.9			70 ± 7.8			63.7 ± 6.1	
Averge sprout number			2			4.3			4			3.3	

However, the cap sizes seem smaller in 80% humidity than the mushrooms grown in 75% humidity. This could be because the two outlier mushrooms grew bigger than expected and raised the average value, although this hypothesis needs to be validated by a bigger sample size. The texture of stalks and the depth of gills are qualitative results, and it can be observed from Figure 6 that in 80% humidity, gills are shallower, and stipes are hairier than 75%.

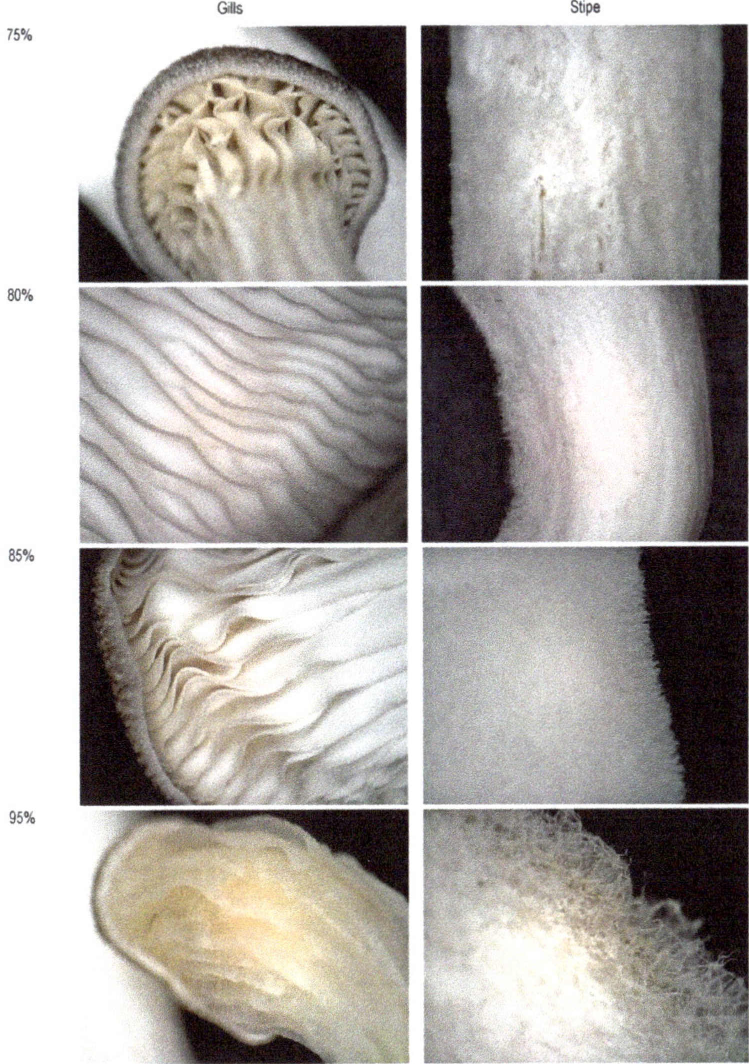

Figure 6. The comparison of gills and stipe under different humilities; images are captured using a Dino-Lite digital microscope at $70\times$ magnification.

4.2. The Results of the CO_2 Experiment

Altering CO_2 levels has a significant effect on the cap size [3]. High CO_2 decreases the cap size and inhibits mushroom maturation. As seen in Figure 7, although there are many sprouts, they elongate without cap formation. Their stalks get longer up to a certain level, as seen in Table 3. However, after a certain level (somewhere between 5000 ppm and 3000 ppm) the stalk length starts to decrease due to the high CO_2 level.

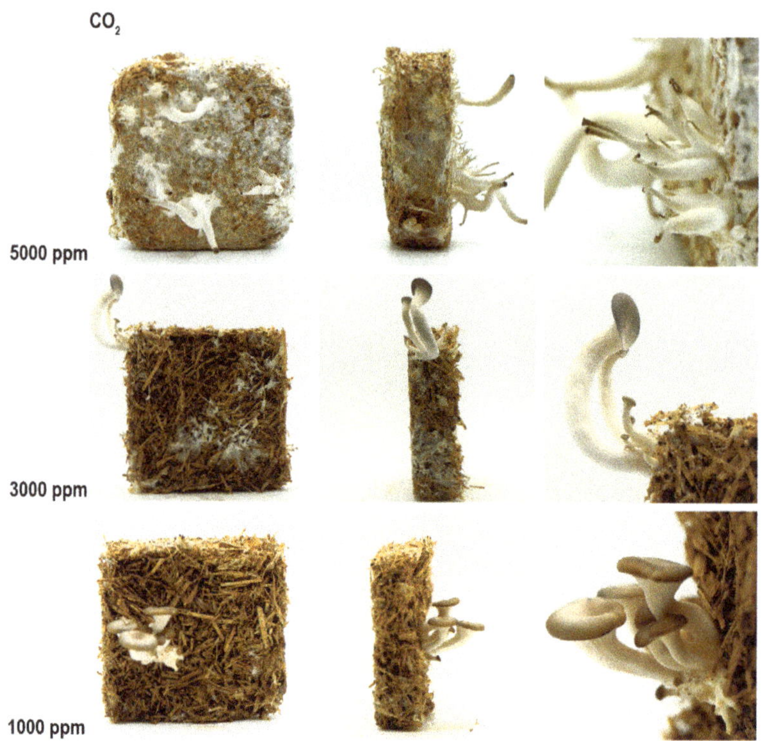

Figure 7. The results of the CO_2 experiment, front (**column 1**), side (**column 2**), and detailed photos (**column 3**).

Table 3. The measurements of the CO_2 experiment.

		5000 ppm				3000 ppm			1000 ppm	
		Rep-1A	Rep-2A	Rep-3A	Rep-1B	Rep-2B	Rep-3B	Rep-1C	Rep-2C	Rep-3C
Cap	Size (cm) Average size	0.4	0.2 0.2 ± 0.1	0.1	1.3	0.8 1 ± 0.2	1	2.4	2.3 2.4 ± 0.1	2.5
	Curvature(degree) Average curv.	31	2 12.7 ± 11.8	5	41	53 44.7 ± 5.3	40	8	7 8 ± 0.8	9
Stalk	Length (cm) Average length	3.8	0.2 2.3 ± 1.5	3	4.5	3.8 4.2 ± 0.3	4.4	2	1.8 2 ± 0.1	2.1
	Curvature(degree) Average curv.	66	79 72.7 ± 5.3	73	44	41 41 ± 1.2	39	53	61 56.3 ± 3.3	55
Averge sprout number		8.3				6			6.6	

4.3. The Results of the Gravity Experiment

As seen in Figure 8, there was a tendency for the fruiting body to grow vertically so the mushroom caps tended towards being parallel with the horizontal plane (Table 4). This led to the stalks being bent from underneath the cap, as they grow away from the tilted plane of the tile towards a vertical direction. As seen in Figure 9, the mushrooms grown at 5000 (high) and 2000 (low) ppm of CO_2 presented the same behavior in terms of orientation.

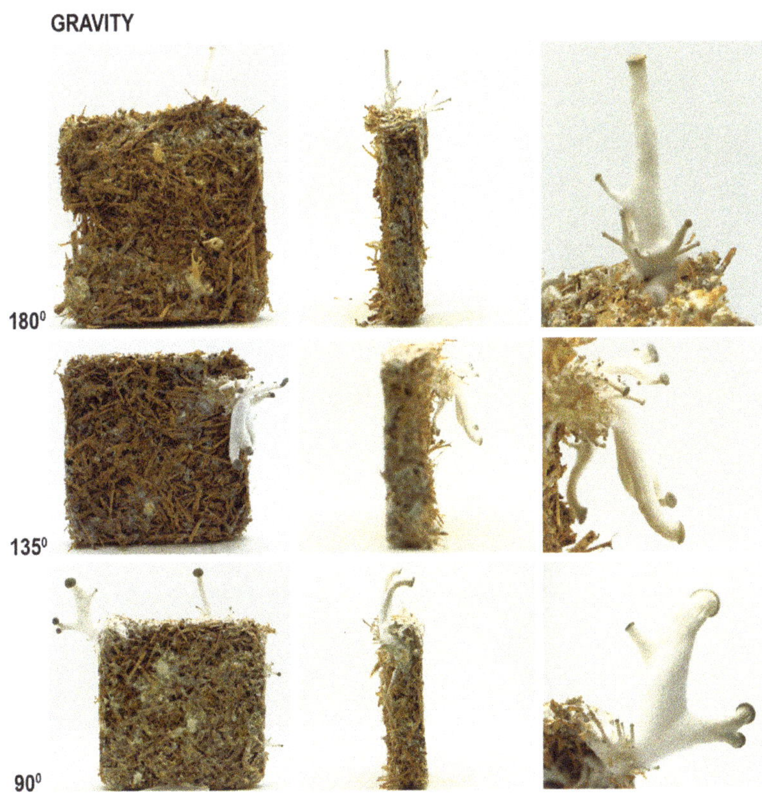

Figure 8. The results of gravity experiment at high CO_2 (5000 ppm), front (**column 1**), side (**column 2**), and detailed (**column 3**) photos.

Table 4. The measurements of the gravity experiment in high CO_2.

		90°			135°			180°		
		Rep-1A	**Rep-2A**	**Rep-3A**	**Rep-1B**	**Rep-2B**	**Rep-3B**	**Rep-1C**	**Rep-2C**	**Rep-3C**
Cap	Curvature(degree)	6	12	0	61	35	16	44	19	35
	Average curv.		6 ± 2.4			37.3 ± 10.6			32.7 ± 10.2	
Stalk	Curvature(degree)	44	51	57	35	23	46	23	36	0
	Average curv.		50.7 ± 3.7			34.7 ± 6.6			19.7 ± 10.5	

In summary, gravity affects the orientation of the caps, which leads the stalks to curve accordingly but there is no significant impact on the size of the mushrooms.

4.4. The Results of the Substrate Amount Experiment

When the mushroom sizes and the number of sprouts is compared, as seen in Figure 10, an increase in cap size, stalk length and sprout number can be observed as the substrate amount increases (see Table 5). The reason for mushroom stalks growing with different curvatures is that they curled as they came out of the hole in the foil wrap.

The variable that has the most effect on the overall size of the mushrooms is the amount of substrates. Although an increase in humidity enlarges them to some extent, the substrate amount is the main determinant. Without sufficient substrates, the mushrooms cannot reach their maturity.

Figure 9. The results of gravity experiment at low CO_2 (200 ppm), front (**column 1**), side (**column 2**), and detailed (**column 3**) photos.

SUBSTRATE AMOUNT

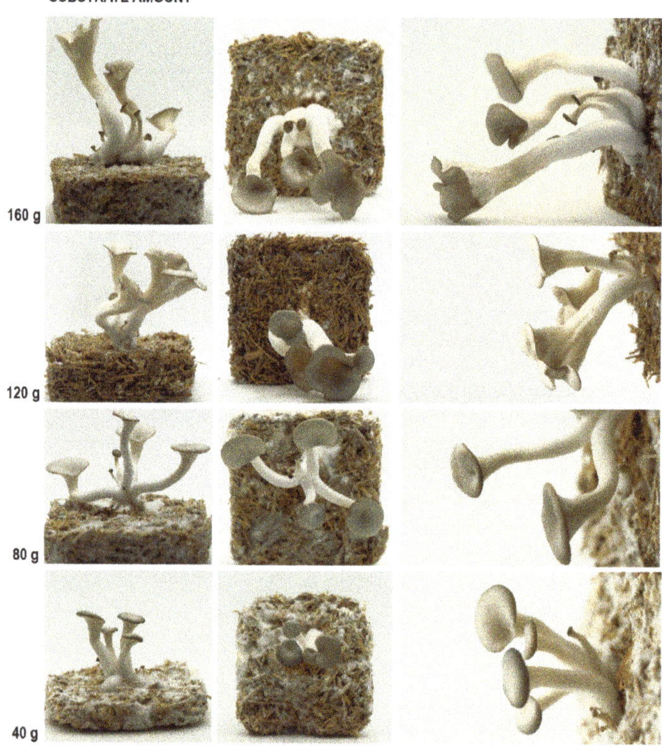

Figure 10. The results of the substrate amount experiment, side (**column 1**), top (**column 2**), and detailed (**column 3**) photos.

Table 5. The measurements of the substrate amount.

		40 g			80 g			120 g			160 g		
		Rep-1A	Rep-2A	Rep-3A	Rep-1B	Rep-2B	Rep-3B	Rep-1C	Rep-1C	Rep-3C	Rep-1D	Rep-1D	Rep-3D
Cap	Size (cm)	1.3	1.2	2.9	2.2	2.3	1.8	3.4	3.7	1.5	3.9	2.5	2.8
	Average size		1.8 ± 2.1			2.1 ± 0.6			2.9 ± 2.7			3.1 ± 0.4	
Stalk	Length (cm)	5.3	2.9	5.1	5.8	6	5.2	5.8	6.7	5.2	6.7	5.4	6.3
	Average length		4.4 ± 1			5.7 ± 0.1			5.9 ± 0.4			6.1 ± 0.5	
Average sprout number		4			6			5			9		

The samples grown in-between conditions exhibit in-between morphologies. As we know from the previous substrate amount experiment, the size of cap edges and the stalks get bigger as the substrate amount increases. All the curvature measurements in the 120 g mixture are somewhere between the 80 g and 160 g substrate amount. Although the sprout number is similar to the 80 g sample, it is less than the 160 g sample.

5. Discussion

The experiments showed that the effect of different variables is often connected to similar affects in terms of mushroom morphology. For example, high humidity and substrate amount will affect mushroom size. And single variables are related to more than one affect. For example, humidity also affects curvature. Despite this, there is a fairly predictable relationship between each of the environmental parameters described in the experiments. Cap curvature is related to humidity, cap size is related to CO_2, stalk bend is related to gravity, and overall mushroom size is related to substrate amount, as seen in Figure 11.

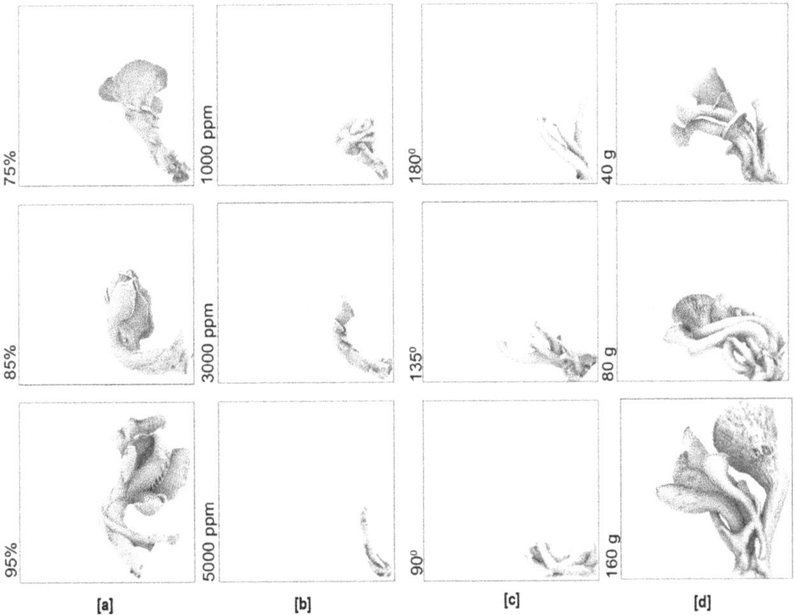

Figure 11. The overlay of the triplicate 3D scans of the mushrooms (at the same scale) grown under the same environmental conditions: The average sizes and the overlayed mushrooms in the (**a**) humidity (**b**) CO_2; (**c**) gravity; (**d**) substrate amount experiment.

The experiments also demonstrate that fungi exhibit linear parametric properties when a single parameter is changed, at least for the limited parameters and a single family of mushrooms tested here. As shown in Figure 12, each parameter change exhibits a distinct trend, although the high degree of variability exhibited in relation to mushroom size and substrate amount should be noted. More replicates are required in the future to achieve a higher significance of results. In all experiments, further replicates would need to be conducted to provide more significant relationships. Nevertheless, using these parameters, it should be possible to predict the morphology of mushrooms given specific parameters—within the range of values tested by the experiments.

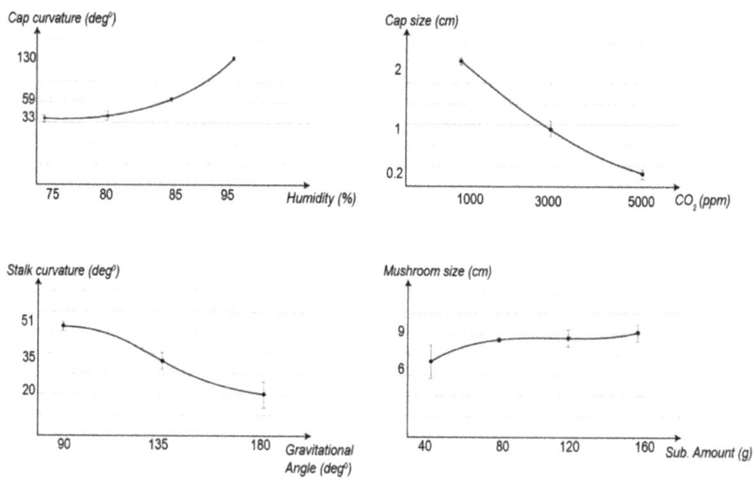

Figure 12. The mushroom growth—variables relationship.

Future research should also explore (1) the critical thresholds where growth is inhibited; (2) tipping points which lead the developmental pathway for the mushrooms to change; or (3) where normal development is critically disrupted where changing the variable no longer affects (or affects as expected) the mushroom morphology. For instance, mushroom growth may not be expected at very low humidity, or mushrooms cannot grow above a certain size even if the amount of nutrients are increased.

6. Conclusions

This paper asked the question: Is the growth of the mushroom fruiting bodies parametric? Or, more precisely: can mushroom morphology be predicted by altering the environmental parameters? As a designer, can we design fungal morphology using a parametric design approach? While we tend to think of biological systems as highly complex and non-linear systems, these albeit limited set of experiments have shown that given defined environmental conditions including factors such as CO_2, humidity, orientation etc. we can see, within the limits of these experiments, linear relationships between environmental parameters and morphology outcomes. This points to the possibility of computational simulations for these systems and for the development of parametric-like software to estimate aspects of biological growth. It is also worth noting, however, that given the small sample size of the experiments (restricted to a single family of edible mushrooms) and the often significant, variation between mushrooms, we also need to recognize that these environmental factors are linked, and that mushroom morphology is highly sensitive to slight variations in conditions. This means that any attempt to model and predict the outcomes of different growth conditions will need to be, to some extent, probabilistic. The next step in this research will be to build such a model.

While the ability to alter the morphology of mushroom growth may be useful in, for example, agricultural contexts, these experiments are practical thought experiments. By trying to take a parametric design concept (which is well discussed in generative and computational design in architecture) and applying it to biological systems, we are revealing both its strength and weakness as a concept. With rapidly growing (literally and metaphorically) interest in the use of biomaterials in design and ideas of harnessing biological fabrication emerging from fields such as engineering living materials, this paper offered an alternative approach to the often gene-centric idea of engineering living organisms. We have shown that the developmental plasticity of mushrooms allows us access to control parameters outside the living cell of the mushroom that in turn remotely influence, rather than control, the material outcomes of mushroom growth. What is true for mushrooms may also be true of other sorts of biological systems.

Future work will need to extend the parameters explored and examine more closely their interrelationships as well as to increase the types of biological systems and processes amenable to change. It will also need to address the challenges of uncertainty of outcomes which are inherent in the design with biological systems.

Author Contributions: Conceptualization, D.O.; methodology, D.O., M.D.-R., R.M. and M.Z; validation, D.O., M.D.-R., R.M. and M.Z; formal analysis, D.O.; investigation, D.O.; resources, D.O.; data curation, D.O.; writing—original draft preparation, D.O.; writing—review and editing, M.D.-R., R.M. and M.Z.; visualization, D.O.; supervision, M.D.-R., R.M. and M.Z.; project administration, D.O., M.D.-R., R.M. and M.Z. All authors have read and agreed to the published version of the manuscript.

Funding: This research was supported by the Hub for Biotechnology in the Built Environment funded by Expanding Excellence in England (E3), Research England.

Institutional Review Board Statement: Not applicable.

Informed Consent Statement: Not applicable.

Data Availability Statement: Not applicable.

Conflicts of Interest: The authors declare no conflict of interest.

References

1. Sherry, A. Enhancing PFASs Attenuation in Coastal Brownfield Soils (EPACS): Enhancing Natural System Attenuation Capacity for a Key Emerging Contaminant. 2016. Available online: https://ebnet.ac.uk/wp-content/uploads/sites/343/2021/02/EBNet-POC2020-Summary-Sheet-funded-projects-v1.pdf (accessed on 14 March 2022).
2. Adamatzky, A.; Ayres, P.; Belotti, G.; Wosten, H. Fungal architecture. *Int. J. Unconv. Comput.* 2019, pp. 1–15. Available online: http://arxiv.org/abs/1912.13262 (accessed on 14 March 2022).
3. Ozkan, D.; Dade-Robertson, M.; Morrow, R.; Zhang, M. Design a living material through bio-digital fabrication, in towards a new, configurable architecture. Proceedings of 39th ECAADE Conference, Novi Sad, Serbia, 8–10 September 2021; pp. 77–84.
4. Ramirez-Figueroa, C. Biomaterial Probe-Design Engagements with Living Systems. Ph.D. Thesis, Newcastle University, Newcastle upon Tyne, UK, 2017.
5. Ozkan, D.; Christagen, B.; Dade-Robertson, M. Demonstrating a Material Making Process through the Cultivation of Mycelium Growth. In *4th International Conference of Biodigital Architecture & Genetics*; Butterworth-Heinemann: Oxford, UK, 2020; pp. 94–103.
6. Jabi, W. *Parametric Design for Architecture*; SAGE Publications: New York, NY, USA, 2013. [CrossRef]
7. Simon, H.A. *The Sciences of the Artificial*, 3rd ed.; The MIT Press: Cambridge, MA, USA, 1969; Volume 11. [CrossRef]
8. Oxman, R. Theory and design in the first digital age. *Des. Stud.* **2006**, *27*, 229–265. [CrossRef]
9. Skipper, M.; Weiss, U.; Gray, N. Plasticity. *Nature* **2010**, *465*, 703. [CrossRef] [PubMed]
10. Dade-Robertson, M. *Living Construction*, 1st ed.; Routledge: New York, NY, USA, 2020. [CrossRef]
11. Carpo, M. Digital indeterminism: The new digital commons and the dissolution of the architectural authorship. In *Architecture in Formation: On the Nature of Information in Digital Architecture*; Routledge: New York, NY, USA, 2013; pp. 48–52.
12. Arnardottir, T.H.; Dade-Robertson, M.; Mitrani, H.; Zhang, M.; Christgen, B. Turbulent casting. In Proceedings of the 40th Annual Conference of the Association of Computer Aided Design in Architecture (ACADIA), Online and Global. 24–30 October 2020; pp. 300–309.
13. Oxman, A.N. *Silk Pavilion-II*. 2021. Available online: https://oxman.com/projects/silk-pavilion-ii (accessed on 14 March 2022).
14. Zhou, J.; Barati, B.; Wu, J.; Scherer, D.; Karana, E. Digital biofabrication to realize the potentials of plant roots for product design. *Bio-Des. Manuf.* **2021**, *4*, 111–122. [CrossRef]
15. Watkinson, S.C.; Boddy, L.; Money, N.P. *The Fungi*, 3rd ed.; Elsevier: Amsterdam, The Netherlands, 2016. [CrossRef]

16. Kaul, H.; Ventikos, Y. On the genealogy of tissue engineering and regenerative medicine. *Tissue Eng. Part B Rev.* **2015**, *21*, 203–217. [CrossRef] [PubMed]
17. Kües, U. Life history and developmental processes in the basidiomycete *Coprinus cinereus*. *Microbiol. Mol. Biol. Rev.* **2000**, *64*, 316–353. [CrossRef] [PubMed]
18. Moore, D.; Gange, A.C.; Gange, E.G.; Boddy, L. Chapter 5 fruit bodies: Their production and development in relation to environment. *Stress Yeast Filam. Fungi* **2008**, *28*, 79–103. [CrossRef]
19. Sakamoto, Y. Influences of environmental factors on fruiting body induction, development and maturation in mushroom-forming fungi. *Fungal Biol. Rev.* **2018**, *32*, 236–248. [CrossRef]
20. Bellettini, M.B.; Fiorda, F.A.; Maieves, H.; Teixeira, G.L.; Ávila, S.; Hornung, P.S.; Júnior, A.M.; Hoffmann-Ribani, R. Factors affecting mushroom *Pleurotus* spp. *Saudi J. Biol. Sci.* **2019**, *26*, 633–646. [CrossRef] [PubMed]
21. Bayer, E.; McIntyre, G. Method for Producing Rapidly Renewable Chitinous Material Using Fungal Fruiting Bodies and Product Made Thereby. U.S. Patent No 8,001,719 B2, 23 August 2011.
22. Mycelia. Mycelium for Professionals. 2020. Available online: https://www.mycelia.be/en/strain-list (accessed on 26 February 2020).
23. Jang, K.-Y.; Jhune, C.-S.; Park, J.-S.; Cho, S.-M.; Weon, H.-Y.; Cheong, J.-C.; Choi, S.-G.; Sung, J.-M. Characterization of fruitbody morphology on various environmental conditions in *Pleurotus ostreatus*. *Mycobiology* **2003**, *31*, 145. [CrossRef]
24. Moore, D. *Fungal Morphogenesis*; 6-Development of Form; Cambridge University Press: Cambridge, MA, USA, 2010. [CrossRef]

MDPI

St. Alban-Anlage 66

4052 Basel

Switzerland

Tel. +41 61 683 77 34

Fax +41 61 302 89 18

www.mdpi.com

Biomimetics Editorial Office

E-mail: biomimetics@mdpi.com

www.mdpi.com/journal/biomimetics